Measurement of Airborne Pollutants

Measurement of Airborne Pollutants

Editor:

Suzanne Couling

BUTTERWORTH
HEINEMANN

Butterworth-Heinemann Ltd
Linacre House, Jordan Hill, Oxford OX2 8DP

A member of the Reed Elsevier group

OXFORD LONDON BOSTON
MUNICH NEW DELHI SINGAPORE SYDNEY
TOKYO TORONTO WELLINGTON

First published 1993

British Library Cataloguing in Publication Data

Measurement of Airborne Pollutants
I. Couling, Suzanne
628.5

ISBN 0 7506 0885 4

Printed and bound in Great Britain

Contents

Preface

The goal of environmentally sustainable development requires sound management of the atmosphere at a local, regional and global scale. The availability and quality of air pollution measurements is clearly critical to the success of this process. Yet these measurements stretch present technical capability to its limits because excellent sensitivity, specificity and robustness are often needed to allow a proper understanding of the condition of the atmosphere, the processes within it, and the impact on it of various emissions. Thus the development of air pollution measurements is central to progress in the formulation of environmental policy, the efficient regulation of emissions, and the satisfactory control of processes which emit pollutants into the atmosphere.

It was against this background that the National Measurement System Policy Unit of the Department of Trade and Industry commissioned Warren Spring Laboratory to undertake a programme of work on air pollution measurements. The results of this programme, which began in October 1989 and finished in September 1992, were presented at the symposium which this volume records.

At the present time there is a rapid expansion in both the number and variety of air pollution measurements being made. The emphasis here is on those which present particular problems and which are required to either provide sounder close-to-ground air quality data, or, to quantify stack and fugitive emissions. The purpose is to report improvements in methodolgy, traceability and harmonization. The effectiveness of sampling is given special prominence because this is generally recognized to be one of the weakest links in the measurement chain for air pollution.

A number of organizations and many individuals have contributed to this volume and the work it describes; Warren Spring Laboratory wishes to record its thanks to them. Among those deserving of special mention are the National Measurement System Policy Unit who commissioned the programme, colleagues at other institutions who have contributed to this volume and to the programme in general, Suzanne Couling—who has brought care and diligence to the role of editor, and Duncan Enright of Butterworth Heinemann Ltd who has patiently advised and encouraged us.

Mark Kibblewhite
Warren Spring Laboratory

Contributors

B R Bushby
Warren Spring Laboratory, Department of Trade and Industry, Gunnels Wood Road, Stevenage, Hertfordshire, SG1 2BX.

P Clayton
Warren Spring Laboratory, Department of Trade and Industry, Gunnels Wood Road, Stevenage, Hertfordshire, SG1 2BX.

J L Cocksedge
Warren Spring Laboratory, Department of Trade and Industry, Gunnels Wood Road, Stevenage, Hertfordshire, SG1 2BX.

P Coleman
Warren Spring Laboratory, Department of Trade and Industry, Gunnels Wood Road, Stevenage, Hertfordshire, SG1 2BX.

J Cox
Warren Spring Laboratory, Department of Trade and Industry, Gunnels Wood Road, Stevenage, Hertfordshire, SG1 2BX.

B J Davis
Warren Spring Laboratory, Department of Trade and Industry, Gunnels Wood Road, Stevenage, Hertfordshire, SG1 2BX.

A L Dennis
Warren Spring Laboratory, Department of Trade and Industry, Gunnels Wood Road, Stevenage, Hertfordshire, SG1 2BX.

G S Driver
Warren Spring Laboratory, Department of Trade and Industry, Gunnels Wood Road, Stevenage, Hertfordshire, SG1 2BX.

S Eaton
Warren Spring Laboratory, Department of Trade and Industry, Gunnels Wood Road, Stevenage, Hertfordshire, SG1 2BX.

I Fängmark
Institute of Environmental Chemistry, University of Umeå, S-901 87 Umeå, Sweden

A R Fernandes
Warren Spring Laboratory, Department of Trade and Industry, Gunnels Wood Road, Stevenage, Hertfordshire, SG1 2BX.

D J Hall
Warren Spring Laboratory, Department of Trade and Industry, Gunnels Wood Road, Stevenage, Hertfordshire, SG1 2BX.

J G Irwin
Warren Spring Laboratory, Department of Trade and Industry, Gunnels Wood Road, Stevenage, Hertfordshire, SG1 2BX.

M Kendall
Warren Spring Laboratory, Department of Trade and Industry, Gunnels Wood Road, Stevenage, Hertfordshire, SG1 2BX.

M G Kibblewhite
Warren Spring Laboratory, Department of Trade and Industry, Gunnels Wood Road, Stevenage, Hertfordshire, SG1 2BX.

J E Lampert
Warren Spring Laboratory, Department of Trade and Industry, Gunnels Wood Road, Stevenage, Hertfordshire, SG1 2BX.

S Marklund
Institute of Environmental Chemistry, University of Umeå, S-901 87 Umeå, Sweden

G W Marsland
Warren Spring Laboratory, Department of Trade and Industry, Gunnels Wood Road, Stevenage, Hertfordshire, SG1 2BX.

C Rappe
Institute of Environmental Chemistry, University of Umeå, S-901 87 Umeå, Sweden

W Rudolf
Umweltbundesamt, Germany

C Schofield
Warren Spring Laboratory, Department of Trade and Industry, Gunnels Wood Road, Stevenage, Hertfordshire, SG1 2BX.

B Sevruk
Swiss Federal Institute of Technology, ETH-Zentrum, 8092 Zurich, Switzerland

K J Stevenson
Warren Spring Laboratory, Department of Trade and Industry, Gunnels Wood Road, Stevenage, Hertfordshire, SG1 2BX.

B H Stone
Warren Spring Laboratory, Department of Trade and Industry, Gunnels Wood Road, Stevenage, Hertfordshire, SG1 2BX.

M Tysklind
Institute of Environmental Chemistry, University of Umeå, S-901 87 Umeå, Sweden

S L Upton
Warren Spring Laboratory, Department of Trade and Industry, Gunnels Wood Road, Stevenage, Hertfordshire, SG1 2BX.

A van der Meulen
National Institute of Public Health and Environmental Protection, Air Research Laboratory, P.O. Box 1-3720 BA Bilthoven, The Netherlands

D S Walker
Warren Spring Laboratory, Department of Trade and Industry, Gunnels Wood Road, Stevenage, Hertfordshire, SG1 2BX.

D S Wallace
Warren Spring Laboratory, Department of Trade and Industry, Gunnels Wood Road, Stevenage, Hertfordshire, SG1 2BX.

J Wilken
Warren Spring Laboratory, Department of Trade and Industry, Gunnels Wood Road, Stevenage, Hertfordshire, SG1 2BX.

M Williams
Warren Spring Laboratory, Department of Trade and Industry, Gunnels Wood Road, Stevenage, Hertfordshire, SG1 2BX.

— Part 1 —

Ambient air quality

— 1 —

Introduction

M. Williams

Over the past decade, environmental issues have become of increasing concern to the public at large and consequently to governments, regulators and pollution control authorities. Air pollution is no exception to this trend, indeed issues such as acidification, global climate change, photochemical pollution and urban air quality have all moved into the forefront of the environmental agenda in recent years.

Even before this increased interest, of course, the need for traceable measurements of high quality has always existed. However with more attention being focused on air pollution issues, and with an increasing amount of legislation on the subject, this need becomes more and more important. The results of air pollution measurement programmes are thus being used in an increasing number of areas, and comparisons with other data and with regulatory limits are being made. These requirements mean that, to an increasing degree, air pollution measurements should be made using methods which are well characterized in terms of precision and accuracy, specificity for the pollutant in question, reliability and limitations. Operational protocols and calibration procedures need to be well defined and documented, and calibration standards need to be traceable to accredited bodies.

Despite the amount of atmospheric measurements which have been undertaken in the past few decades, much remains to be done in this area. New pollutants emerge for which new measurement methods are developed (or vice versa!), or new techniques are developed for measuring 'traditional' pollutants more accurately, more efficiently or at lower levels than was previously possible. The present 'workshop' was convened to present some of the results of a programme of research, funded by the National Measurement System Policy Unit of the UK Department of Trade and Industry, designed to ensure that airborne pollutant measurements can be made with the traceability and harmonization needed to bring them within the National Measurement System.

Following an introductory keynote paper from Dr Werner Rudolf, three papers are presented which deal with operational evaluations of emerging techniques for ambient measurements of airborne particles and for low levels of nitrogen dioxide. Calibration techniques for automatic analyses or for gas cylinders obtained from commercial suppliers are considered in Stewart Eaton's paper.

Fundamental issues in the measurement of acid deposition are covered by three papers. The first two deal with the measurement of precipitation, checking gauge performance and analysing and improving the aerodynamic design of precipitation collectors. The third deals with the use of simulated solutions for quality control of the chemical analysis of rainwater samples.

— 2 —

Continuous air pollution measurements on a national and world-wide scale

W Rudolf

2.1 Abstract

The content of this paper is mainly concerned with quality assurance of air pollution networks operated in Germany, other countries within the Commission of European Communities (CEC) and those within the frame of UNEP/WHO (GEMS). Quality assurance is regulated by national and international guidelines. The national programmes of CEC member countries have to be in line with the regulations of CEC, where guidelines exist, i.e. for SO_2, black smoke, total suspended particles (TSP), NO_2 and lead. Since the GEMS programme is under the supervision of both UNEP and WHO, either WHO guidelines or ISO guidelines regulate the quality assurance programmes.

2.2 Measuring networks in Germany

In Germany there are 13 fully installed air pollution networks. The so-called 'New Länder' are in the process of putting five other networks into operation within the next 2 years. From these 18 networks, 17 are under the responsibility of the Länder Governments and one is under the responsibility of the Federal Environmental Agency. These networks operate more than 400 different stations, the bulk of which are equipped with continuously operating monitors for the measurement of SO_2, CO, NO, NO_2, TSP, O_3 and total hydrocarbons (excluding methane).

The task of this monitoring is to assess air pollution to fulfil German and CEC requirements for the surveillance and abatement of air pollution in the long term and for the purposes of smog alerts. During a smog episode, of elevated SO_2 concentrations in winter or O_3 in summer, the normally independent networks exchange their data for the use in an early smog alert system. This, together with a meteorological forecasting system, aims to provide early information on transboundary transport of pollutants and a better understanding of the processes of air pollution chemistry during such transport. The resulting smog charts of measured and forecast concentrations are distributed automatically, via a computer system, back to the operators of the networks. Intensive co-operation with the neighbours of Germany (BeNeLux, France, Austria, Switzerland and the eastern neighbours Poland and CSFR) will hopefully lead to a combination of these networks for monitoring transboundary air pollution. The larger scale of photochemical smog formation with its complex chemical reactions suggests extending such an early alert system to the whole of Europe including the UK. In particular, since the channel tunnel will possibly be ready in 1993, transboundary and air pollution transport will reach new dimensions. An example of the type of information such a system can create is shown in Figure 2.1. It is obvious that such an information system can be of considerable help to the national authorities in deciding appropriate measures for minimizing smog situations.

2.3 Trends in air pollution concentration in Germany

When discussing the possible connection of networks, the cost-benefit relation has to be considered. One aspect to be considered is the trend of the most important air pollutants, such as suspended particles SO_2, CO, NO_x and O_3, during the recent years.

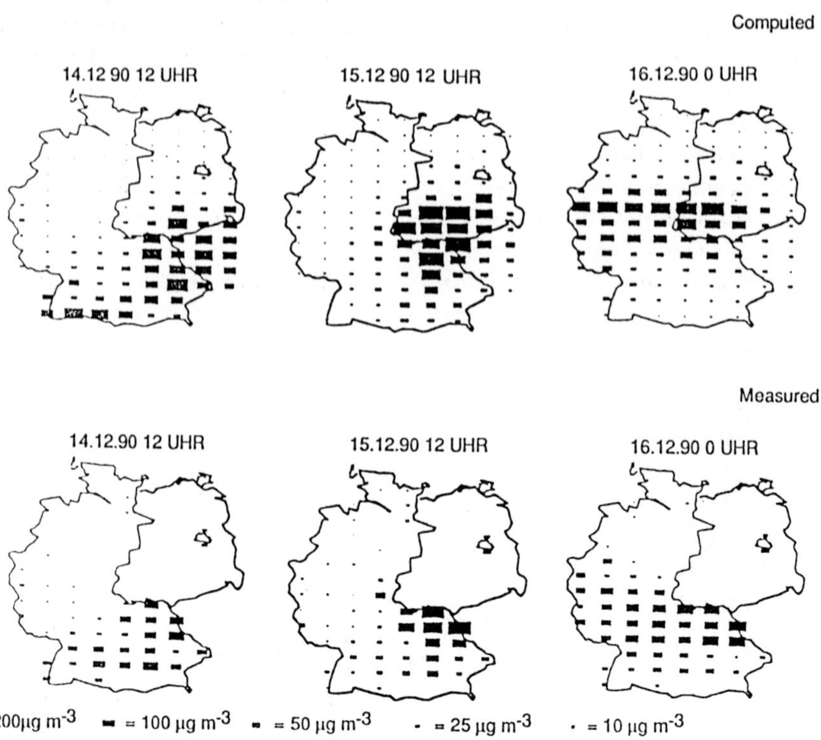

Figure 2.1 *Computed and measured SO_2 concentrations, Germany, December 1990*

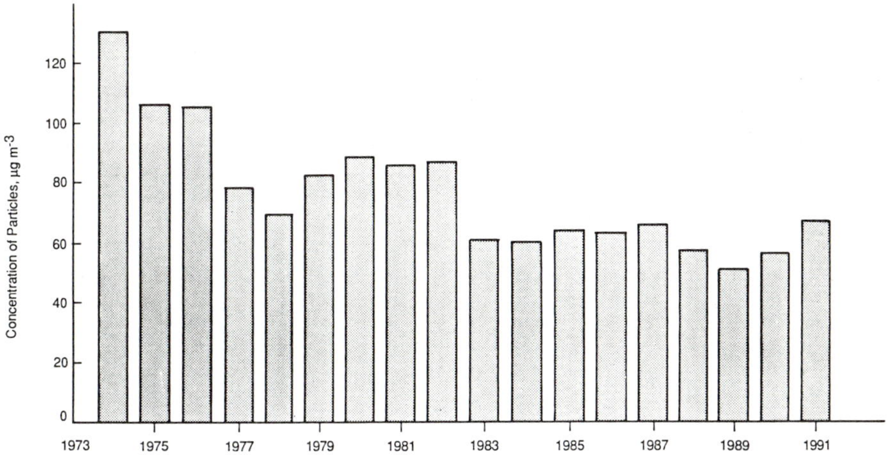

Figure 2.2 *Trend of suspended particles. Federal environmental agency office and habiting area*

Figure 2.2 shows a pronounced decrease in the concentration of suspended particles over the last 10 years at the measuring site at Pilotstation Frankfurt/M. Similar results have been obtained from most of Germany.

The trend of SO_2 at Pilotstation Frankfurt/M over the last 30 years is shown in Figure 2.3 indicating a downward trend. The concentration in 1990 was about 30% of the value in the early 1960s. This pattern is the same in all parts of Western Germany, and even in the former DDR SO_2 concentrations have started to decrease.

CO is one of the most important components related to traffic emissions. The measurements in a downtown street canyon of Frankfurt/M., shown in Figure 2.4, show a downward trend by a factor of two over the last 15 years. Not all German stations show such a clear trend. The reduced CO-emission-factor of private cars is partly compensated by an increase in traffic density.

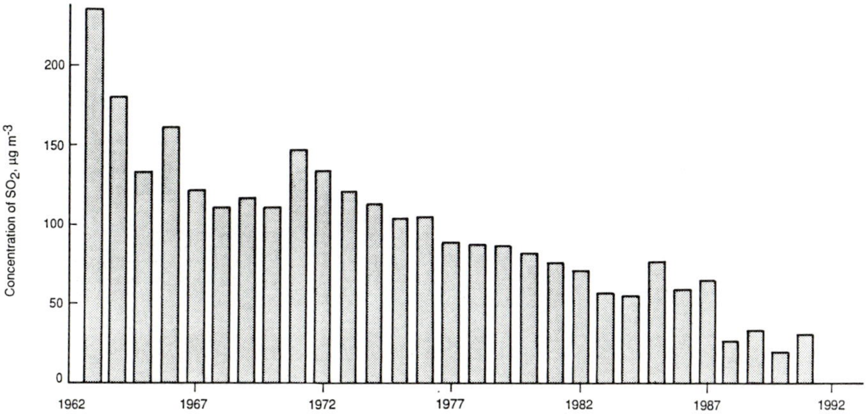

Figure 2.3 *SO_2 trend. Federal environmental agency office and habiting area*

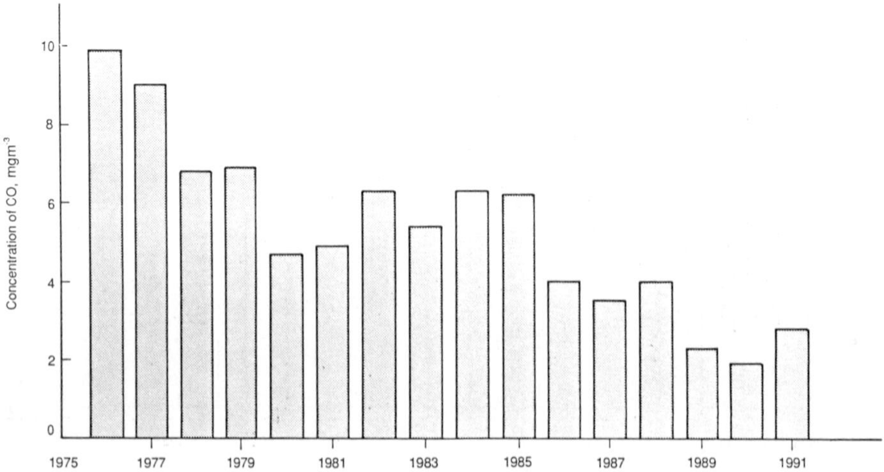

Figure 2.4 *CO trend for Frankfurt/M city. Downtown street canyon, measuring height 3.5 m*

Figure 2.5 shows that the trend of NO_x concentrations (sum of NO and NO_2) has a dramatic increase from the beginning of the measurements in 1961 until 1976. The following 15 years show a more stable concentration level. The indicated downward trend in recent years may not be statistically sound; other German measuring sites show stable or slightly increasing concentrations due to the fact that the impact of catalysts has been compensated for by the increasing traffic density.

The yearly average concentrations of O_3, shown in Figure 2.6, do not indicate a clear trend. The picture is still dominated by the high concentrations in 1976 when the whole of Europe including the UK suffered from O_3 concentrations up to 400 $\mu g\,m^{-3}$.

These trends indicate that components of great importance in the past may not need to be monitored to the same extent in the future, while others may

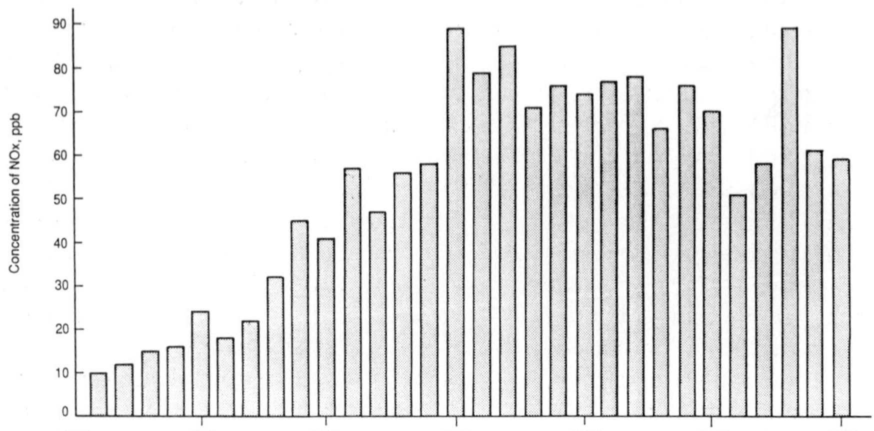

Figure 2.5 *NO_x trend for Frankfurt/M Westend. Living and office area*

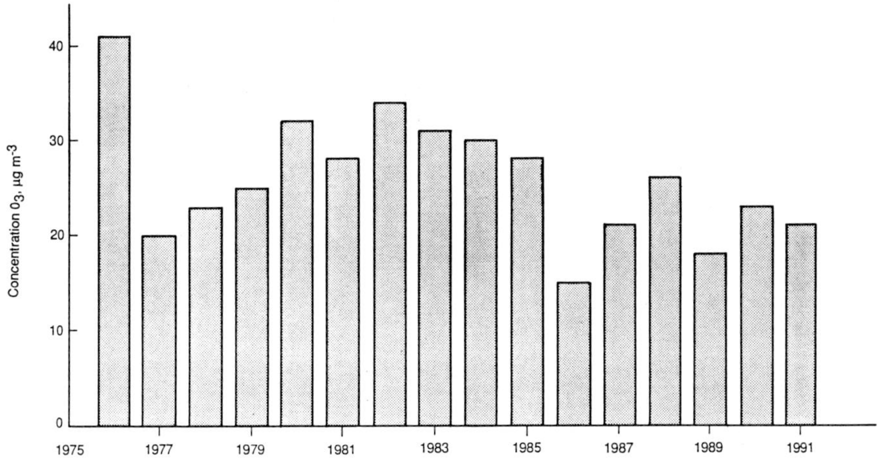

Figure 2.6 *Ozone trend for Frankfurt/M Westend. Living and office area*

claim more attention. So focusing on the right component is one problem. The necessity for quality assurance and harmonization of measuring methods is another aspect.

2.4 Quality assurance

It is widely accepted that station siting, quality and comparability of instruments, calibration, data acquisition and data control are the most important criteria in quality assurance. To touch on the problem of siting, we should remember the discussion in the Articles of SO_2 and NO_2 Directives of the CEC. In particular, the estimation of the risk with respect to human exposure to NO_2 was intensively discussed in Germany and other CEC member states. Figure 2.7 shows the vertical distribution of NO and NO_2 in a downtown street canyon. While it is obvious that the so-called primary traffic components, which are directly emitted from the exhaust, such as NO, CO, diesel-soot and benzene, have a sharp gradient with height, the secondary component NO_2 shows a much smoother gradient. This makes it possible to give a relatively broad recommended range for the height for measuring NO_2 in street canyons; a measuring height up to 3.5 m, corresponding to the suggested height of the sampling inlets of German measuring equipment, seems to be reasonable. The primary components may be more problematic and a sampling height lower than 2 m may be advisable. This strongly affects network design and different components may require different networks. Practical aspects such as vandalism or other disturbances to the sampling device are discussed as well.

The possibility of making measurements at street level is demonstrated by the sampling inlet operated at the street canyon measurement site of the Pilotstation Frankfurt/M. At a distance of 2 m from the traffic, a pole of 3.5 m length has been erected right on the sidewalk. The instruments, installed

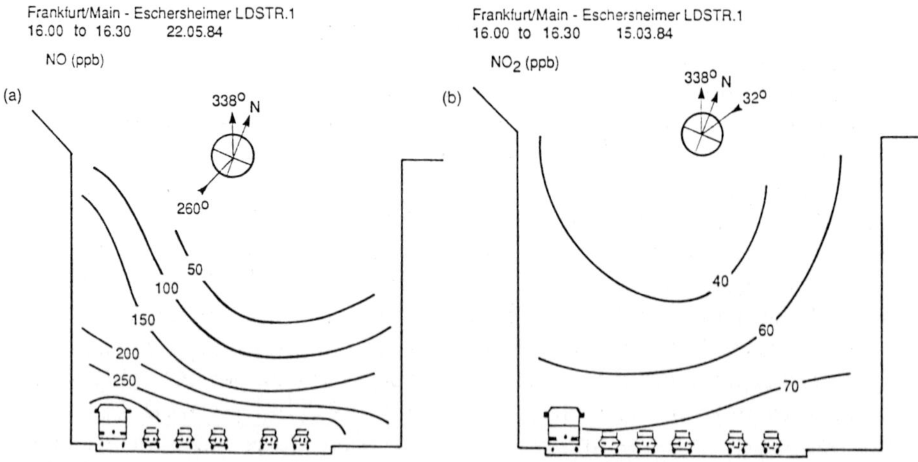

Figure 2.7 *Iso-lines of (a) NO and (b) NO_2 concentrations in a downtown street canyon*

in a cellar below, sample through tubing that passes up through the pavement and into the pole. At a height of 1.5 m up the pole, there is a net with a sufficiently wide mesh size to let the air pass through to be sucked via the tubing into the instruments. People do not see the inlet and in 8 years of operation no disturbance or vandalism has been reported.

2.5 Comparability of measurements

In Directive 80/779/CEC (Directive on SO_2 and TSP) two important definitions have been created with respect to the assessment of comparability of the measurements.

1. Satisfactory correlation: accepted performance test of instruments, with resect to minimum requirements.
2. Reasonable stable relationship: accepted as the comparability of national measurements with the defined comparison method, which for SO_2 is UV-fluorescence, the calibration of which must be traceable back to the tetra chloro mercurate (TCM) method. For NO_2, it is the chemiluminescence with ozone method, the calibration of which must be traceable back to NO_2-permeation.

During the SO_2 quality assurance programmes of CEC, six countries decided to use the comparison method (Belgium, Denmark, France, Ireland, Luxemburg, United Kingdom). Later on, Germany followed with a special programme at selected sites. In addition, one of the experiments compared the total acidity method with the comparison method described above. The results were not quite satisfactory and different from site to site. As a rule, at low concentrations the total acidity method overestimated compared to the comparison method, while at concentrations near the 98% threshold, there was a tendency to underestimate.

A second programme to test the performance characteristics of monitors was accepted by all member states. Nevertheless, the test procedures suggested by CEC were refused by France.

Performance tests and minimum requirements have been carried out in Germany for more than 10 years for NO, NO_2, CO, TSP, O_3 and NMHC, as well as for SO_2, in accordance with German test procedures and defined minimum requirements. A lot of instruments are approved by the combined vote of our Länder and the Federal Government. At Umweltbundesamt Pilotstation, the 15th test will be available soon. The Commission is informed about test reports and the announcements of approval. Because a lot of the procedures used in Germany were included in the proposed CEC test procedure, and because the German test procedure is equal to or more stringent than the CEC test, Germany asked the Commission in 1989 for general acceptance of the German test results until the CEC procedure is accepted by all member states.

2.6 Calibration

In the CEC Directives for SO_2 and NO_2, reference instruments are defined which are not basic methods and therefore have to be calibrated against an accepted primary standard or the reference method. For SO_2 the accepted reference method is the TCM method when used under laboratory conditions. Primary standards are the permeation method, gravimetric method and static volumetric dilution. Laboratory comparisons of the permeation and static volumetric dilution methods, the results of intercalibrations of SO_2 by the Commission, and stability and certification test by BCR are discussed below.

In 1981 a comparison between laboratories was carried out with SO_2 calibration gases in pressure cylinders. The TCM method was used in parallel to check the stability of the SO_2 calibration gases. Table 2.1 shows the results which indicate particularly good cylinder stability but deviations between the analyses of participating laboratories of more than $\pm$ 10%.

In 1985 a further CEC experiment was carried out with SO_2 cylinders. The programme included member states: Italy, France Germany and Greece. The cylinders were transported by a CEC lorry to the national reference laboratories, who arranged for their circulation to the participating national networks. The results (see Figure 2.8) show, with some exceptions, reasonable agreement between the 'German Länder Laboratories' but the cylinders were clearly unstable compared to the experiment in 1981.

The next CEC experiment in 1988, combined the possibilities and experiences of the previous experiments and was designed to test the accuracy of networks by direct control of designated measuring stations. For that reason a reference instrument together with two SO_2 cylinders were carried around together. The results are shown in Figure 2.9. The most

Table 2.1 CEC SO_2 comparison in 1981

Laboratory	*Date*	*Permeation tube*	*Cylinder I*	*Cylinder II*	*Chem. anal.*
IHE/SCK	07.07.81	–	306	324	–
IHE/SCK	29.07.81	–	304	320	–
IHE/SCK	06.08.81	–	307	321	–
London	20.08.81	286	293	305	293
Dublin	27.08.81	246[x]	290	302	294
Riso	03.09.81	330	330	358	295
Frankfurt	10.09.81	272[x]	324	334	297
Athens	17.09.81	250	257	267	300
Roma	24.09.81	248[xx]	270	285	312
Ispra	01.10.81	293[xx]	336	348	310
IHE Brussel	13.10.81	314	304	315	316
Luxemburg	16.10.81	282	282	295	321
Bilthoven	21.10.81	306	306	317	310
Paris	05.11.81	352	310	320	343
SCK Mol	09.11.81	342	309	321	340

The measured SO_2 concentration for both cylinders are represented graphically in Figures 2.8 and 2.9.

[x]In two cases the SO_2 concentration of the permeation device was underestimated, probably due to an accidental dilution caused by an insufficient difference in the flow aspired by the analyser and the flow delivered by the permeation device.

[xx]A discrepancy between the measured concentration of the permeation device and the pressure bottles was also noticed in two other laboratories where the measurements were compromised by some technical problems.

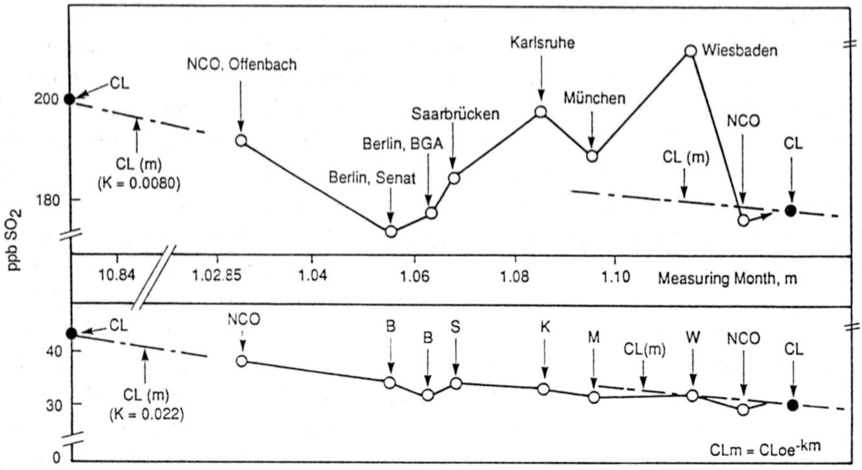

Figure 2.8 *SO_2 intercomparison by the CEC in 1985, German results. Filled circles; CL measurement: empty circles; NLAB measurement: dashed line; CL(m) versus measuring month*

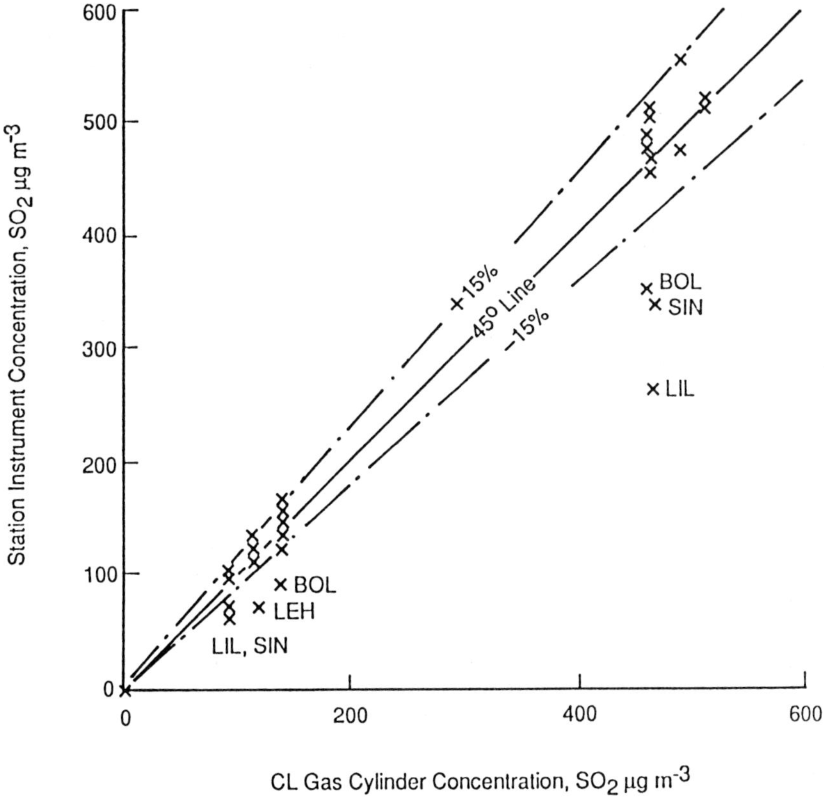

Figure 2.9 *SO_2 test cyclinder measurements*

important finding from this experiment was that the severest mistakes were caused during sampling. Therefore an atmospheric generator is now under construction which can include the whole sampling line in the calibration. As a last step in CEC activities, a detailed and sophisticated manual is being prepared by experts from the member countries. It is hoped that this will help to avoid a lot of mistakes and lead to an increasing harmonization of SO_2 measurements. Nevertheless, written guides such as the manual mentioned above must be supported by personal contact between technicians and scientists. A good opportunity for this is participation in ring experiments on a regular basis. Also, the acceptance of a harmonization cascade is very important; CEC is no longer able, if one looks at the existing geographical spread which will increase with the incorporation of eastern European countries, to carry out such experiments with a mobile unit. The measuring car is no longer a suitable vehicle. One has to think about the use of airplanes. From past experiences it is recommended that the national reference institutes should convene on a regular basis at laboratories, designated because of their special capability, so that standards can be transferred to the national reference laboratories by trained staff. Further steps of QA are then in the hands of the national authorities.

2.7 Particulate matter

As long as particles are sampled and measured, questions must be asked about the inlet geometry and the particle cut-off size. In 1984, under contract to CEC, field and wind-tunnel measurements were carried out. The results of the field experiment have shown deviations between different samplers of more than 30%. Therefore, in addition to other activities during the 1988 SO_2 experiment, such as sending around filters loaded with particles, measurements at selected sites were carried out. The results, shown in Figure 2.10, are quite reasonable despite the fact that some stations show deviations which are not acceptable. The most important question now is the construction of a reference sampler with a cut-off of 10 μm. After vigorous

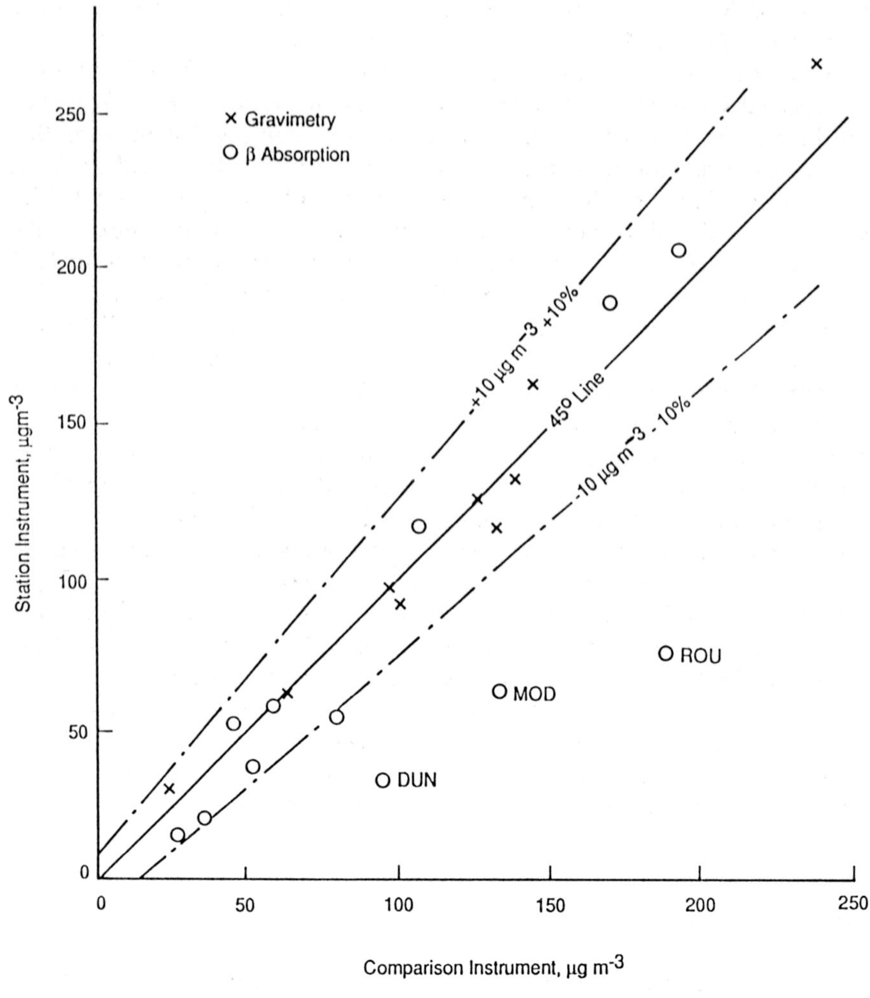

Figure 2.10 *Intercomparison of suspended particle monitors: ambient concentration measurements made by gravimetry and beta-absorption techniques*

discussion, this cut-off size has been defined to be the most acceptable for the surveillance of human health. The standardization of the method will be completed within CEN. However, bearing in mind the decreasing concentrations shown in Section 2.3, one has to pay attention not only to the 10 μm cut-off but also to the practical detection limit of TSP monitors when operated under field conditions.

2.8 Nitrogen dioxide

The defined calibration method in the CEC 85/203/CEC Directive is the permeation method. Other methods for the preparation of primary or secondary calibration gases are: the Saltzman method; static volumetric dilution of NO and titration with O_3; and gravimetric determined static dilution of NO_2. In May 1992 at Essen, Landesanstalt für Immssionsschutz, the CEC will carry out a comparison experiment at which eight national reference labs will take part. In fact, NO_x-harmonization began in 1978 when a technician and myself drove a measuring car, equipped with a dilution system, a permeation tube and NO cylinders, to Ispra, Rome and Toulous, and later to IHE, Belgium, the laboratory in MOL, Belgium, and RIVM, Netherlands. The deviations of NO_2 standards, with the exception of the laboratory in Toulous which used a filter method, were within 10%. In 1985 cylinders were sent around (organizers were WSL, MOL and UBA) accompanied by visits of the UBA measuring car to MOL, RIVM and Ispra. The result was that the reference laboratories using permeation and static injection were within $\pm$ 5% while the laboratories using the Saltzman method tended to be lower by up to 15%. This led, in Germany, to the instigation of an *ad hoc* group which investigated some permeation tubes. There was a strong indication that, especially during the beginning of the permeation period, other components in addition to NO_2 were permeated. In the meantime, a reinvestigation of static injection, permeation and Saltzman methods at Ispra supported by IR-spectroscopy has shown only 0.5% N_2O_2 in the pure liquid NO_2 of permeation tubes and in the cylinders of pure NO. Permeation and static injection agreed to within 3% showing no systematic deviation. The Saltzman method showed a systematic deviation of between 3 and 5%, which was smaller than former experiments. In 1992 at UBA Pilotstation gravimetric determined NO_2 injection in the 5 ppm range was started. A comparison against NO calibration gases prepared by static volumetric injection has shown promising agreement in the range of 3%.

2.9 Conclusion

The importance of the quality assurance programmes described above is a world-wide issue going beyond national borders and beyond the borders of CEC member states. The improving results should encourage the CEC and

Table 2.2 SO_2 and TSP concentrations of 17 'Megacities' study of GEMS within UNEP

	SO_2 $\bar{x}$ $\mu g\, m^{-3}$	SO_2 98% $\mu g\, m^{-3}$	TSP $\bar{x}$ $\mu g\, m^{-3}$	TSP 98% $\mu g\, m^{-3}$	
Bangkok	15–20	60–25	100–380	200–800	
Buenos Air.			60–240	300–1000	
Bombay	40–20	220–40	100–180	350–450	
Beijing	20–150		250–500		
Cairo	?		600–750		
Calcutta	100–40	250–100	380–420	1000–600	
Delhi	10–90	50–300	250–500	400–1200	
Karachi			250–400		CO $\bar{x}$ = 49 ppm
London	100–40				
Manila	50–100		50–250		
Mexico	60–160		200–600		O_3 very high
Moscow	3–15		230–100		SO_2 very low
Rio	40–200		200–80		
Sao Paulo	50–250	100–350	80–180	200–400	
Seoul	150–240		200–150		
Shanghai	20–100	150–400	200–330	400–1000	

member states to export their experience especially into developing countries. The necessity of such a transfer is emphasized by the results of the so-called Megacity Study within the GEMS programme of UNEP/WHO. To demonstrate the ongoing activities, Table 2.2 shows a summary of the results of measured SO_2 and TSP values in cities all over the world with more than 10 million inhabitants.

Harmonized and quality assured measurements of air pollution in such cities are important to protect people against destruction of their environment, caused not least by the industries of developed countries which have gone into those areas to avoid the regulations and thresholds existing in their home countries.

—3—

Ambient test procedure to demonstrate reference equivalency of measurement methods for fine suspended particulate matter up to 10 μm for compliance monitoring

A van der Meulen

3.1 Introduction

According to Article 1(4) of Directive 89/427/EEC (CEC, 1989), the European Commission shall submit a proposal to revise Directive 80/779/EEC on sulphur dioxide (SO2) and suspended particulate matter (SPM) (CEC, 1980). Among others, such a proposal should deal with a reference method to determine SPM.

The *Executive Summary on Studies of the Commission of the European Communities on the Development of a SPM Reference Method* (van Elzakker *et al.*, 1990 and Payrissat *et al.*, 1991) summarize the work done by EC to develop an appropriate SPM reference method according to the requirements of the aforementioned Article 1(4). Originally, EC intended to propose two reference methods: one for inhalable SPM, and one for the thoracic fraction of SPM. However, there turned out to be no instruments available to meet all of the practical requirements for measuring inhalable and thoracic particulates simultaneously (van der Meulen *et al.*, 1990). As a consequence, under point 6 of the 'Executive Summary' the following is proposed:

1. To harmonize the SPM monitoring methodology in EC, an appropriate EC reference method should be established for PM10, that is for fine SPM with particle size up to some 10 μm, according to the CEN/ISO (1991) thoracic particle size fraction convention.
2. The EC Member States have to use the reference method under routine conditions for the survey of ambient air quality, or methods which are equivalent to the reference method (PM10 reference equivalence methods).
3. Member States have to take the necessary steps to ensure that equivalence methods are tested by an appropriate EC PM10 reference equivalence procedure.

This paper aims to assist in the development of such an EC-PM10 reference equivalence procedure. This involves a discussion on the comparison of wind-tunnel and ambient procedures. Wind-tunnel methodology can demonstrate the reference equivalence of the sampling performance of the candidate sampler, whereas ambient field testing of co-located reference and candidate samplers can demonstrate compliance over a wide range of ambient situations. The proposal to set up an ambient reference equivalence procedure is also discussed.

3.2 Wind-tunnel methodology for sampler testing

The pertinent regulations of the US Federal Register (1987) prescribe performance specifications and wind tunnel test procedures applicable to candidate methods for PM10. Key elements are:

1. Ideal sampling performance based upon the ISO/CEN thoracic particle size convention (ISO/TC, 1991), i.e. smooth, S-shaped sampling effectiveness curve dropping from near 100% to near 0% as the particle size increases from 1 to some 30 μm, with a 50% cutpoint (D50) of 10 $\pm$ 1 μm aerodynamic diameter, and geometric standard deviation (σ) of 1.5.
2. Expected mass concentrations calculated from the sampling effectiveness curves for three wind speeds must be within $\pm$ 10% of that calculated for the ideal sampler.

The tests for the sampling performance of a candidate PM10 sampler are to be conducted in a wind-tunnel test facility. Particles up to 20 μm, i.e. liquid particles (5, 5, 10, 14, and 20 μm) and solid test particles (20 μm), and winds within a commonly encountered ambient range (2, 8 and 24 km hr^{-1}) should be employed in such tests.

Now, the following EC findings (van Elzakker *et al.*, 1990; van der Meulen *et al.*, 1990; Laskus *et al.*, 1989) are of some importance:

1. Systematic differences up to 30% were consistently observed in the ambient field experiments between high and low volume PM10 instruments, both approved by the US EPA wind-tunnel procedure.
2. Wind-tunnel results regarding the same SPM and PM10 device(s), obtained in different European wind-tunnel facilities, do show considerable differences in sampling performance.
3. Wind-tunnel results could not be validated by ambient field tests; comparing observed particle size distributions and the wind-tunnel-determined sampling effectiveness gave considerable deviations.

Note, that the applicability of wind-tunnel results (obtained in isotropically turbulent flow) to ambient situations (where a wide range of anisotropic and non-stationary turbulence is observed) has been questioned before (Benarie, 1977). Hence, it is thought that an EC-PM10 reference equivalence procedure should be based on elaborate ambient field testing, instead of a wind-tunnel methodology.

3.3 Proposed ambient field test procedure

Above all, the EC-PM10 reference equivalence procedure should be a practical one, enabling the EC Member States to assess their candidate PM10 sampling systems under ambient conditions. Specifically, the procedure has to focus only on the equivalence of the PM10 inlet characteristics. It shall not deal with the equivalency of the various ways to analyse the collected particulate matter (e.g. gravimetric, β-dust or TEOM). In this way, the procedure can be seen as the complete ambient analogy of the US-EPA wind-tunnel testing methodology. For, the US-EPA procedure focuses only on the sampling performance of the PM10 inlet as such, and does not deal with the complete procedure to determine the ambient particulates. The reference equivalence procedure basically should cover the aspects discussed below.

3.3.1 PM10 reference device

As pointed out in the Introduction (Section 3.1), the PM10 reference device shall be based upon the CEN/ISO thoracic particle size fraction convention (1991). Basically, the so-called WRAC (wide range aerosol classifier) can be used as reference sampler both as PM10 (and also SPM) reference sampler, and to characterize the size fractionation of the ambient particulate matter between fine particles, $< 10\ \mu m$, and coarse particles, $> 10\ \mu m$.

The WRAC system has been amply tested within the context of the aforementioned EC programme to develop a reference sampler for suspended particulates (van Elzakker *et al.*, 1990). The WRAC shows the most optimal sampling performance as compared to all other present particulate sampling systems. However, most regrettably, the practicability of the WRAC turned out to be a major drawback. Therefore, in order to recommend a practical way of working, the feasibility of the High Volume Sampling-PM10 as reference sampler should be considered, since the HVS-PM10 performs very well compared to the PM10 results of the WRAC, in combination with conventional high volume sampling to estimate the PM10 contribution to SPM. Note that the HVS-PM10 inlet has been approved by US-EPA.

3.3.2 Candidate PM10 sampling instrument

Preferably, the candidate sampling method is based on discrimination and selection of PM10 particles by inertial separation in a specially shaped inlet. For the purpose of reference equivalence testing a complete sampling system should consist of: PM10 inlet + dust collection filter + regulated flow device; that is, a PM10 inlet followed by conventional filtration of a measured volume of sampled air, and determination of the net weight gain of the filter using standard gravimetric procedures.

Basically, the candidate PM10-inlet is tested by co-located sampling with the reference instrument; subsequently, the filters are weighed off-line in the laboratory. This is a time-consuming, laborious procedure, hence, it might be useful to develop a direct-response ambient test procedure to check the cut-off characteristics the PM10-inlet. To this end, the aerodynamic particle sizer (APS) could be used, sampling isokinetically from the connecting tubing between the PM10 inlet and the collecting filter-holder. In this way, the fine particle spectrum can be used as 'tracer'. When the candidate sampler is tested with and without the PM10 cut-off device, the ratio of the concomitant APS spectra yields the cut-off characteristics of the PM10 device. Moreover, the cut-off characteristics could be tested over a wide range of ambient conditions.

3.3.3 Compliance: calculated

Compliance can be judged by comparing the calculated concentrations starting from observed particle size distributions and the appropriate sampling characteristics of the candidate sampler with the CEN/ISO target. The sampling characteristics of the candidate sampler should preferably be available from first principles, or alternatively from wind-tunnel experiments. The expected concentration C can be calculated from (van der Meulen *et al.*, 1987):

$$C = A_i \times C_i$$

where C_i is the concentration level ($\mu g\,m^{-3}$) of size range i; A_i is the sampling efficiency factor of size range i; n is the number of size ranges. The C_i values are taken from measurements considered to be the 'true' values such as from the aforementioned WRAC. The A_i values compare the sampling characteristics of the candidate sampler and the CEN/ISO thoracic convention respectively. In this way, one could get an idea of the optimum compliance of the candidate sampler vs. the ISO/CEN target, or in other words, what is the best agreement to be expected when comparing measured concentrations from the candidate and reference sampler. The sensitivity of PM10 sampling efficiency of the candidate sampler to small cut-off fluctuations can also be judged.

3.3.4 Compliance: measured

Comparative data from the candidate and the reference method should be obtained in characteristic situations covering a wide range of relevant ambient situations within EC to demonstrate the reference equivalence of the candidate PM10 sampler. Conditions to be tested include: low and high concentrations; mainly fine particles or mainly coarse particles; low and high winds and low and high humidity.

3.3.5 Statistical procedure to show equivalence of comparative data

Regarding the definition of the reference equivalence, or, 'how good is good enough for equivalence', it seems worthwhile to consider the following points:

1. different levels of equivalence (qualitatively: from 'very good' to 'acceptable') do require a concomitant set (i.e. minimum amount) of comparative data;
2. subsequently, the EC directive has to specify the inaccuracy of PM10 measurements.

Such an approach does not suffer from the drawback of one and only one predetermined level of equivalence. For, as a rule, the user is only interested in the final answer: compliance or not, i.e. is the total deviation from the target response within the predetermined limits (van der Wiel and van der Meulen, 1988). The total deviation of 'real' measurement values from the 'true' values consists of a random (statistical) part covered by imprecision, and a systematic part called bias. Among others, bias depends on functional characteristics, describing the deterministic aspects of the measuring process, like sensitivity or calibration (i.e. deviation from the CEN/ISO target), and operational characteristics like logistics, maintenance, electric power, etc.

The advantages of such a procedure are obvious:

1. it facilitates equal stringency of different measuring methods (it is the final answer that counts);
2. the relative importance of various performance characteristics is reflected by the operational conditions (e.g. temperature controlled measurement stations or not);
3. the error budgeting of accuracy related characteristics can be left to the discretion of the users.

This approach is contrary to the conventional procedure which specifies requirements for all of the various characteristics separately; this can be unnecessarily restrictive.

3.3.6 Reference equivalence

According to the scheme thus outlined, reference equivalence is to be obtained in any and/or all of the following proposed categories:

Background:	fine particle fraction dominating, making up over some 70–80 % of SPM.

Urban:	increased concentrations, coarse particle fraction up to some 50%.
Industrial (average):	average-to-high concentrations, fine and coarse particle fractions are in the same order of magnitude (50%).
Industrial (heavy):	high concentrations, coarse particle fraction well over 50%.
Extreme:	high concentrations, (very) large coarse particle fraction, and high winds.

3.4 References

Benarie M M (1977). Continuous sampling of urban suspended matter over large grainsize spectrum: validity of the HiVol sampler and other results. *Atmospheric Environment,* **11**, 527.

CEN/ISO (1991). Air quality—particle size fraction definitions for health related sampling. ISO/TC 146/SC 2+3 *ad hoc* WG & CEN/TC 137/WG 3 (1991).

ISO (1983). Air quality—particle size fraction definitions for health related sampling, Technical report ISO/TR 7708-1983.

Commission of the European Communities (1989). Council Directive 89/427/EEC (21 June 1989) amending Directive 80/779/EEC on air quality limit values and guide values for sulphur dioxide and suspended particulates. O.J. L 201/53, 14.07.1989.

Commission of the European Communities (1980). Council Directive 80/779/EEC (15 July 1980). Air quality limit values and guide values for sulphur dioxide and suspended particulates. O.J. L 229, 30.08.1980, 30-48.

Laskus L, Bake D, Konig R, van der Meulen A, van Elzakker B G (1989). Measurements of particle size distribution and dust concentration. Staub-Reinhaltung der Luft 49 395-400.

Payrissat M, Sandroni S, Hecq P (1991). The European harmonization programme on: the revised reference method for suspended particulates, the monitoring network design, and prospectives. *Pollution Atmospherique,* 1991, 404–417.

US Federal Register (1987). Ambient monitoring reference and equivalent methods. United States Environmental Protection Agency. Federal Register 40 CFR Part 53. 1st July 1987.

van Elzakker B G, Hall D, Holländer W, Mark D, van der Meulen A, Laskus L, Upton S, Vincent J H, Zierock K H, (1990). Executive Summary on Studies of the Commission of the European Communities on the Development of a SPM Reference Method. EC, December 1990.

van der Meulen A, van Elzakker B G, Mark D, and Zierock K H (1990). Improvements in the design and operation of a sampler for the measurement of suspended particulate matter in the ambient atmosphere. RIVM report no. 222403001 (1990).

van der Meulen A, van Elzakker B G and van den Hooff G N (1987). PM10 : Results of a one year monitoring survey in the Netherlands. *JAPCA,* **37**, 812-818.

van de Wiel H J and van der Meulen A (1988). Framing minimum performance requirements. In: Air quality in Europe. Proceedings of the International EC Workshop on Harmonisation of the Technical Implementation of the EEC Air Quality Directives 80/779 (SO2 & SPM), 82/884 (Lead), 85/203 (NO2). Lyon (France) 1988

— 4 —

Evaluation of monitoring systems for ambient aerosols

J E Lampert

4.1 Abstract

Warren Spring Laboratory is conducting a programme to evaluate a number of measurement systems for ambient aerosols. This work will intercompare the different systems and recommend their suitability for future monitoring applications by government and others.

This report describes the operational details of the programme and presents initial results from laboratory and field trials conducted at Warren Spring Laboratory. The work has been carried out on behalf of the Department of Trade and Industry under the auspices of the Standards and Quality Measurement Advisory Committee (SQMAC), and forms part of a larger programme designed to incorporate airborne emission measurements into the UK National Measurement Scheme.

4.2 Introduction

There is increasing interest in the measurement of aerosol concentrations in the atmosphere, to assess their impact on the environment and, in particular, to investigate any harmful health effects. The monitoring of aerosol in the form of black smoke has been carried out routinely in the UK for many years as part of the European Commission (EC) Smoke/Sulphur Dioxide Directive monitoring programme. The existing British Standard for black smoke is based on a non-gravimetric reflectance method. However, the measurement of particle concentrations by gravimetric methods is becoming increasingly preferred by regulatory bodies such as the EC. Proven techniques suitable for unattended operation in air quality monitoring networks have yet to be established in the UK for the measurement of aerosol by gravimetric methods.

Warren Spring Laboratory (WSL) is therefore currently undertaking a programme designed to evaluate and compare a range of aerosol samplers. Their suitability for use in this application is being assessed and, in particular, systems for the continuous monitoring of thoracic particles by means of gravimetric methods are being identified and evaluated. Since the efficiency of sampling inlet heads has been studied and reported on elsewhere (Barrett *et al.*, 1985, vanOsdell, 1991), this programme has concentrated on comparing the operational performance of the sampling systems. This work forms part of a larger programme designed to incorporate airborne emission measurements into the UK National Measurement Scheme. This paper describes some of the monitors included in the tests, and our experiences of operating them in the field. Initial results are also presented.

4.3 Samplers

4.3.1 Candidate samplers

All of the gravimetric systems under test operate by drawing a measured volume of air through a filter paper. The mass of particulate matter collected is either measured manually at the end of the sampling period, or automatically as the sampling proceeds. The main samplers included in the study were as follows:

Manual methods: Conventional high volume samplers (the GMW/Sierra and the Ecotech)

Automatic samplers: The tapered element oscillating microbalance (R&P TEOM 1400) and the β-ray absorption sampler (Environnement 100M β-ray)

The following are also being tested as a subsidiary part of the programme: the automatic high volume sampler (Digitel DHA 80), the Institute of Occupational Medicine (IOM) ambient sampler and two automatic black smoke monitors. The various systems were calibrated in the laboratory and then transferred to the west field of WSL for extended intercomparison tests. WSL is situated in a semi-rural area at the edge of a light industrial estate. The samples were positioned ~ 50 m from a busy motorway.

4.3.2 Particulate sampling standards

Aerosol, as considered in these tests, can be described by three main health related fractions.

Inhalable: covers the whole range of particle sizes that can potentially be inhaled by humans.

Thoracic: covers particles which can penetrate into the thoracic region of the respiratory tract.

Respirable: covers particles which can penetrate deep into the lungs.

The International Standards Organisation/Comite European de Normalisation (ISO/CEN) has recently specified new conventions for all three

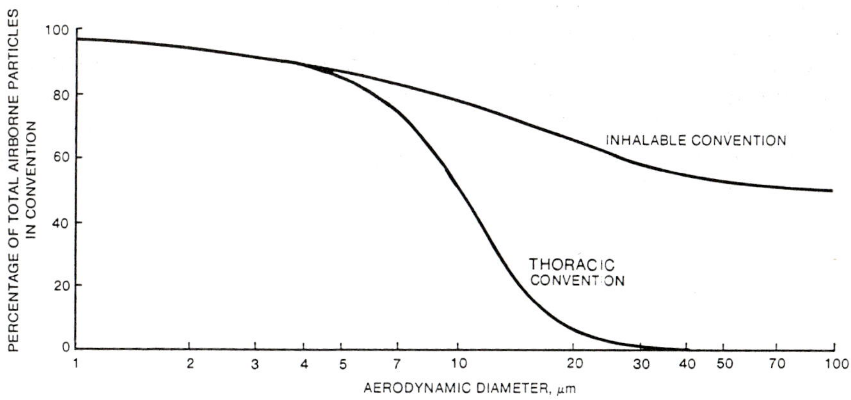

Figure 4.1 *The inhalable and thoracic conventions as percentages of total airborne particles*

aerosol fractions. Figure 4.1 shows a plot of sampler collection efficiency required to meet these conventions. The US PM10 standard is designed to collect particles with an aerodynamic diameter below 10 microns and is meant to comply with the ISO thoracic convention for a specified representative particle size distribution.

Figure 4.2 shows the efficiency of a typical PM10 sample inlet, as determined in the WSL wind-tunnel. This indicates that the PM10 inlet

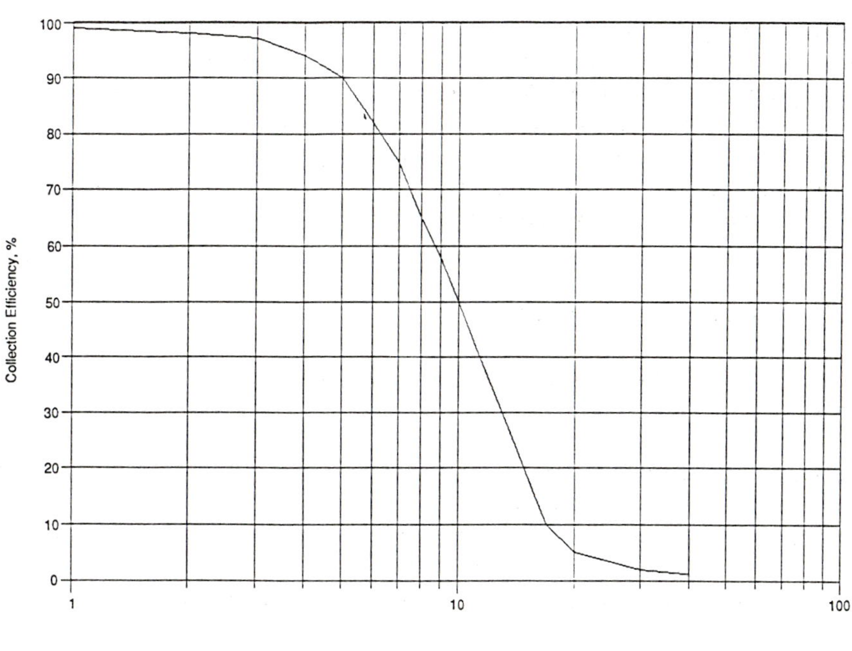

Figure 4.2 *Typical PM10 sampling head efficiency*

encompasses respirable particles but has a cut-off below the upper thoracic range. Where available, the PM10 inlet was used for the samplers in this evaluation programme because it is well tested and suitable for a range of analysers.

4.3.3 Description of the samplers

4.3.3.1 High volume samplers (Figure 4.3)

Conventional high volume samplers consist of a blower motor which draws air through a sampling head and a large glass fibre filter. At the end of the sampling period the mass of particulate matter collected on the filter is weighed manually. The particulate aerosol concentration in the atmosphere is given by:

$$C = \frac{W_2 - W_1}{F_c \times T}$$

where W_1, W_2 are the filter weights before and after exposure, T is the exposure time and F_c is the flow corrected to STP.

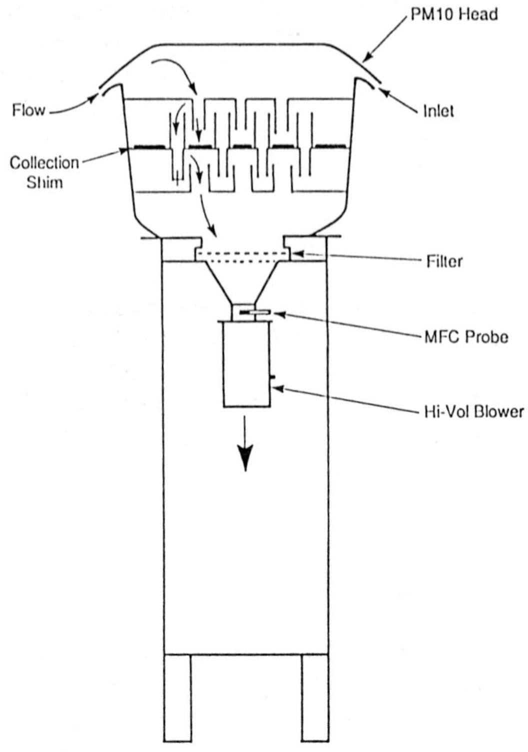

Figure 4.3 *High volume sampler*

General Metal Works /Sierra high volume sampler

The GMW high volume sampler consists of an air blower providing a flow of up to ~100 $m^3\,hr^{-1}$ through a large glass fibre filter. The flow through the system is monitored using a thermal sensor which is mounted above the filter holder. The sensor is connected to electronic circuitry which adjusts the blower motor to maintain a constant flow. The flow is recorded using a borden-gauge unit connected to an in-built circular chart. The sampler was operated with a Sierra Andersen 1200 PM10 sampling head at a flow of approximately 68 $m^3\,hr^{-1}$ (40 $ft^3\,min^{-1}$).

Ecotech 2000 high volume sampler

The Ecotech high volume sampler consists of an air blower providing a variable flow of up to ~80 $m^3\,hr^{-1}$ through a large glass fibre filter. The flow is adjusted to the required constant value and is maintained automatically by a pressure control valve. The set flow rate or total flow are continuously displayed. The sampler was operated with the same type of sampling head as used with the GMW, at a flow of 68 $m^3\,hr^{-1}$.

4.3.3.2 Tapered element oscillating microbalance (TEOM) (Figure 4.4)

The tapered element oscillating microbalance system consists of a filter cartridge, located on top of an oscillating crystal element, and an air pump controlled by a mass flow controller unit. As particulate matter accumulates on the filter, the frequency of vibration of the tapered crystal element changes. The particulate mass, M, is proportional to the change in frequency, of vibration of the crystal element, f_0, f_1:

$$M = k\left(\frac{1}{f_0^2} - \frac{1}{f_1^2}\right)$$

A computer unit continuously calculates the mass concentration from the particulate mass and the flow through the filter, k. The sampler was operated with an R & P PM10 sampling head which is closely modelled on the Sierra 246B head. The total flow through the sampling head of 1 $m^3\,hr^{-1}$ (16.7 $l\,min^{-1}$) is split into a sample flow of 3 $l\,min^{-1}$ through the filter, and a bypass pass flow of 13.7 $l\,min^{-1}$. Mass concentration and total filter loading were recorded as analogue voltages.

4.3.3.3 β-ray Environnement 100M (Figure 4.5)

The Environnement 100M consists of a beta source, a glass fibre band filter and a pump. A promethium 147 source is fitted into the sampling line directly above the filter, and a Geiger Muller type detector is used to measure the change in the absorption of beta radiation as particles are deposited. The absorption of the source radiation is proportional to the filter

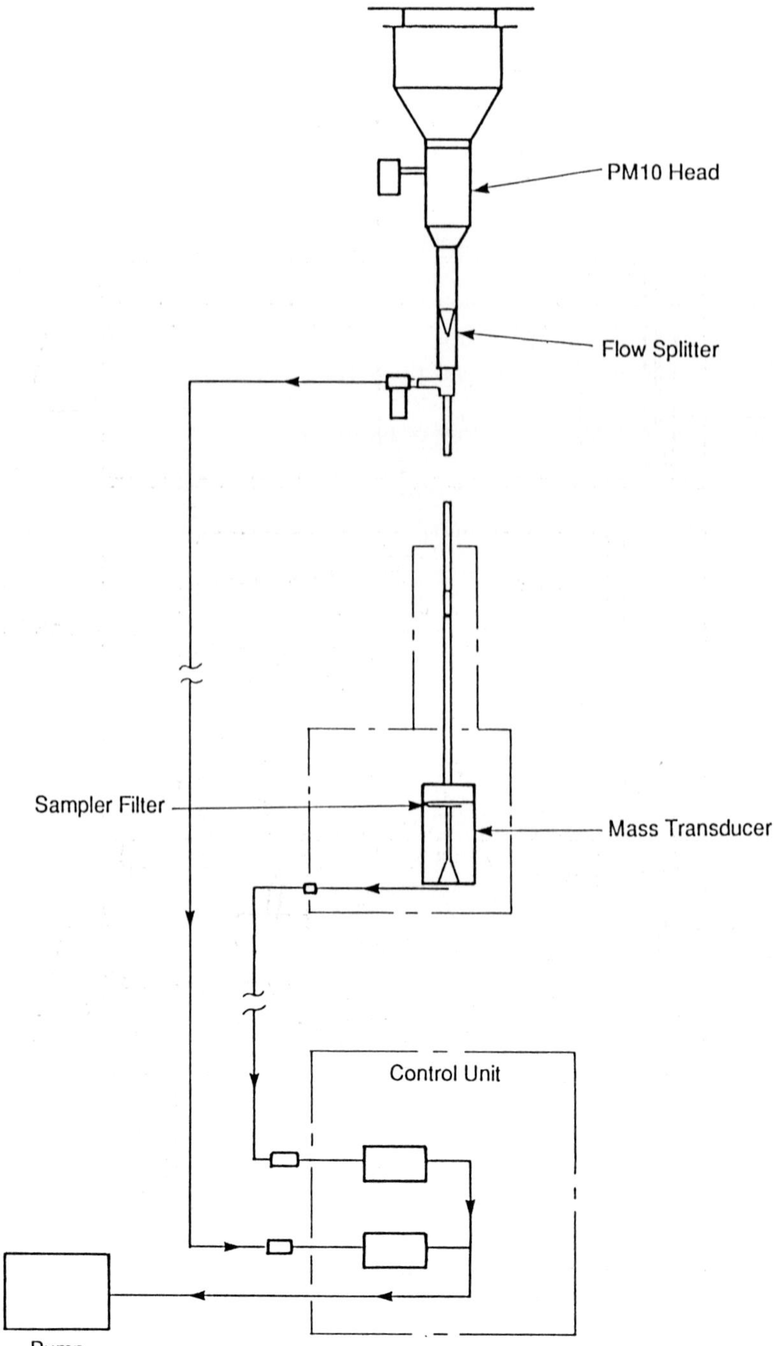

Figure 4.4 *Tapered element oscillating microbalance*

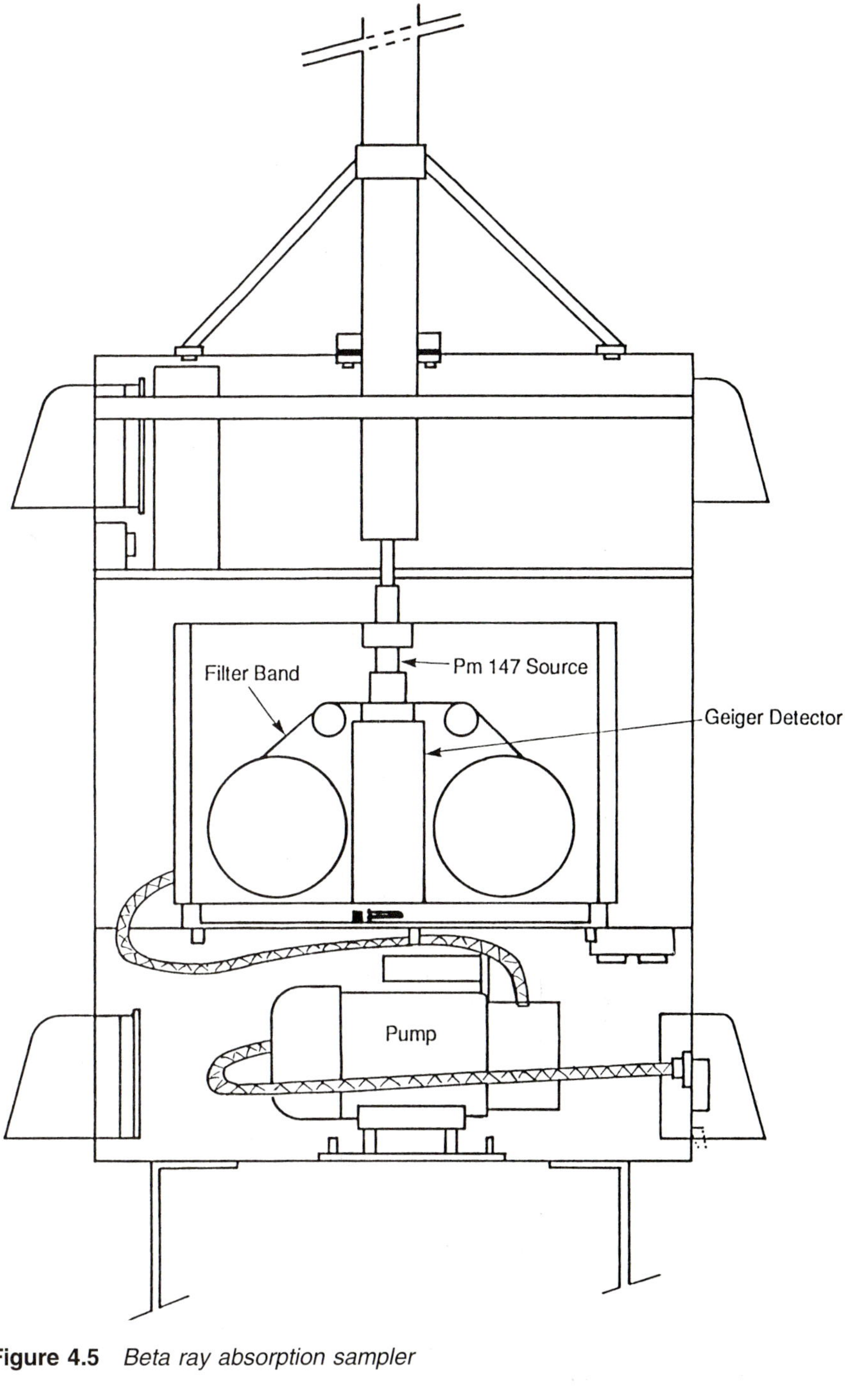

Figure 4.5 *Beta ray absorption sampler*

loading and is reasonably independent of chemical composition. The absorption, *I*, is therefore proportional to total mass:

$$\frac{I}{I_0} = e^{\mu m}$$

Where m is the mass loading, μ is the absorption coefficient and I_0 is the absorption for a blank filter.

The sampler can be set to operate from 1/2 up to 24-hour cycles with intermediate averages if selected. The sampler will automatically take a measurement and feed the filter tape on if the filter reaches a predetermined level. This ensures that the filter does not become overloaded. Any intermediate results are stored and the final average concentration is calculated automatically at the end of the normal sampling period. The analyser was operated with a Sierra Andersen 246B PM10 sampling head at a flow of 1 $m^3 hr^{-1}$.

4.3.3.4 Institute of Occupational Medicine sampler

This sampler was developed by the IOM at the request of the European Commission (Mark *et al.*, 1990) and was designed to accommodate two interchangeable sampling heads to match the ISO/CEN inhalable and thoracic conventions. It is intended as a possible reference sampler against which equivalent monitors can be compared. The sampler consists of a rotating sampling head connected to a pump and flow controller system. The air flow rate is maintained at approximately 70 $l\,min^{-1}$ using a pressure transducer operated compensation valve.

4.3.3.5 Digitel automatic high volume sampler

The Digitel DHA-80 is an automatic high volume sampler with a cassette filter loading system allowing up to 15 samples to be exposed. The sample flow is controlled by a rotameter, incorporating a light beam system, which detects the position of the float and regulates the blower motor to maintain a constant flow. Sampling times and filter changing times are programmable and can be operated directly or remotely. An in-built printer automatically records sample changing times, motor loading and other significant events. The sampler uses circular glass fibre or paper filters of 150 mm diameter.

4.4 Laboratory calibration

Laboratory tests were undertaken to calibrate the sample flow rates, and other parameters of the samplers as necessary, to correct any major faults and to establish familiarity with the measurement systems. It was not possible to control temperature and humidity within the test area, but temperatures and pressures were noted at the time of calibration of each sampler. The samplers were assembled according to the manufacturers instruction manuals, run for a few days to ensure correct operation, and then air flows and other relevant parameters were measured. The operation of the flow measuring devices used was referenced to a bubble flow meter traceable to an accredited standard.

The GMW/Sierra high-volume sampler was calibrated using a specially designed manometer (BGI calibrator) which enables calibration and adjustment of the built-in flow recorder chart. The BGI calibrator consists of a variable orifice, designed to fit in place of the filter mounting, and a water manometer. The pressure difference in inches of water is measured by the manometer as the orifice size is varied. The flow recorded on the GMW graph is plotted against the flow recorded on the calibration sheet for a given water pressure. Thus a calibration curve of actual versus indicated flow was obtained. The thermal flow controller was then used to set the sampler to the correct flow in the field.

The Ecotech 2000 high-volume was calibrated with the same BGI manometer calibrator as used for the GMW instrument and the same procedures were followed. The flow was adjusted to $\sim 68 \pm 1\ m^3 hr^{-1}$ to meet the requirements of the sampling head.

The calibration of the TEOM consisted of measuring the flows through the mass flow controllers and calculating the mass balance calibration constant. The sample flow was measured with a film flow meter, which has an accuracy of $\pm$ 2%, and the bypass flow was measured with a UGI dry gas meter which has an accuracy of $\pm$ 5%. The calibration constant was calculated by measuring the oscillation frequency of the balance with and without a filter in place. Leaks tests, as specified in the manual, were performed and no leaks were indicated. All calibration factors were within the manufacturer's specification.

The relation between mass concentration and voltage for the β-ray Environnement 100M is given in the manual as:

$$M = \frac{V_1 - 100}{2.5}$$

where V_1 is the absorption voltage for the exposed filter (mV) and 100 mV is the nominal voltage for a blank filter. The monitor is set to give an absorption voltage of $\sim 2.5\ mV \mu g^{-1}$; this is checked during calibration. The microprocessor of the analyser is set to adjust the analogue output to the flow rate and sampling time used. The flow measured at the outlet on the analyser, as received, was found to be high and required adjustment. The flow was adjusted to $16.7\ l\,min^{-1}$ as measured at the outlet of the sample pump. The absorption factor of the source was calculated using a standard foil and found to be as listed on the manufacturer's calibration sheet.

The (IOM) sampler was set up in the laboratory and allowed to stabilize for 24 hours. The flow was measured using the UGI gas meter and the flow controller gradually adjusted until a stable flow of approximately $70\ l\,min^{-1}$ was established.

4.5 Field trials

The GMW and Ecotech high volumes were set up in the field site and the flow rates adjusted to approximately 68 $m^3 hr^{-1}$. For the GMW, flow is monitored with a Borden gauge unit which records onto a circular paper chart. For the Ecotech, the flow is continually displayed on the panel LED display. The display was adjusted to zero with no sample flow, and the flow then set to 68.5 $m^3 hr^{-1}$ by means of the manometer calibrator. The monitors were then adjusted to display the set flow. The high volume samples were collected using standard 25 x 20 cm (8 x 10 in) glass fibre filters. Batches of clean glass fibre filters were removed from the packet, cut to size if necessary, and conditioned for at least 24 hours at 20 °C and 40% humidity in an environmentally controlled chamber. The filters were then inspected for damage, numbered, and weighed on a Mettler AC 100 balance housed in the chamber. At the end of the 24-hour sampling period, the exposed filters were removed and conditioned in the environmental chamber for 24 hours before reweighing. The instrument sampling heads were cleaned and the collector plates regreased before each new sample was collected.

The TEOM sensor units and sampling head were set up in the field in the purpose-built shelter supplied by the manufacturer. Data were collected as 15-minute average voltages on a central computer via a telemetry link. The sample filter cartridge was changed when the indicated sample loading exceeded 90%. The PM10 sampling head was cleaned at the same time. The bypass flow inlet filter was changed every second time the sample filter was changed.

The β-ray Environnement 100M was set up and the sample flow checked using a calibrated rotameter and adjusted to 16.7 $l\,min^{-1}$. The β-ray monitor can be set to operate at a range of cycling times between 1/2 and 24 hours. So a 2-hour sampling period was chosen, to provide a reasonable comparison with the TEOM monitor. The monitor sampled for 2 hours and then determined the absorption of the particulate matter collected. This absorption value is output as a constant analogue voltage during the next sampling period. Thus the analogue voltage always represents the previous two-hourly sample. Data were logged as 15-minute average voltages and collected on a central computer via a telemetry link. The PM10 sampling head was cleaned whenever the PM10 head used with the TEOM was cleaned.

The IOM sample flow was checked and set to 70 $l\,min^{-1}$ with a dry gas meter. Daily samples were taken using the inhalable sampling head fitted with 47-mm glass fibre filters.

The Digitel DHA-80 high volume sampler was set to operate at flow of 667 $l\,min^{-1}$ (40 $m^3 hr^{-1}$) as indicated by the rotameter float. The sampler was loaded with 14 filter samples which were changed automatically every 24 hours.

4.6 Operational performance

Both high volume samplers have operated reasonably well so far. The GMW sampler operated satisfactorily during the period of the tests with no major faults. One problem with this sampler, however, was that the blower unit brushes required replacement after approximately one month of operation. The main problem experienced with the Ecotech sampler was malfunctioning of the microprocessor LED digital display unit. This made it difficult to monitor the instrument flow and therefore, an assumed flow had to be used for some calculations.

There were some initial problems with the TEOM monitor tested, but it operated reasonably successfully in the field. The main operational problem found was that the control unit did not apparently indicate the correct filter loading. This problem is being investigated by the manufacturer. The filters should last for up to a month, although, during periods of high ambient concentrations, it was found that they became fully loaded in less than a week. The manufacturers have now modified the instrument so that, by altering the ratio of the sample flow to the bypass flow, a sample flow of 1 $l\,min^{-1}$ can be used. This modification, which is intended to increase filter life, is currently being tested.

The Environnement β-ray sampler operated very successfully in the field. Unfortunately, however, there was a major component failure and the monitor had to be returned to the manufacturer for repair which means that there is currently only a limited data set for this sampler.

There were operational problems, due to the flow control system, with the particular IOM monitor used, which is a prototype, and therefore only a few samples have currently been taken.

The Digitel DHA-80 high volume sampler operated successfully in the field.

4.7 Results

The following are some provisional data, which have not been fully ratified, and may be subject to change.

The PM10 high volume samplers were chosen as the reference standard because they represent the simplest sampling technique, and particle concentrations are determined independently of the sampler (by weighing). The two conventional high volume samplers gave reasonable agreement for the daily average concentrations as would be expected; see Figure 4.6.

Daily average data values for the TEOM are in reasonable agreement with the high volume samplers except for a couple of points where the high volume data were significantly higher than those for the TEOM; see Figure 4.7. The reasons for this are not fully understood at present, but it was noted

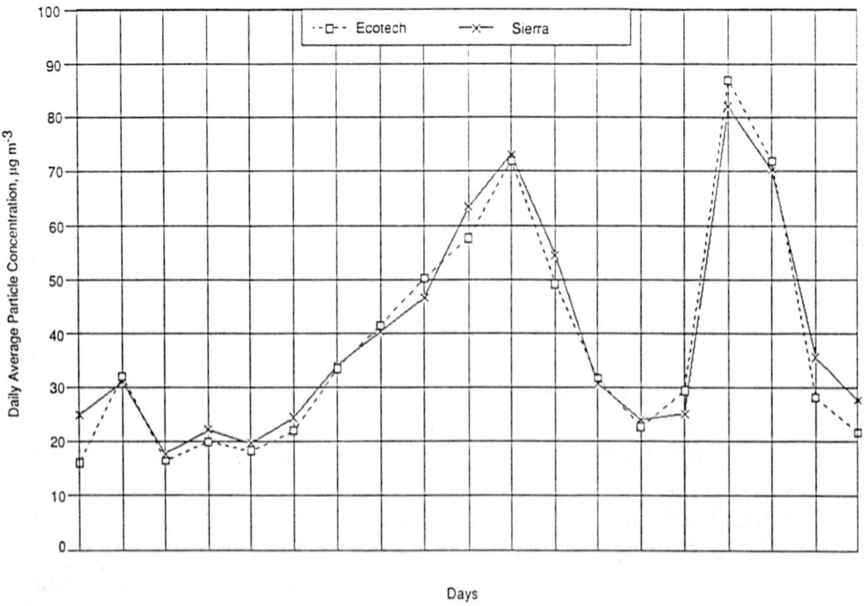

Figure 4.6 *Daily average particle concentration as measured by the Sierra and the Ecotech high volume samplers*

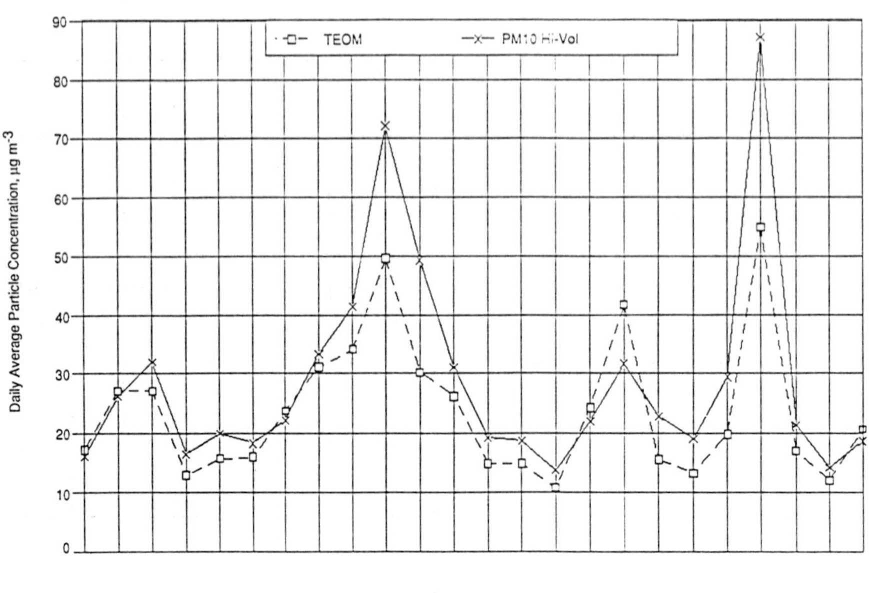

Figure 4.7 *Daily average particle concentrations measured by the TEOM compared with the results from the PM10*

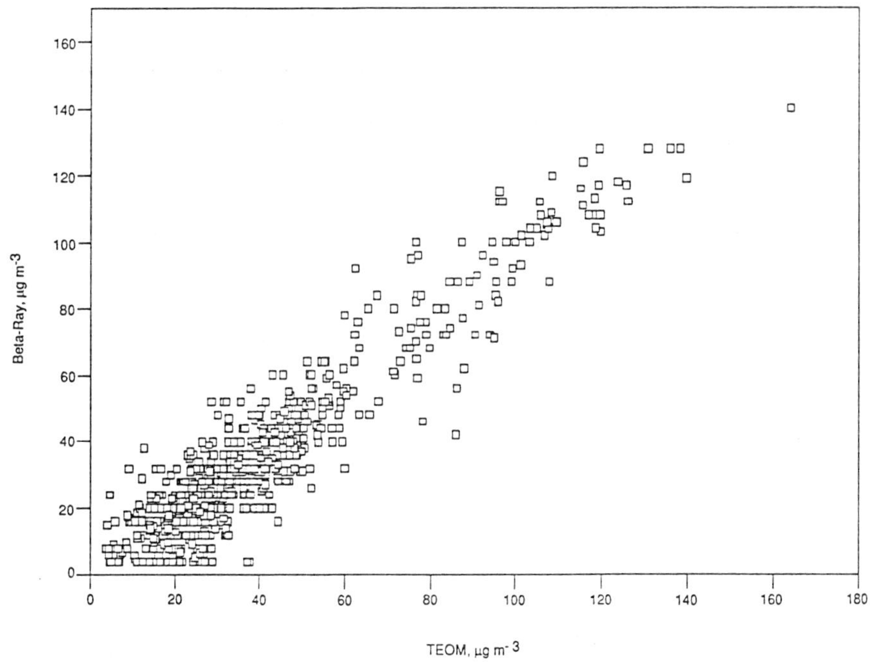

Figure 4.8 *Scatter plot of particle concentration measured by the TEOM against the concentration measure by the Beta-ray monitor*

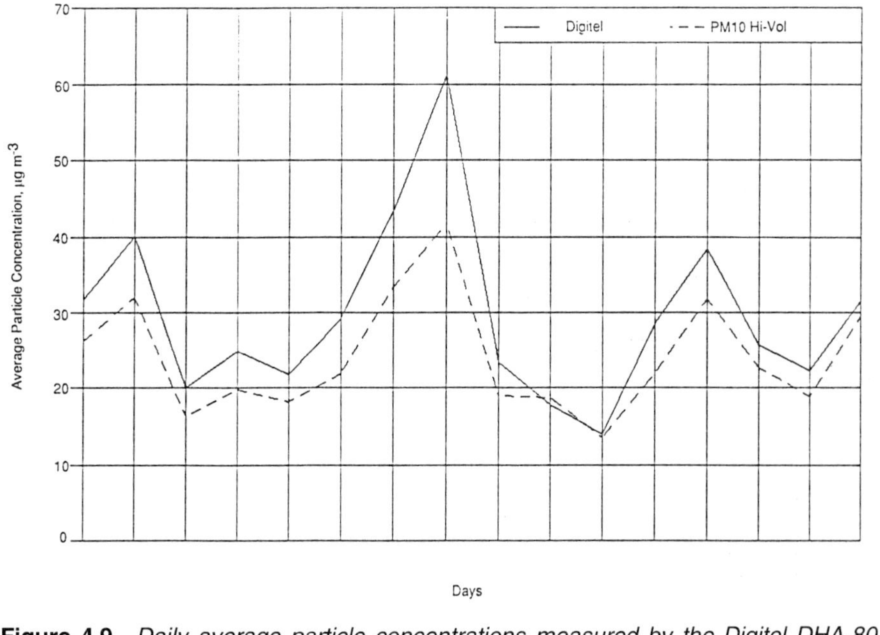

Figure 4.9 *Daily average particle concentrations measured by the Digitel DHA-80 compared with concentrations measured by the PM10 for the period February to April 1992*

that for the days in question, there were periods of dense damp fog which may have affected one or other of the samplers.

Because of the failure of the β-ray monitor, there are no reliable comparative data for the β-ray Environnement 100M and the high volume samplers. There are some intercomparison data with the TEOM, although the monitors were not in the west field site during the period when these data were collected. The data show good agreement except at low concentrations, see Figure 4.8. Recorded β-ray absorption values were at, or below, the nominal zero when ambient concentrations recorded by the TEOM were of the order of 10 $\mu g\,m^{-3}$. This indicates that a 2-hour exposure time for the β-ray was insufficient for low ambient concentrations.

The Digitel DHA-80 automatic high volume sampler showed the same trends as the PM10 high volume samplers, but with higher overall concentration, see Figure 4.9. This is consistent with the different sampling heads used on the two monitors. The Digitel also showed a period when concentrations were much higher than those recorded by the TEOM, again, for reasons which have to be investigated further.

There are very limited data from the IOM sampler and due to instrument problems they cannot be considered reliable. However, concentrations measured with the IOM sampler appear to be generally higher than those obtained with the PM10 samplers.

Table 4.1 Advantages and disadvantages of ambient aerosol samplers

Monitor type	*Advantage*	*Disadvantage*
Hi-Vol	Less expensive than continuous systems Simple to operate Provides sufficient sample for chemical analysis	Daily samples Problem to change filters at weekends and at remote locations Filters need conditioning and accurately weighing Requires experienced staff
TEOM	Continuous data enables rapid detection of pollution episodes Remote data collection Filter loading can be monitored	Expensive Requires regular replacement of sample and other filters Computer required for instrument control
Beta-ray	Remote data collection possible Requires minimal attention Filters will not overload	Lower sensitivity than TEOM Source covered by UK Health and Safety regulations making relocation difficult Finite source life (2–3 years)

4.8 Conclusion

This intercomparison study is not yet complete and no final conclusions have been reached about the various systems, however it is possible to consider their various advantages and disadvantages at this stage; these are summarized in Table 4.1

4.9 References

Barrett C F, Ralph M O and Upton S L (1985). Wind Tunnel Measurements of the Inlet Efficiency of Four Samplers of Suspended Particulate Material. Stevenage: Warren Spring Laboratory, Report No. LR 545 (AP)M.

VanOsdell D W (1991). Wind Tunnel Test Report No. 29A - Test of the Rupprecht and Pataschnick TEOM PM10 sampler inlet at 2 and 24 km/h. Research Triangle Institute, Triangle Park, North Carolina.

Mark D, Vincent J H, Aitken R J, Botham R A, Lynch G, van Elzakker B G, van der Meulen A and Zierock K H (1990). Measurement of Suspended Particulate Matter in the Ambient Atmosphere. Institute of Occupational Medicine Report, Edinburgh, Report No. TM/90/14.

— 5 —

Evaluation of ambient monitoring techniques for low levels of nitrogen dioxide

J Wilken, G S Driver and K J Stevenson

5.1 Introduction

There is increasing interest in the monitoring of low levels of nitrogen oxides, particularly within the framework of the United Nations Economic Commission for Europe (UNECE) NO_x Protocol, which seeks to limit emissions of oxides of nitrogen and to establish exposure limits (critical loads and critical levels) for the whole of Europe. (Throughout this paper, the term NO_x refers to the sum of nitric oxide (NO) and nitrogen dioxide (NO_2).) In order to determine deposition of NO_2 to soil and vegetation, accurate measurements of ambient levels in rural areas are required. Such data are also necessary for the verification of mathematical models developed to assess the problems of photochemical pollutants and in the determination of the long range transport of these pollutants. Current techniques used for measuring NO_2 in urban areas are generally not sensitive enough to detect the sub-ppb concentrations typically found in remote rural areas, and they are also known to suffer from interferences from other atmospheric species. High sensitivity NO_2 measurement techniques have recently become commercially available, however these have not yet been thoroughly evaluated for continuous operation in the field.

The overall objective of this project was therefore to evaluate and compare three new low-level NO_2 analysers, each employing a different measurement principle, with a view to selecting a preferred method for use in remote rural areas. The instruments under evaluation were a gas phase chemiluminescent analyser (Teco 42), a liquid phase luminol chemiluminescent NO_2 analyser (Luminox), and a differential optical absorption spectrometer (Opsis). These analysers were also compared with a conventional chemiluminescent analyser (ML 8840) of the type commonly used in urban monitoring networks. For convenience only, the analysers will be referred to by their manufacturer's trade name. Use of a trade name does not in any way imply endorsement of the product. The analysers were tested both in the laboratory and in the field to assess each instrument's measurement capability in terms of data quality and operational performance. For successful deployment in rural locations the instruments must have sufficient sensitivity to ensure that a high percentage of ambient data are above the instrument's detection limit, and also the instrument must be robust and reliable enough to operate in a remote unattended location for long periods of time.

This paper reports the results of two consecutive instrument intercomparisons, the first of which was based at Warren Spring Laboratory in Stevenage, and the second in a remote location in northern Scotland. A discussion of results is given, with regard to selecting the most appropriate and cost effective instrument for use in a rural NO_2 monitoring network.

5.2 Experimental

5.2.1 Description of the analysers

All the instruments under evaluation were operated to the manufacturers' recommendations, in their commercially available form and without any significant modifications. Table 5.1 contains information on the instruments manufacturers, UK suppliers and price at time of purchase.

Table 5.1 Instrument details

Instrument	*Manufacturer*	*UK supplier*	*Approx cost*
Teco 42	Thermo Environmental Instruments Inc, USA	Thermo Electron Ltd	£12,000 (1989/90)
(Teco 42S)	"	"	£16,500 (1991)
ML 8840	Monitor Labs. USA	Hobré Instruments (UK) Ltd	£10,000 (1991)
Luminox	Scintrex, Canada	Quantitech Ltd	£9,000 (1989/90)
Opsis	Opsis AB, Sweden	Environmental Technologies Ltd	£80,000 (1989/90)

5.2.1.1 Gas phase chemiluminescent analyser

The gas phase chemiluminescent analyser evaluated was a Thermo Environmental Instrument Model 42 analyser (Teco 42). This instrument, like conventional chemiluminescent analysers uses the chemiluminescent reaction between nitric oxide (NO) and ozone (O_3) to determine levels of total nitrogen oxides and nitric oxide (NO), with nitrogen dioxide (NO_2) being derived by subtraction. The Teco 42 is a microprocessor-controlled analyser with a single reaction chamber enabling NO and NO_x measurements to be made sequentially at 10-second intervals. Nitrogen dioxide must first be converted into NO before determination using this technique; this is accomplished by the use of a molybdenum converter heated to ~ 325°C. The major disadvantage of this technique arises from the use of this converter. In addition to NO_2, several other nitrogen containing species, e.g. peroxyacetyl nitrate (PAN), HNO_2, HNO_3 and organic nitrates, can also be reduced to NO in the converter. The net effect is an overestimation of ambient NO_2. The effect of these interferences in urban and more polluted rural areas is usually small in relation to the NO_2 concentrations being measured, however, for quantitative measurements in areas where the levels of NO_2 are low, they may be significant. To make corrections, quantitative determinations of the instrument's response to each interfering species, together with independent measurements of the concentration of each species in ambient air are required. Further corrections may be necessary if the conversion of NO_2 to NO is not 100% efficient, as is often the case. Thus, NO_x concentrations determined by this method must be considered to be not only the sum of NO and NO_2, but the total of all oxides of nitrogen species often referred to as NO_y.

The advantages, however, of the chemiluminescent technique are high sensitivity and linearity over several orders of magnitude. The detection limit for the Teco 42, the signal equivalent to twice the noise level, is 500 ppt (parts per trillion) as specified by the manufacturer. However, laboratory based tests at WSL have shown this to be as low as 180 ppt for the 300-second time constant setting (see Table 5.2).

Table 5.2 WSL tested and manufacturer's instrument specifications

Specification	*Teco 42*		*Luminox*		*Opsis*	
Manufacturer/WSL tested	Man.	WSL	Man.	WSL	Man.	WSL
Noise (ppt)	250	80	5	2	100	350
Minimum detection limit (ppt)	500	160	10	4	200	700
Zero Drift	0.5 ppb 24 hours	0.6 ppb 12 hours	–	0.014 ppb 12 hours	1 ppb 1 month	–

5.2.1.2 The liquid phase luminol NO_2 analyser

The liquid phase luminol NO_2 detector employed in this study was a model LMA-3 Luminox manufactured by Scintrex/Unisearch Limited in Canada. The detection technique is based on the chemiluminescent reaction between NO_2 and luminol (5-amino-2,3-dihydro-1,4-phthalazinedione) in solution. Unlike gas phase chemiluminescent analysers, this technique utilizes a gas/liquid phase interaction which is specific to NO_2, without the necessity to convert to NO. Sample air is drawn into the reaction cell, where any NO_2 in the sample reacts on the surface of a wick soaked with luminol solution. The intensity of the resulting chemiluminescence is recorded by a photomultiplier tube.

Advantages of this instrument are its small size and hence portability, its high sensitivity and selectivity, and fast response time (see Table 5.2). The minimum detection limit quoted by the manufacturer is 10 ppt and the response time is of the order of a few seconds. The disadvantages of this technique however, are its non-linear response to NO_2 concentrations below 3 ppb, and its positive response to other species which oxidize the luminol solution, in particular ozone and PAN. A chemical scrubbing device has been developed by the manufacturer in an attempt to eliminate the interference by O_3. However, other workers (Kelly *et al.*, 1990) have shown that this device also removed up to 50% of the NO_2 from the sample air. Hence, this device was not used on the Luminox analyser during this study. A significant disadvantage of this instrument is its need for ongoing maintenance. Some minor modifications were made to the Luminox analyser in order to make it more suitable for operation at a remote field site. Larger capacity luminol solution reservoirs were fitted, to allow about five months operation between refills. The peristaltic feed pump was replaced with a more durable metal geared version available from the manufacturer. This pump also incorporates a leak detector which automatically stops the pump in the event of luminol leakage from split peristaltic pump tubing.

In essence, careful calibration, routine maintenance and corrections for interferences and non-linearity are required in order to obtain accurate measurements of NO_2 when using this method.

5.2.1.3 The differential absorption spectrometer

The DOAS (differential optical absorption spectrometer) instrument used in this study was manufactured by the Swedish Company OPSIS AB.

DOAS is a remote sensing spectral analysis technique based on the principle of optical absorption. Different molecules absorb light at various characteristic wavelengths. The magnitude of the absorption is dependent on the concentration of the molecules present. The method is therefore capable of measuring several different pollutant gases conveniently in one system. The DOAS analyser used here was configured to measure O_3 and SO_2 as well as NO_2.

The basic instrumental arrangement consists of a light source, a receiver and a spectrometer/computer unit. The light source is in the form of a powerful high pressure xenon lamp which emits a beam of visible and UV light. The beam is directed over a path length of between 100 m and 1 km to a receiver. Here the light is collected and transmitted through a fibre optic cable to the spectrometer/analyser unit which must be housed in a sheltered location. Using this long-path arrangement, Opsis provides a measure of the average pollutant concentration over the length of the beam.

There are two configurations of lamp and receiver commonly used for Opsis. One is a double-ended system where the lamp and receiver are separated by a distance of several hundred metres. The alternative single-ended system utilizes a combined lamp and receiver at one end of the measurement path and a retroreflector at the other end. The retroreflector consists of a cubic arrangement of mirrors constructed to return the light beam along exactly the same path. The requirement for mains power at only one end of the system was considered an advantage, especially in remote rural locations, hence a single ended Opsis system was purchased for evaluation purposes.

Advantages of the Opsis technique include high selectivity and sensitivity to NO_2. The instrument used in this study was a high sensitivity analyser with a moveable diffraction grating, specifically designed for use in remote background monitoring. The specified minimum detection limit of this analyser is 0.4 $\mu g\,m^{-3}$ ($\equiv$ 0.2 ppb, parts per billion) for NO_2, although instrument tests at WSL showed a minimum detection limit of closer to 1 ppb (see Table 5.2). Long-path measurements allow a large measurement area to be covered with the advantage of no direct contact with the sample gas. A fast sampling time can be achieved, in the order of 0.5 seconds. However, if Opsis is being used to measure several gases, as was the case here, then measurements are made sequentially and at less frequent intervals. Different lamps, fibre optic cables and/or path lengths may be necessary for the determination of some pollutant species such as NO. The information content reported by the analyser is high, since for each pollutant concentration, the margin of error for each measurement (the deviation), and the level of light transmission are also recorded. This information was used to check analyser performance and is essential for data validation.

A disadvantage of this technique, as with all optical remote sensing techniques, is that the signal is lost or degraded during low visibility weather conditions such as fog.

5.2.1.4 Conventional gas phase chemiluminescent NO_x analyser

Routine measurements of ambient NO_x at the Stevenage site were made using a conventional Monitor Labs Model 8840 Nitrogen Oxides Analyser. This analyser, like the Teco 42, utilizes the gas phase reaction of NO with ozone to produce the excited NO_2^* molecules from which light is subsequently emitted and detected. The ML 8840 has two reaction chambers and photomultiplier tubes, enabling NO_x and NO to be measured simultaneously, unlike the Teco 42 which employs a switching mechanism. This instrument also uses a molybdenum converter and hence potentially suffers the same interference from other nitrogen species. The analyser's minimum detectable concentration of 2 ppb limits its use to urban NO_2 monitoring.

5.2.2 Description of sites

Two intercomparison exercises were conducted, the first was held at Warren Spring Laboratory in Stevenage, followed by a second more extensive exercise under more realistic field operating conditions at a remote site in Scotland. The site locations are described below.

5.2.2.1 Stevenage monitoring site

A preliminary evaluation of the instruments was undertaken at WSL in Stevenage. At the laboratory, technical support is readily available and data can be compared with that obtained from monitoring instruments routinely operated at the Stevenage site as part of the UK National Air Quality Monitoring Network. The Stevenage monitoring site is situated in a suburban area in close proximity ($\sim$75 m) to a major motorway.

The Luminox and Teco 42 analysers were housed on the monitoring site alongside the Monitor Labs Model 8840 chemiluminescent NO_x analyser used for routine urban monitoring. All analyser sample lines were connected to a common sampling manifold which was located at a height of 8 m above ground level. The Opsis transmitter/receiver head was attached to an outer wall of the laboratory, with the retroreflector mounted on a post at a distance of 122 m away. Figure 5.1 shows the location of Opsis with respect to the sampling inlet of the other analysers housed on the Stevenage site. All instruments were operated at the site between January and June 1990 although the Opsis analyser remained in operation until September.

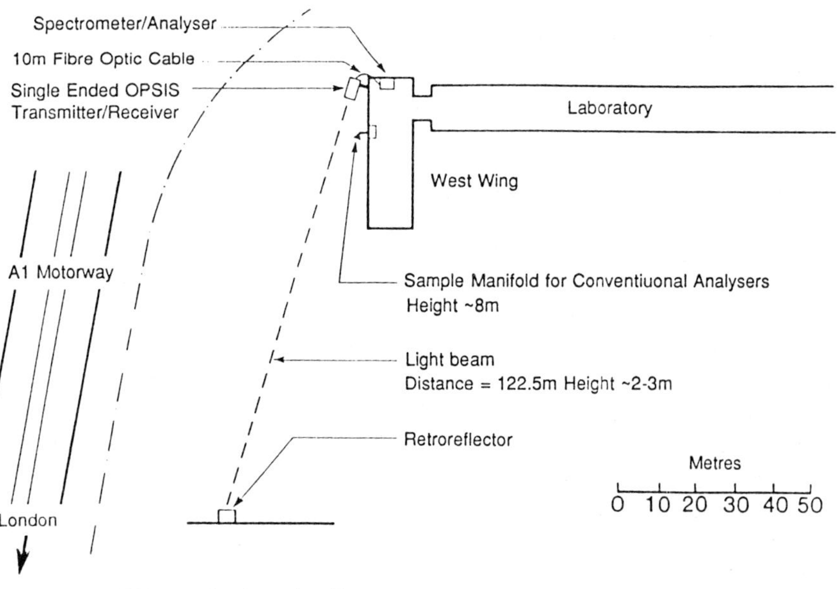

Figure 5.1 *WSL monitoring site, Stevenage*

5.2.2.2 Strath Vaich monitoring site

WSL have an established air pollution monitoring site in a remote part of northern Scotland, near to the Strath Vaich Dam. The site is located on exposed and slightly elevated (270 m above sea level) moorland. There are negligible sources of local pollution and the site is prone to prevailing westerly winds that bring in clean air from the Atlantic. Typical annual average NO_x concentrations are around 1 ppb making this a suitable remote background location at which to install and test the monitoring capabilities of the new instrumentation. The intercomparison exercise ran between November 1990 and August 1991, however, the Luminox instrument was removed from March to June for use elsewhere.

The Luminox and Teco 42 instruments were installed in the site hut and connected to the existing sample manifold. The Opsis transmitter/receiver head was mounted on a specially constructed solid steel platform alongside the monitoring hut. The retroreflector was mounted on the wall of a tower by the dam, at a distance of ~300 m away from the Opsis head (see Figure 5.2). Once Opsis was installed, the Ordnance Survey were commissioned to accurately measure the path length.

5.2.3 Data acquisition

The Teco 42 and Luminox analysers were interfaced with an existing data telemetry and processing system. The analysers' analogue outputs were connected to a data logger which recorded and averaged 15 minutes of instantaneous readings taken at 10-second intervals. Further data reduction and data scaling were accomplished using specially developed software. This included calculating pollutant concentrations from the raw 15-minute averages using the relevant calibration factors of each instrument, and in the

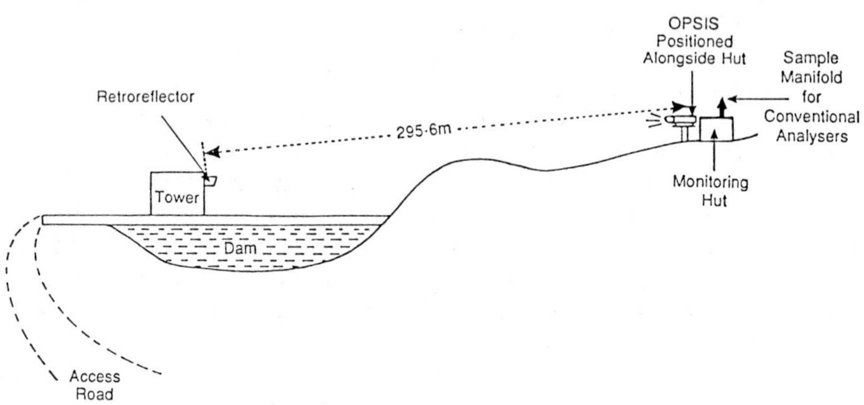

Figure 5.2 *Strath Vaich monitoring site, Scotland*

case of the Luminox instrument, applying any necessary linearization factors.

During the Stevenage intercomparison, the Luminox analyser was operated on a running range of 100 ppb NO_2. For determination of NO_x and NO, the Teco 42 and ML 8840 analysers were operated on running ranges of 1 and 2 ppm respectively.

At Strath Vaich, all the analysers were operated on a range of 100 ppb for NO_2, NO_x and NO, corresponding to a full-scale output of 1 V. The site logger was configured to accept a 1 V input on each channel with a maximum resolution of 1 mV, hence, the minimum detectable change in concentration was limited by the logger resolution to $\sim$0.1 ppb for the Teco 42 analyser and 0.4 ppb for the Luminox analyser.

The Opsis instrument has its own data logging and storage software. Instantaneous NO_2 measurement values together with the corresponding values of the standard deviation and light level, were automatically stored on its own computer hard disk. The Opsis internal modem and customized communications software (VIS/COM 500) were used for data transmission.

All measurement values recorded by Opsis were stored in units of $\mu g\, m^{-3}$ corrected to 273 K and 101.13 Pa. These concentrations were later converted to ppb using the correction factor of 2.05 $\mu g\, m^{-3}$ NO_2 = 1 ppb.

5.2.4 Instrument calibration methodologies

Instrument calibrations were conducted at approximately monthly intervals throughout the field intercomparisons tests. Calibrations were performed on site using both NO and NO_2 transfer standard cylinders which had been previously checked against primary standards. Low-level NO_2 cylinders were both difficult to obtain from manufacturers, and proved unreliable due to cylinder instability. Large discrepancies between the manufacturer's calibration certificates and WSL's checks against primary standards were found. These cylinders were therefore recalibrated either directly before or after being used in the field. Zero air was generated on site by passing air through chemical scrubbing agents.

Calibration requirements and methodologies were different for each analyser. Single-point calibrations were performed on the Teco 42, by administering NO and NO_2 span gases from transfer standard cylinders of known concentrations, $\sim$500 ppb, to the sample inlet of the analyser. Calibration of the Luminox analyser was more complicated for the following reasons. Firstly, the Luminox analyser has a maximum full-scale range of only 200 ppb, therefore special low-level transfer cylinders of $<$100 ppb NO_2 were required for performing single point calibrations. Secondly, in order to accurately quantify the non-linear response of the Luminox analyser, a multi-point calibration down to sub-ppb levels had to be performed on site. This was accomplished by dynamic dilution of a 500 ppb NO_2 cylinder with zero air. Mass flow controllers and a soap film meter were used to accurately control and measure the flows, thus enabling NO_2 concentrations in the range 0.1 to 60 ppb to be generated on site. The Luminox zero response was obtained either from sampling scrubbed room air, or by switching off the Luminox air pump, and allowing the chemiluminescent signal to decay away. The zero response in either case was identical.

The Luminox response in the linear region was defined by the slope and intercept obtained from the linear 'best fit' to all data points above 3 ppb obtained by multi-point calibrations. The non-linear response below 3 ppb was defined by fitting a series of linear approximations to the data points in this region. Linearization of the 15-minute average NO_2 data, using a combination of the slopes and intercepts calculated from the calibration curve, was performed automatically using data processing software. Because of the complexity involved, multi-point calibrations were performed on site only at the start and end of the field intercomparison exercises, however, the Luminox response in the linear region was checked at regular intervals using a low-level NO_2 transfer cylinder.

The Opsis analyser is already calibrated by the manufacturer for each individual gas species being determined. The manufacturer's calibration factors are stored in the Opsis computer and used to automatically scale the data, hence measurements are presented directly in concentration units. It is possible to check these calibration factors by performing zero and span gas calibrations at regular intervals. Opsis calibration techniques are fairly complex and require the use of additional equipment such as a zero lamp and corresponding power supply, various calibration cells, extra fibre optic cables and very high concentration span gas cylinders. It is recommended by the manufacturer that zero point calibrations should be performed on a monthly basis when using Opsis to measure ambient concentrations at very low levels, i.e. near its detection limit. A zero point calibration was therefore performed on site by directly connecting a fibre optic cable from a zero lamp source to the spectrometer/analyser. Since there was no absorption due to pollutant species in the light path, the spectrum recorded by the analyser corresponded to a zero reading. Each new zero reading was stored in the computer and used to calculate the ambient concentrations. A zero calibration was considered acceptable, as defined by the manufacturers, if the value of a measurement made with only the zero lamp in place was close to zero, that is within a few $\mu g\,m^{-3}$, and not greater than three times the standard deviation.

In practice, on-site evaluation of the analyser's zero point required several attempts before an acceptable zero calibration was recorded. As seen in Section 5.3.2.4, ambient NO_2 data at Strath Vaich derived using these 'acceptable' zero calibrations, often resulted in large negative NO_2 concentrations being reported. These negative concentrations indicate either an interference in the NO_2 spectrum recorded, or subtraction of an acceptable zero spectrum which was in fact unsatisfactory. Careful inspection of the ambient data collected over several days following a zero point calibration proved to be the only effective way of determining the success of the zero point calibration when measuring very low levels of pollutant.

Performing an Opsis span calibration involved determining the light absorption due to the flow of a gas of known concentration through a calibration cell of known path length. The product of the span gas concentration and the length of the calibration cell, must be of the same order of magnitude as the pollutant concentrations in ambient air over a path length of ~500 m. Thus, for a calibration cell of path length 0.01 m, a span gas concentration of approximately 1000 ppm was used.

Several attempts to perform NO_2 span gas calibrations in the field gave unsatisfactory results in terms of both accuracy and reproducibility. The span gas concentration as determined by Opsis showed large variations of ~15% between individual readings during a calibration, and ~20% between separate calibration events. As a result no effort was made to scale the data using these calibration results and all Opsis data presented in this report are based on the manufacturer's original span calibration only.

Daily automatic zero and span checks, using permeation tube calibrators, were provided for both the Teco 42 and Luminox analysers. This facility provided an overall performance check and rapid fault detection system for the analysers. Similar autocalibration facilities for the Opsis analyser have not yet been commercially developed.

5.2.5 Data processing and validation

In order to obtain accurate and precise ambient measurements, the raw data from the analysers must be processed and validated. For the Teco 42 and Luminox instruments this involved using existing software to calculate the pollutant concentrations from the raw 15-minute averaged instrument voltages. The zero and span factors used in the calculations were determined from the site calibrations or from the autocalibration zero. Other calculations, including the linearization of the Luminox response and the difference calculations for data channels, i.e. $NO_2 = (NO_x - NO)$, were also performed. In order to reduce the data sets to a manageable size, hourly average concentrations were used for the intercomparison. For the Teco 42 and Luminox data, hourly average pollutant concentrations were calculated for those hours comprising not less than three 15-minute averages.

The Opsis instrument was configured such that approximately twelve ambient NO_2 measurements were collected in 1 hour. Opsis automatically filters out individual measurements which are not to be used in calculating the hourly average values, according to the following criteria:

1. data flagged to indicate that the spectrum had not been evaluated successfully;
2. negative concentrations with values greater than twice the standard deviation.

The minimum number of measurements used to determine the hourly average value is not specified by Opsis, therefore, in theory, only one

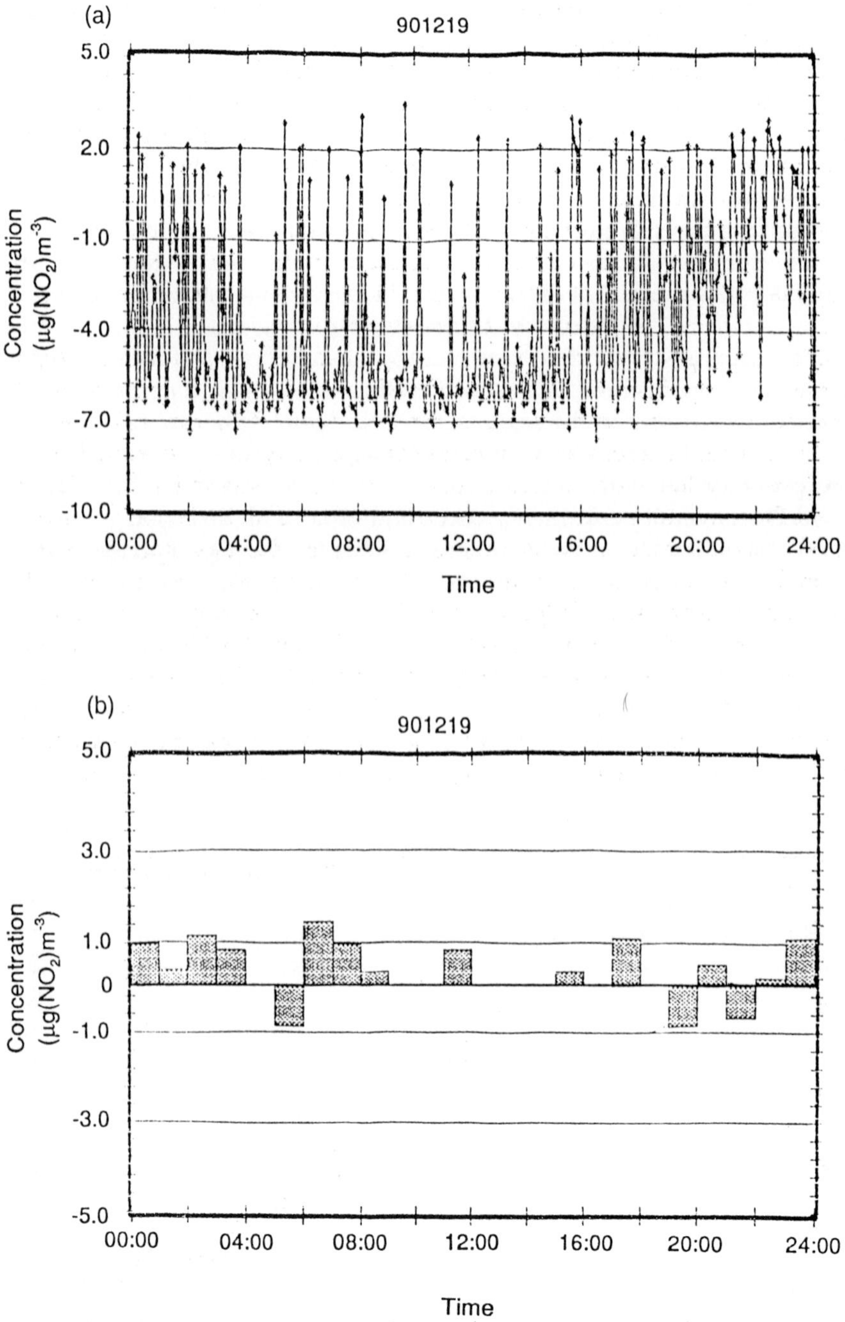

Figure 5.3 *(a) Instantaneous NO_2 Opsis data, integration time = 0 minutes.*
(b) Hourly average Opsis NO_2 data, integration time = 60 minutes.

measurement corresponding to a 2-minute ambient sampling period could be used to determine the hourly average. Figure 5.3 shows the effect on the 24-hour data capture when collecting hourly average Opsis data as opposed to all measurements. Several data gaps appear in the hourly average data set where very negative measurements have been automatically rejected. It is not therefore, possible to correct the ambient data based on the lowest hourly average ambient data recorded, since the largest negative values, corresponding to the lowest ambient concentrations, will have been rejected from the hourly average determinations. This highlights the importance of being able to discern and obtain an acceptable zero point calibration, thereby avoiding the consequence of negative concentrations.

In addition to pollutant concentrations, Opsis also records the level of light transmission together with an estimate of the uncertainty (or deviation) associated with each ambient measurement. This additional information must be carefully inspected in order to eliminate any data associated with periods of very low light levels or high deviations. Validation criteria, in terms of the maximum permissable deviation and minimum light level, were selected according to several factors including the gas species being measured, the ambient concentrations being recorded, and the overall operational performance of Opsis. Figure 5.4 shows the range of light levels and deviations recorded by Opsis at Stevenage during May 1990. In May, the light levels were seen to be generally high and stable, i.e. between 50 and 65%, and the deviations small, i.e. 0.6-1 ppb. On average, the deviations were less than 10% of the ambient concentrations recorded. For comparison, data recorded at Strath Vaich in February 1991 are also plotted in Figure 5.4.

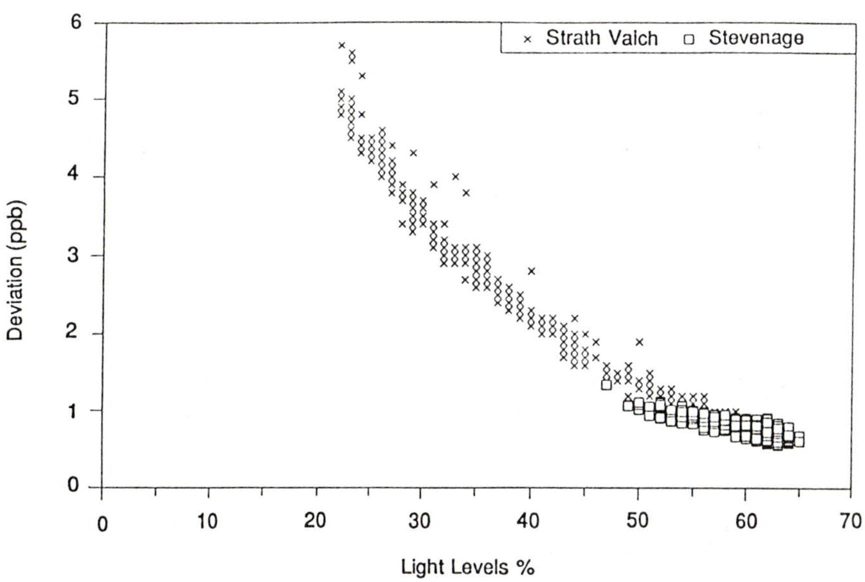

Figure 5.4 *Comparison of deviation against light level at Stevenage, May 1990, and at Strath Vaich, February 1991.*

During this period the Opsis performance parameters were significantly different, showing reduced and fluctuating light levels in the range 20-60% and larger deviations of 1 to 6 ppb. The ambient NO_2 concentrations were however, very low in comparison to the magnitude of the deviations. Hence, average deviations of over 165% of the ambient concentrations were often recorded, indicating a high degree of uncertainty associated with each measurement. Figure 5.4 clearly illustrates that careful inspection of the data is required in order to select appropriate data criteria. For the Stevenage intercomparison, all Opsis data with corresponding light levels above 50% and deviations below 1.5 ppb were considered valid. At Strath Vaich however, a less stringent set of validation criteria had to be adopted, to accommodate the overall poorer operational performance of Opsis, yet still exclude any very poor measurements; in this case light levels above 30% and deviations below 3 ppb were considered valid.

5.2.6 Accuracy and precision of measurements

The accuracy and precision of the measurements are important factors in determining the level of agreement between the analysers used in the intercomparison studies. The absolute accuracy of a measurement depends not only on the inherent characteristics of the instrument, such as response stability, but also on many other parameters encompassing the entire measurement chain. Factors to be considered when determining the overall accuracy are the efficacy of the primary calibration methods, the resultant accuracy of the transfer standards and the frequency and method of field calibration. The accuracy of the WSL primary calibration methods for NO and NO_2 have been shown to be of the order of 4.8% for the static dilution method and 6.6% for the transfer cylinders (Eaton and Stevenson, 1993).

The additional uncertainty in the analyser calibration due to instrument drift can be estimated from the variation (2σ) in instrument response between individual site calibrations. For the Teco 42 this was found to be 7.4%. The overall accuracy of the Teco 42 measurements can thus be derived from the above estimates in the equation:

$$\begin{aligned}\text{Accuracy} &= \{(4.8)^2 + (6.6)^2 + (7.4)^2\}^{0.5} \\ &= \pm 11.0\%\end{aligned}$$

For NO_2 measurements, a further source of error arising from the inefficiency of the molybdenum converter must be considered. The converter efficiency of the Teco 42 was tested and found to be 98%, resulting in an additional error of –2%, and an overall instrument accuracy for NO_2 measurements of +11.0 to –11.2%.

A different calibration method was used for the Luminox analyser. Therefore an additional uncertainty of 3.7%, associated with performing a dynamic dilution of the NO_2 calibration cylinder, must be added to the uncertainties associated with the primary and transfer standards. The response curve used to scale the Luminox data was determined from a combination of the multi-point and single-point field calibrations. The uncertainty arising from variations in the field calibrations can therefore be estimated from the error associated with the line of best fit to all the calibration data. In the linear response region of the calibration curve, the error in determining the slope was estimated to be 5.3%, however, in the non-linear region, below 3 ppb, the fewer and more widely scattered data points gave rise to a larger uncertainty estimate of 15%. The estimated overall accuracy of the Luminox NO_2 measurements above 3 ppb was therefore 10.4%, and for measurements at levels below 3 ppb the accuracy was 17.5%. For simplicity, a single estimate of the accuracy of the Luminox measurements of 14% has been arbitrarily selected from the means of the above uncertainties. Note however that for both the Teco 42 and Luminox measurements, the possible uncertainty due to interfering species has not been formally evaluated and hence cannot be included in the accuracy estimate.

The overall accuracy of the Opsis measurements could not be similarly derived, since the span calibrations performed on site were inconsistent and therefore not used to scale the data. The Opsis analyser does however, give an estimate of the uncertainty (or deviation) for each individual measurement and this was, on average, ~10% for ambient NO_2 measurements at Stevenage. When determining very low levels of ambient NO_2 at Strath Vaich the errors were significantly larger. Average deviations of ~3 ppb were recorded which were typically of the same order of magnitude or higher than the ambient NO_2 concentrations.

The precision of an analyser, is primarily related to the instrument stability, and is defined here in terms of the repeatability of measured concentrations at ambient levels. A measure of instrument precision can be obtained from the long-term variation in base line response during operation in the field. For the Teco 42 and Luminox analysers, precision at low levels was determined from the variation in response to the daily autocalibration zeros over one month; any instability in the zero gas source is also necessarily included in the determination. The precision of the Teco 42 and Luminox measurements were found to be ± 1 and ± 0.8 ppb respectively, at the 2σ level.

Since autocalibration facilities were not available for the Opsis, a similar calculation of precision cannot be performed, however a monthly base line stability of $\pm 2\ \mu g\, m^{-3}$ (1 ppb) NO_2 is specified by the manufacturer.

The ML 8840 chemiluminescent NO_x analyser as used in the UK NO_2 Directive monitoring network has been shown to have an accuracy of +10.4% to −11.1%, and a precision of ± 3.5 ppb for NO_2 measurements (Bower *et al.*, 1989). Table 5.3 shows a summary of the estimated accuracy and precision of the NO_2 measurements for each analyser.

Table 5.3 Estimated accuracy and precision of NO_2 analysers

Analyser	*Accuracy*	*Precision*
Teco 42	+ 11.0 → − 11.2%	± 1 ppb
ML 8840	+ 10.4 → − 11.1%	± 3.5 ppb
Luminox	± 14% (Avg)	± 0.8 ppb
Opsis	± 10%	± 1 ppb

5.3 Results and discussion

The aims of this project were to evaluate the operational performance of the analysers, in terms of suitability for use at remote unattended sites, and to compare the actual air quality measurements obtained.

Operational performance will be discussed first, followed by a detailed intercomparison of the results from the various instruments.

5.3.1 Operational performance of the analysers

5.3.1.1 Operational performance at Stevenage

The analysers were selected for evaluation on the basis that they were all commercially available instruments intended for routine operation rather than research or prototype instruments.

During the initial evaluation at Stevenage the Teco 42 proved to be the most reliable of all the analysers and required the least maintenance/modifications. However, an intermittent fault with the microprocessor control unit was observed. Occasionally, following a power cut at the site, the Teco 42 would fail to restart again properly unless the analyser was manually switched off and on. This problem was eliminated by plugging a mains power surge protector unit in line with the analyser. Although easily rectified, the implications of this problem are serious when operating the analyser in a remote location, where mains power is often unstable.

The majority of operational problems displayed by the Luminox analyser were due to the peristaltic pump. The original pump supplied with the analyser failed on several occasions and was replaced for a more durable metal geared version. When left to operate unattended the Luminox analyser ran continuously for varying periods of time ranging from one month to a few days, before damage to the peristaltic pump tubing resulted in instrument breakdown. On average, the maximum period of unattended operation was 1-2 weeks, after which the split tubing had to be replaced. It was therefore necessary for a local site operator to make routine fortnightly visits to change the luminol pump tubing. Alternatively, the need for frequent routine maintenance could potentially be eliminated by installing a better quality peristaltic pump which inflicted less wear and tear on the tubing during operation or a different pumping system.

There were a wide variety of problems with the Opsis analyser when it was installed at Stevenage. These problems generally fell into the following categories: (1) optics and beam alignment; (2) computer hardware/software; and (3) other component faults. Initial Opsis data showed large diurnal fluctuations in light levels resulting in either poor quality data, or no data at all. Several attempts were made over a period of 3 months to realign and/or refocus the beam, including changing mirrors, but with limited success. It was finally deduced that the heating effect from the lamp was causing a torsion along the optical bench inside the transmitter/receiver head. This meant that a complete redesign of the optical components inside the Opsis head was necessary. The redesigned version of the transmitter/receiver head was subsequently installed at the site in April, however an improvement in the stability of the light levels was not seen until a few weeks later when the 300 W lamp was exchanged for a smaller 150 W version. The major implication of these problems was realization that all the Opsis data would have to be carefully screened in conjunction with the corresponding deviation and light level data, to ensure valid measurements were being made.

Although dramatic improvements in the overall performance of the analyser were achieved with these modifications, other intermittent problems prevented any extended period of continuous ambient measurements being made. In most cases the problems that developed could only be fixed by the experienced engineers from Opsis in Sweden.

In conclusion, it was apparent that the single-ended Opsis system under evaluation was still in its 'prototype' stage of development, requiring frequent maintenance/modifications and careful inspection of the data in order to ensure satisfactory performance.

5.3.1.2 Operational performance at Strath Vaich

The Opsis and Teco 42 were installed at Strath Vaich in October 1990, and were operated at the site until August 1991. The Luminox analyser was not installed until late November 1990 and was temporarily removed from the site in mid March, for use in a European NO_2 intercomparison programme in Germany. An important measure of an instrument's performance and reliability in the field is given by the data capture statistics recorded during the intercomparison study. Data capture figures for the Teco 42 and Opsis instruments were 82% and 27% respectively, for the 5-month period 1/11/90 to 31/3/91. Data capture for the Luminox analyser for period 22/11/90 to 15/3/91 was 71%.

During 10 months in operation at Strath Vaich, the Teco 42 showed no significant instrument malfunctions. A small (7%) decrease in instrument sensitivity was observed between calibrations, and this was found to be caused by a partially blocked internal zero/span solenoid valve. The valve was bypassed without interruption to ambient measurements and was later replaced. Lost data was mainly as a result of logger/telemetry problems

(~10%), together with site visits, calibrations and operator error (~8%). The Teco 42 again proved to be the most reliable of all the analysers tested, and appeared to be suitably engineered for long-term unattended operation in the field.

The Luminox analyser continued to demonstrate problems with the peristaltic pump tubing, even though it was replaced every fortnight. The analyser's air pump ceased operation as a result of being flooded with solution. This occurred due to problems of electrical interference in the mains power at the site, which initiated a series of continuous 'backflush' cycles, hence flooding the system. Approximately 20% of lost data was due to problems with pump tubing and a further 10% due to the logger/telemetry faults. Clearly, the Luminox instrument requires regular routine maintenance, making it unsuitable for use in an unattended location. However, the Luminox analyser has the benefit of being of simple operational design, thus making it possible for non-technical staff to undertake most routine maintenance and solve commonly occurring problems.

Many of the Opsis problems experienced during the Stevenage trials recurred at Strath Vaich. For example, low and fluctuating light levels continued throughout the winter period, resulting in intermittent periods of good and poor data quality.

On several occasions, the computer 'hung-up' and it was necessary for the local site operator to manually re-boot the system, however, power supply problems at the site may have been responsible in some cases. A variety of new problems arising from faults with the components of the analyser/spectrometer itself were also observed. Again many of the problems required the expertise of an Opsis engineer both to trouble-shoot and repair.

Opsis data capture over 5 months (November – March) prior to data validation was 39%, however once validated, eliminating very poor quality data with high deviations and low light levels (see Section 5.2.4), the data capture figure was reduced to 27%. Of the 73% of lost data, approximately half were due to the problems with beam misalignment and the rest were equally attributable to problems with the computer and other components.

The remoteness of the site, and hence reduced speed at which problems could be fixed, is also reflected in the low data capture figure. However, on a more fundamental level, the low data capture figure reflects the unsuitability of this complex 'prototype' analyser for long-term use in a remote unattended location.

5.3.2 Intercomparison of results

5.3.2.1 Data intercomparison methodology

Analyses were performed on the Teco 42, Luminox and Opsis data sets in order to examine how well a given pair of instruments tracked the changes in ambient NO_2 concentrations, and to compare the differences between the measurements. To quantify the level of agreement between given pairs of instruments, a linear regression analysis of one instrument's results against another was performed. The parameters obtained from these analyses include the slope of the regression line (m), the intercept (c), the correlation coefficient (R), and the error σ_m associated with the slope and the standard error (σ_y) of the y estimate. The correlation coefficient (R), provides an indication as to whether the two instruments are measuring the same parameter; a value of 1 would indicate that the methods are perfectly correlated. The intercept provides a measure of any constant offset existing between the respective data sets, while the slope provides a measure of any proportional bias existing between the data sets, dependent on the magnitude of the NO_2 concentration such as calibration differences, or interference effects that are dependent on NO_2 concentrations.

The standard deviation ($\pm 2\sigma$) criterion was used as a measure of the significance of the difference between the results from respective instruments. If the regression slope $\pm\ 2\sigma_m$ does not include the slope of one, then it can be concluded that the indicated bias is statistically significant at the 5% level. The standard error of the y estimate (σ_y) gives an estimate of the standard deviation of the distribution of residuals about the regression line, and hence serves as a measure of precision. For example, the greater the value of σ_y the larger the scatter about the line and the less precise the measurement technique relative to the chosen reference technique.

In order to make a comparison between the analysers, one instrument must be chosen as the reference method. Since the Teco 42 analyser used the more conventional chemiluminescent technique, it was chosen as the reference system against which to compare the other results. This does not however, imply that the results from the Teco 42 were necessarily better than those from the other analysers.

5.3.2.2 Results of the intercomparison at Stevenage

The Teco 42, Luminox and Opsis analysers were installed at the Stevenage site in January 1990, however during the first few months of operation several teething problems were experienced with the all the analysers, see Section 5.3.1.1. For this reason the intercomparison analyses were performed on just 2 months data (58 days during April and May) when the analysers were considered to be operating to the best of their ability and the data capture was sufficiently high in all cases (>90%). Data from the analysers were also compared with a ML 8840 chemiluminescent NO_x analyser, already in use at the Stevenage site for urban NO_2 monitoring.

The time series of hourly average NO_2 concentrations measured by each of the Teco 42, Luminox, Opsis and ML 8840 analysers, during April – May 1990 are shown in Figure 5.6. A more detailed graph of just six days of simultaneous measurements (May 23-28) is shown Figure 5.5. Generally the instrument responses track very well during periods of both high and low concentration, although the ML 8840 data appear overall, systematically

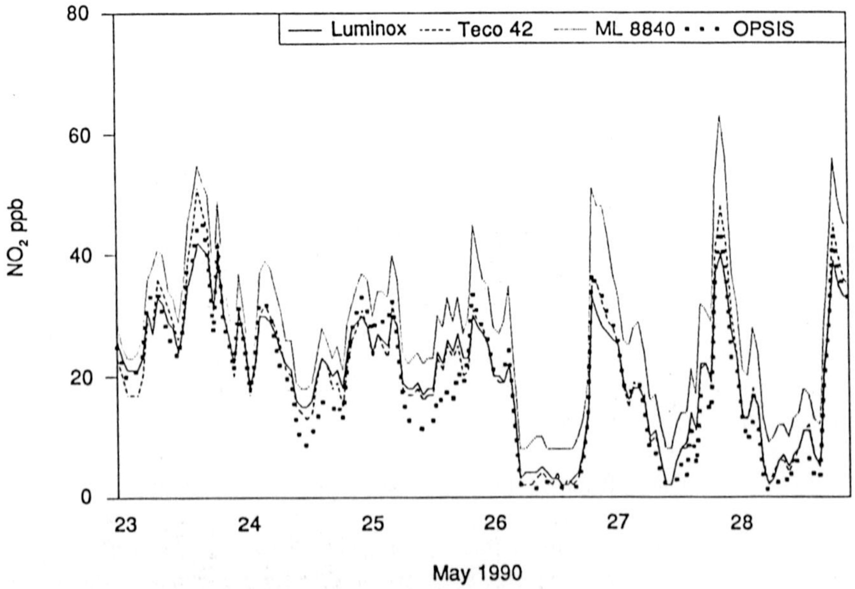

Figure 5.5 *Hourly average NO_2 concentrations, Stevenage, May 23–28 1990*

Table 5.4 Hourly average NO_2 data statistics: Stevenage, April – May 1990

	Teco 42 *NO_2 (ppb)*	*ML 8840* *NO_2 (ppb)*	*Luminox* *NO_2 (ppb)*	*Opsis* *NO_2 (ppb)*
Average	20.9	26.9	22.1	20.3
Minimum	1.0	2.0	2.0	0.3
Maximum	72.0	81.0	60.0	59.5
Std. Dev	11.5	11.9	10.2	11.5
No. Hours	1016	1016	1016	1016

higher. Table 5.3 lists the average, maximum and minimum values recorded by each analyser for 1016 hours (42 days) of coincident NO_2 measurements. As seen in Table 5.4, the average NO_2 concentration recorded by the ML 8840 (27 ppb) was ~30% higher than the average concentrations recorded by the other analysers. The average NO_2 concentration recorded by the Opsis, Luminox and Teco 42 analysers agreed to within 8%. The range of ambient NO_2 concentrations measured extended from below 1 ppb to above 80 ppb. The frequency distribution of ambient concentrations recorded by each instrument are shown in Figure 5.7. Values in the class interval 10 ppb, correspond to the number of hours with a concentration level of greater or equal to 0 but less than 10 ppb. These graphs show that the majority of ambient NO_2 measurements were between 10 and 30 ppb NO_2.

The results from the instruments have been intercompared to determine whether the differences between the methods are statistically significant. Figure 5.8 shows scatter plots of the ML 8840, Luminox and Opsis NO_2 data against the Teco 42 data. The solid line on each graph is the linear regression

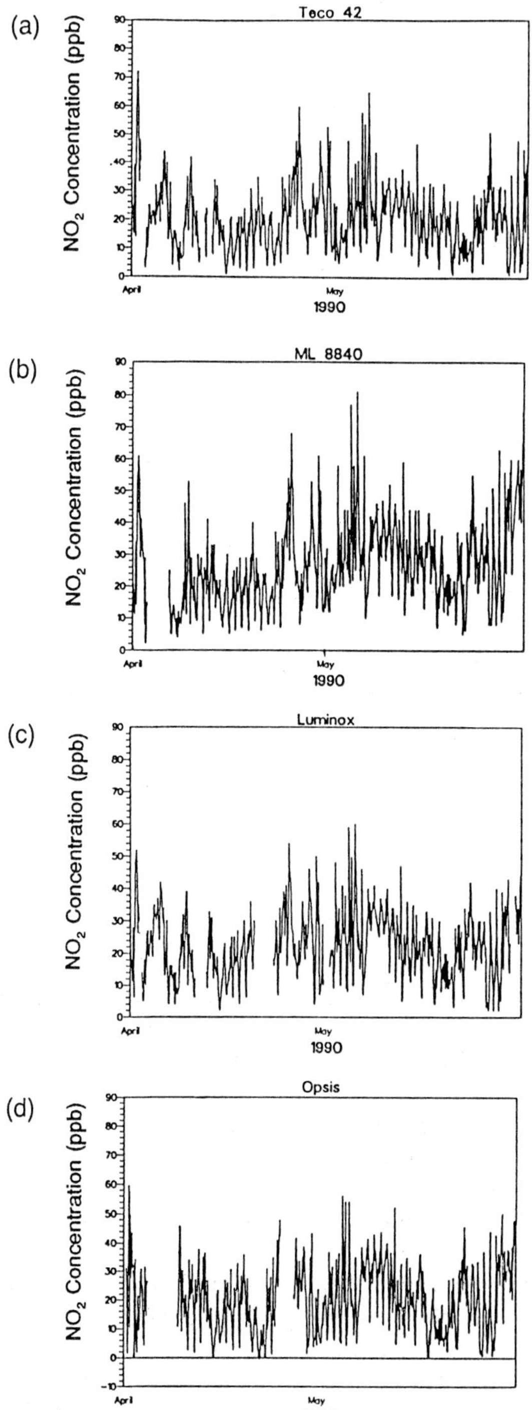

Figure 5.6 *Hourly average NO_2 time series plots for (a) Teco; (b) ML 8840; (c) Luminox; (d) Opsis. Stevenage, April–May 1990*

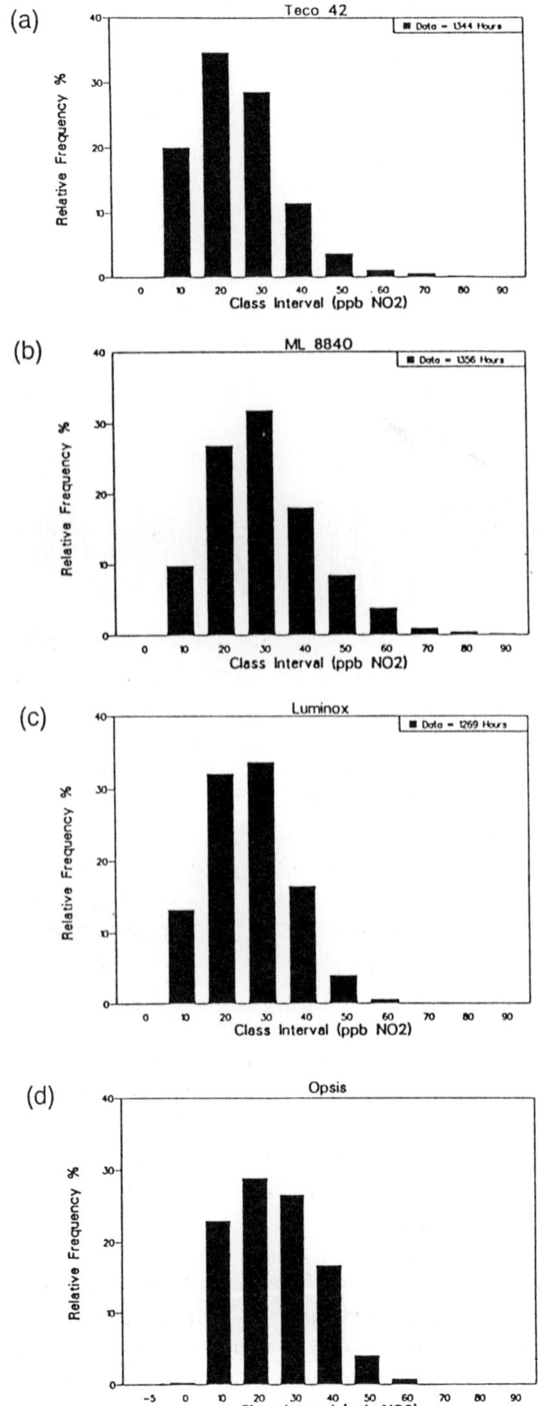

Figure 5.7 *Frequency distributions of NO_2 concentrations for (a) Teco; (b) ML 8840; (c) Luminox; (d) Opsis. Stevenage, April–May 1990*

Table 5.5 Results of the Stevenage NO_2 linear regression analyses

Parameters	*ML 8840 vs Teco 42*	*Luminox vs Teco 42*	*Opsis vs Teco 42*
Slope (m)	0.988	0.856	0.936
σ_m	0.012	0.007	0.013
*	Yes	No	No
Intercept	5.53	4.11	0.473
σ_y	4.88	2.91	4.94
N	1258	1228	1169
R	0.92	0.96	0.90

N = Number of hourly average values in the data set.
R = Correlation coefficient.
σ_m = Error in the slope.
σ_y = Standard error of *y* estimate.
* = Indication of whether the $\pm 2\sigma$ interval for the slope contains the value of 1.

line, while the dotted line is the 1:1 line, for reference. Table 5.5 gives the results of the linear regression analysis for each set of data. As seen from this table, all instruments were highly correlated with the Teco 42, giving correlation coefficients of $\geq$0.9 in all cases. Agreement between the analysers, as determined from the regression slope, was found to be within 1% for the ML 8840, 14% for Luminox and 6% for Opsis. Using the $\pm 2\sigma$ criteria as discussed earlier, the results of the Teco 42 and ML 8840 regression produced a slope that was not statistically different from 1 at the 5% level, indicating excellent agreement between the two analysers employing the same chemiluminescent detection principle. Comparison of the Teco 42 with the Opsis and Luminox data sets however, show that differences in the regression slopes are significant at the 5% level. Figure 5.8 shows that at the high end of the NO_2 measurement range, both the Luminox and Opsis analysers reported, on average, lower NO_2 concentrations than the Teco 42. At low NO_2 levels however, the Luminox and ML 8840 recorded consistently higher NO_2 concentrations than the Teco 42, giving rise to regression slopes with large positive intercepts of 4.1 and 5.5 ppb respectively. The Opsis analyser showed relatively good agreement with the Teco 42 at zero, with a resulting intercept of 0.8 ppb.

Although statistically significant proportional biases of 6% and 14% were detected in both the Opsis and Luminox data sets, the actual level of agreement between the analysers must also take into account the accuracy at which the NO_2 measurements were made. The estimated accuracy of the analysers, as discussed in Section 5.2.5, was no worse than 15% for any of the instruments. The level of agreement between the analysers, as determined from the regression slopes was therefore, within the estimated accuracy of all the analysers.

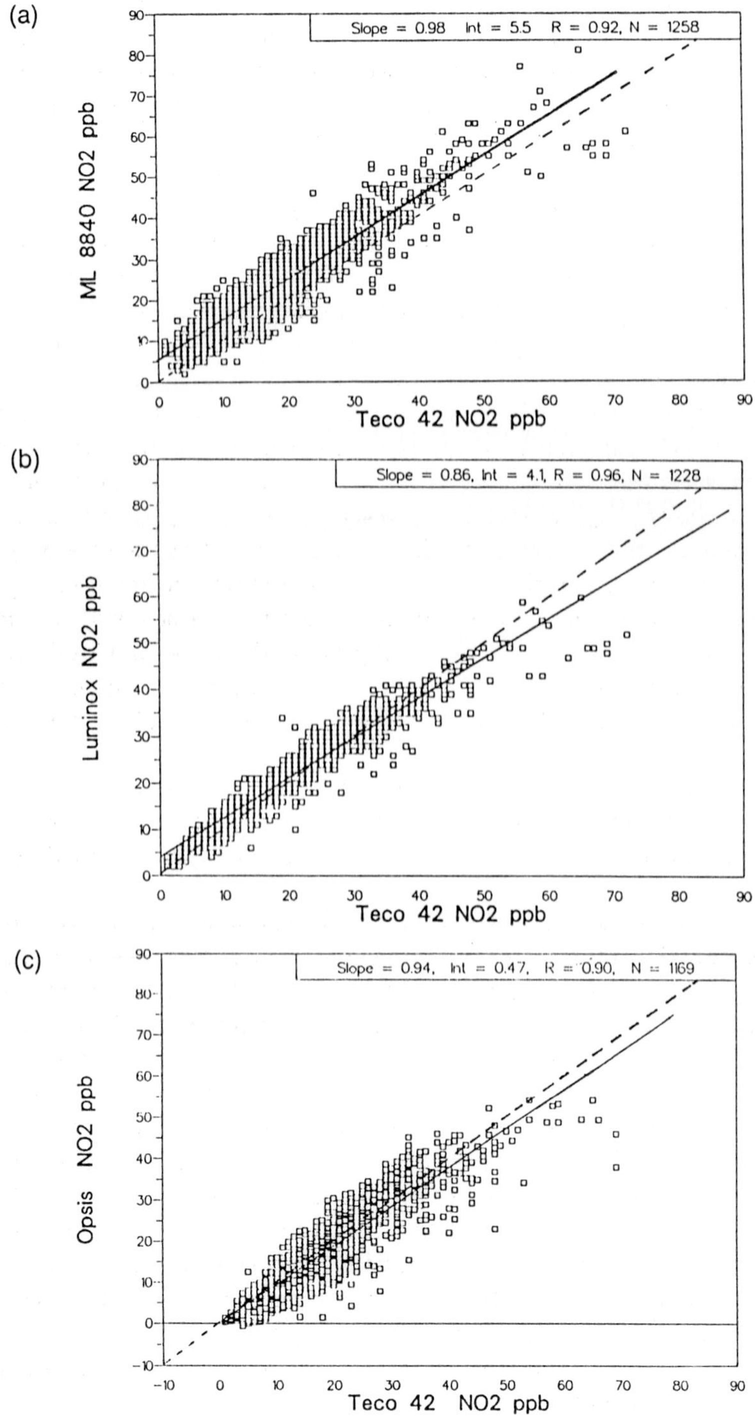

Figure 5.8 *Linear regression graphs of (a) ML 8840 against Teco 42; (b) Luminox against Teco 42; (c) Opsis against Teco 42. Stevenage, April–May 1990*

Similarly the level of agreement at low NO_2 concentrations can only be expected to be within the precision ($\pm 2\sigma$) of the instruments (see Section 5.2.5). Only the Opsis analyser showed acceptable agreement at zero, within the limits of precision, whereas the Luminox and ML analysers showed significant positive offsets. The ML 8840 is the least sensitive of the instruments tested and is not therefore capable of accurately or precisely determining NO_2 at low levels, thus poor agreement at zero may result. The increase in base line response of the Luminox analyser may be due to a positive interference effect from ambient ozone. The average ozone concentration recorded at Stevenage during the same sampling period in April to May, was 27 ppb. The Luminox response to ozone and PAN were quantified during laboratory based tests at WSL and in Germany, and found to be approximately 3.6% for ozone and 17% for PAN. The Luminox analyser is therefore, only 40 times more sensitive to NO_2 than to ozone, and ambient O_3 concentrations of $\sim$30 ppb would give rise to a signal equivalent to $\sim$1 ppb of NO_2. The effect of ozone interference may be very significant at the remote Strath Vaich site where ozone concentrations of $\sim$30 ppb are observed throughout the year and NO_2 concentrations are typically of the order of 1 ppb. The Luminox interference due to ozone as determined by WSL, is approximately ten times larger than the value previously reported by other workers (Shiff *et al*., 1986), however this interference effect is likely to depend on the analyser configuration and the luminol solution in use at the time.

In order to accurately determine Luminox NO_2 measurements, simultaneous ambient ozone (and PAN) data are required. The Luminox measurements reported in this paper have not been corrected for these interferences.

Another way to detect any constant offset existing between the analysers, is to examine the numerical difference between the NO_2 concentrations measured by one analyser to those measured by the chosen reference analyser. Figure 5.9 shows the difference between the Teco 42 measurements and the data from the other analysers, plotted as a function of time. It can be seen from Figure 5.9 that the differences for the Luminox and Opsis analysers are generally uniformly distributed about zero with time. In contrast, the ML 8840 data show an offset, that appears to increase as the sampling period progresses, indicating either a potential drift in instrument response, or a change in zero calibration. The magnitude of these differences in relation to the ambient concentrations levels at which they were recorded, are shown in Figure 5.10. The differences remain fairly constant over a range of NO_2 concentrations up to 50 ppb, although there is an indication that the differences generally increase at higher concentrations. Figures 5.10(a) and 5.10(b) show that at the lower end of the concentration range, the bulk of the ML 8840 and Luminox data sets are consistently 5 to 10 ppb larger than the Teco 42 measurements.

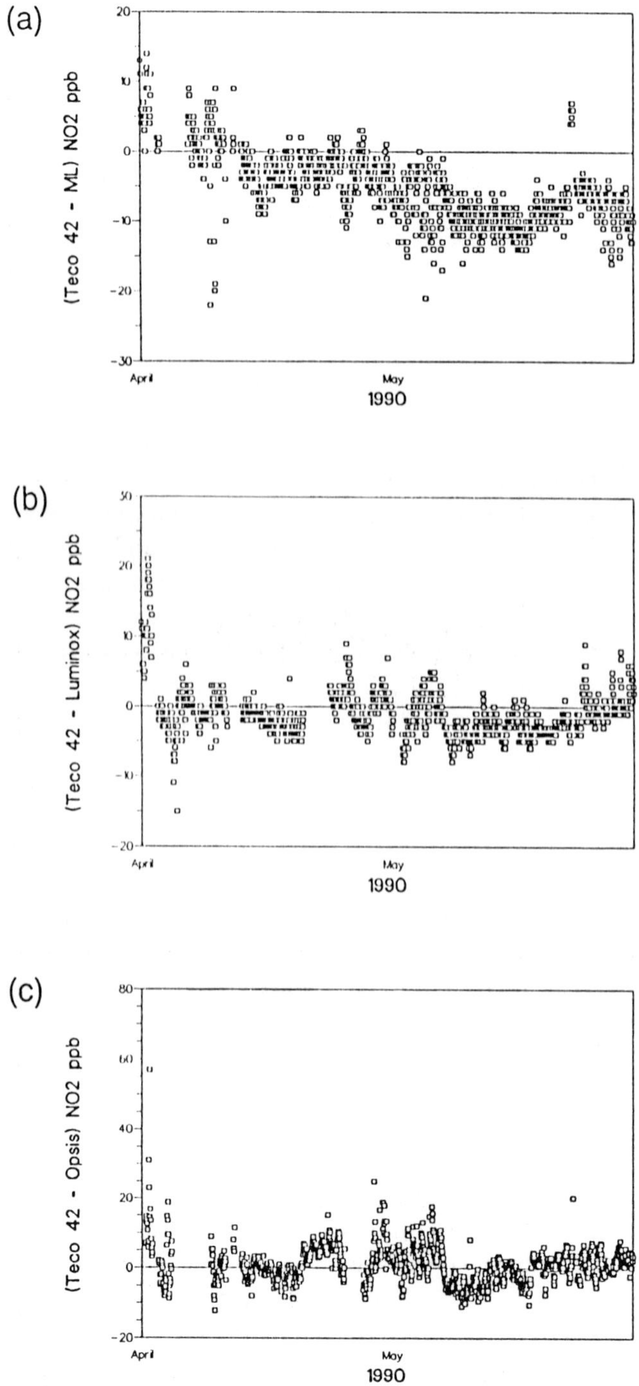

Figure 5.9 *The variation with time of the differences in NO_2 concentration between (a) Teco 42 and ML 8840 data; (b) Teco 42 and Luminox data; (c) Teco 42 and Opsis data. Stevenage, April–May 1990*

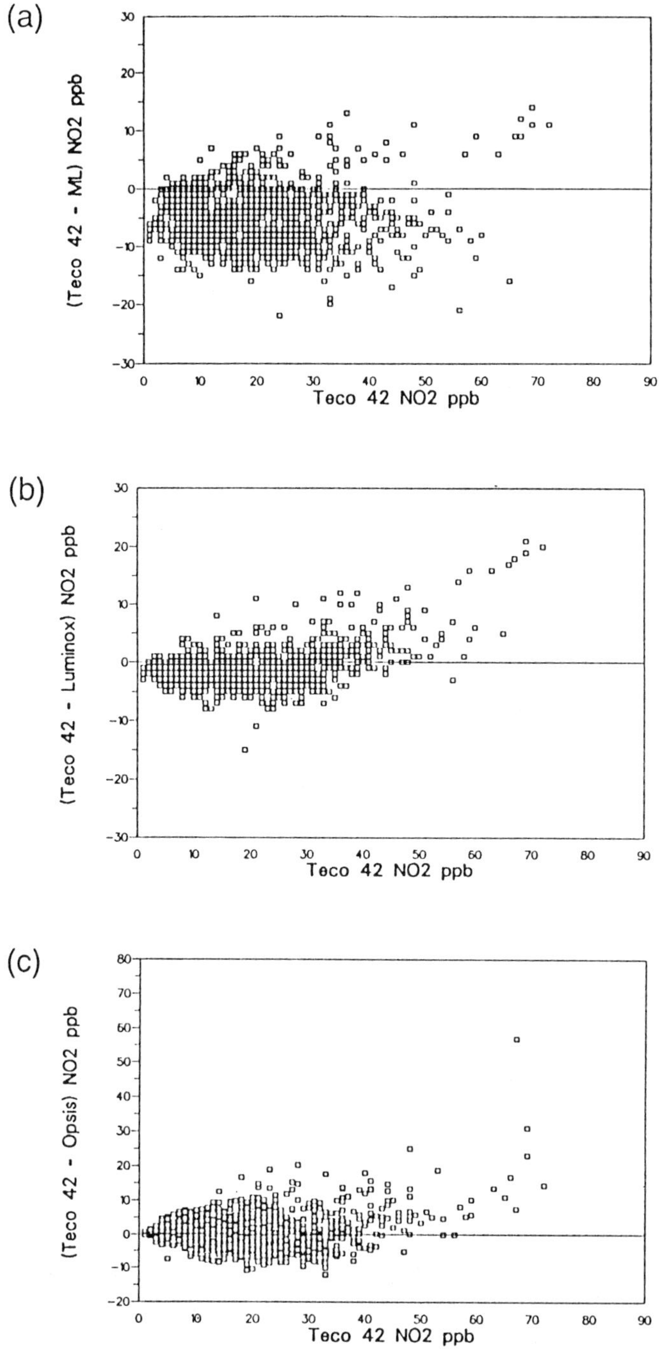

Figure 5.10 *The variation with NO_2 concentration of the differences in NO_2 concentration between (a) Teco 42 and ML 8840 data; (b) Teco 42 and Luminox data; (c) Teco 42 and Opsis data. Stevenage, April–May 1990*

5.3.2.3 Results of the intercomparison at Strath Vaich

The NO_2 analysers were installed at Strath Vaich in November 1990 for a period of 10 months. During this time however, there were long periods when either one or more of the instruments were broken, or had for some reason been removed from the site. Consequently, only the data collected between November 1990 to March 1991 have been used in the intercomparison analyses. Figure 5.11 shows the time series plots of the hourly average NO_2 concentrations recorded by each analyser for this period. It is immediately obvious from Figure 5.11 that, in contrast to the fairly complete Teco 42 data set, the Luminox, and to even a greater extent, the Opsis data set, are both fragmented and of poor quality. Figure 5.12 shows in more detail, a weekly time series plot of simultaneous NO_2 measurements between January 29 and February 4 1991. As seen in Figure 5.12, the instruments generally track one another reasonably well, however, both the Luminox and Opsis measurements appear lower than the Teco 42 values, with Opsis recording negative NO_2 concentrations at the lowest levels. The relative distributions of ambient NO_2 concentrations reported by each analyser from November to March are shown in Figure 5.13. The Teco 42 data show ambient NO_2 concentrations ranging from below 1 ppb to above 25 ppb, with a significant fraction (~70%) being in the range of 0.1 to 1 ppb. In contrast the Luminox data show a larger proportion of measurements at higher NO_2 concentrations with over 80% of the data at levels above 1 ppb. Over 50% of the Opsis measurements were reported as negative NO_2 concentrations. This is probably due to the use of an incorrect zero calibration spectrum. It was not feasible to scale the data, by setting the lowest ambient hourly average value to zero, because of the way in which the Opsis software eliminates some negative data when calculating hourly average concentrations (see Section 5.2.4).

Table 5.6 Results of the Strath Vaich linear regression analysis

Parameters	*Luminox vs Teco 42*	*Opsis vs Teco 42*
Slope (m)	0.784	0.623
σ_m	0.008	0.104
*	No	No
Intercept	0.801	–0.702
σ_y	1.098	1.22
N	1606	913
R	0.92	0.89

N = Number of hourly average values in the data set.
R = Correlation coefficient.
σ_m = Error in the slope.
σ_i = Error in the intercept.
* = Indication of whether the $\pm$ 2σ interval for the slope contains the value of 1.

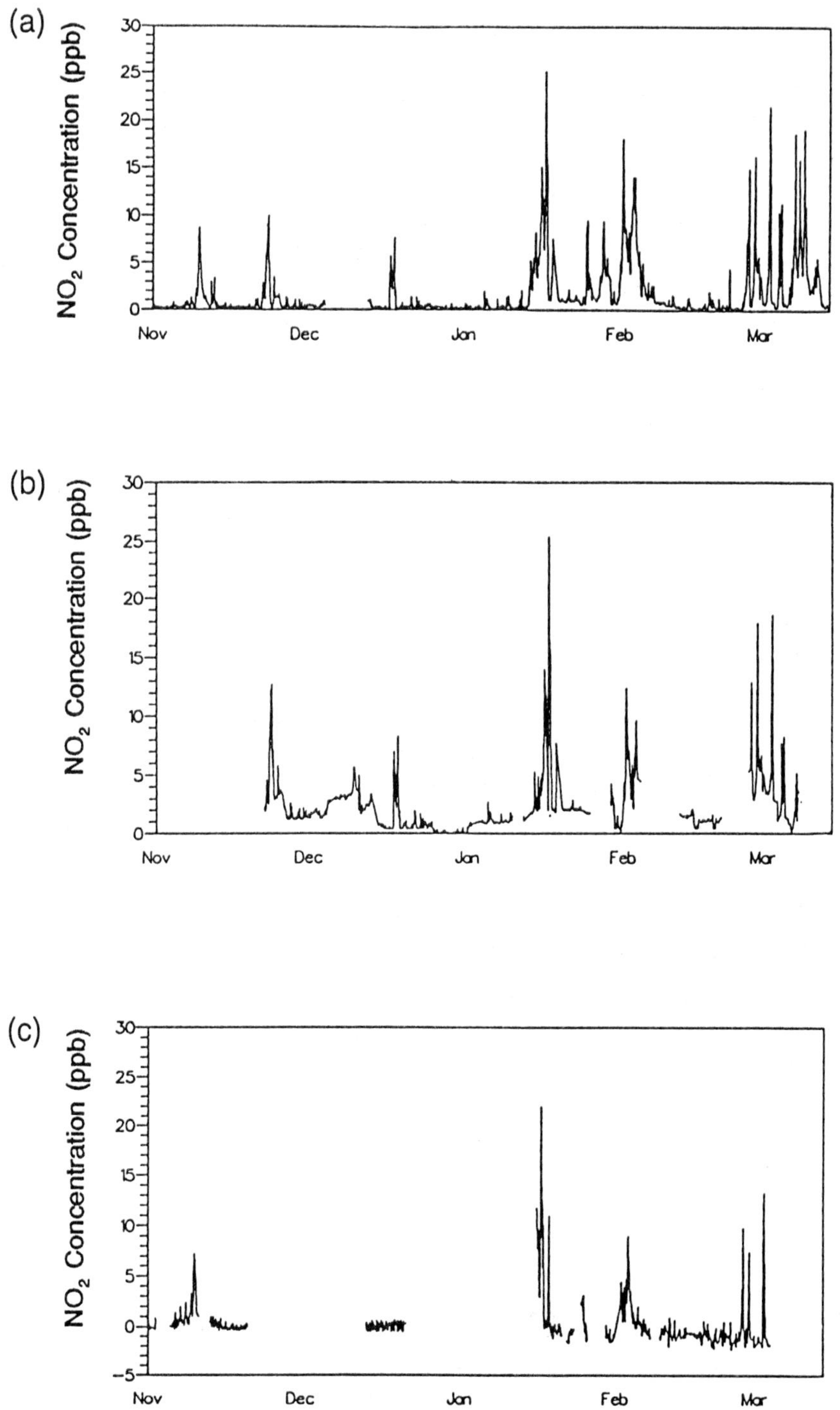

Figure 5.11 *Hourly average NO_2 time series graphs for (a) Teco 42; (b) Luminox; (c) Opsis. Strath Vaich, November 1990–March 1991*

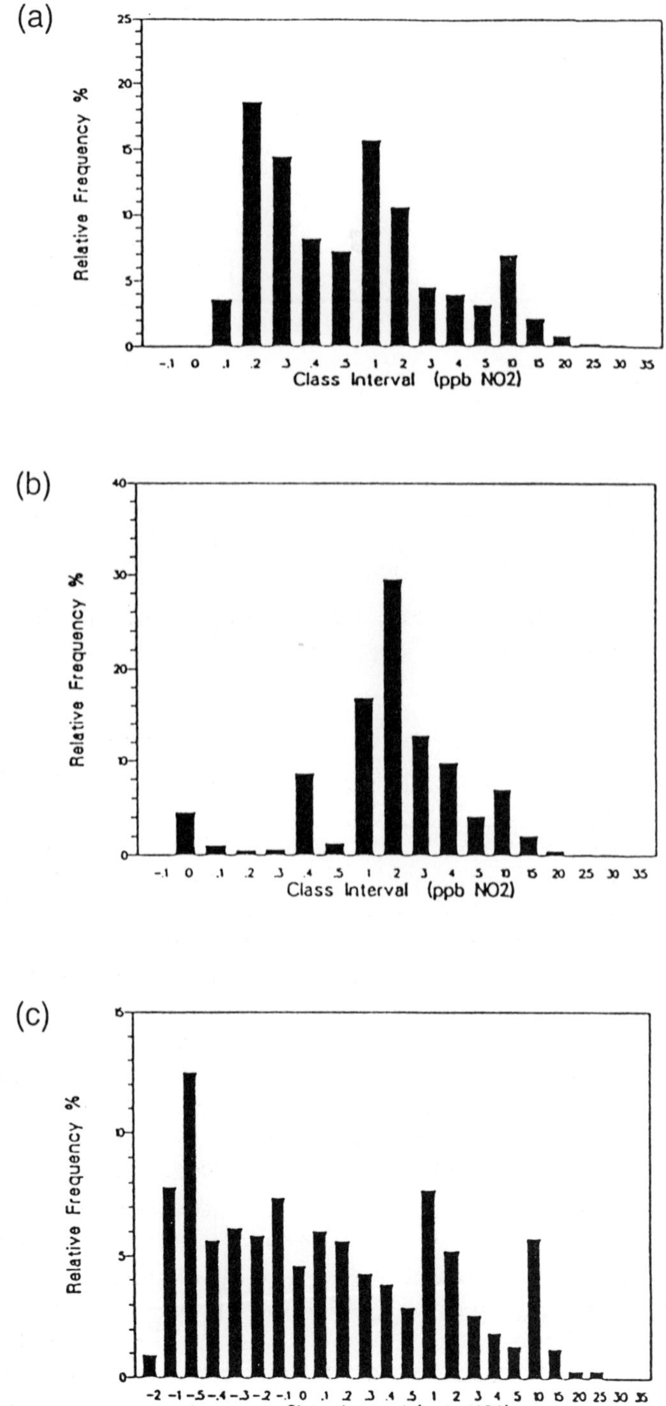

Figure 5.12 *Frequency distributions of NO_2 concentrations for (a) Teco 42; (b) Luminox; (c) Opsis. Strath Vaich, November 1990–March 1991*

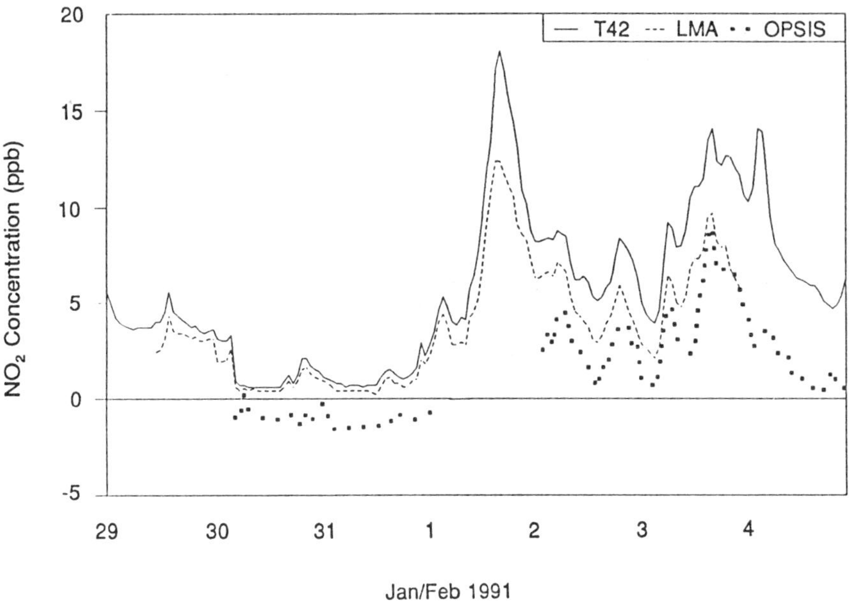

Figure 5.13 *Hourly average NO_2 data for the Teco 42, Luminox and Opsis. Strath Vaich, 29 January 1990–4 February 1991*

Comparisons of the Luminox and Opsis NO_2 measurements against the Teco 42 data are shown in Figure 5.14, and the parameters from the corresponding linear regression analyses are given in Table 5.6. The correlation coefficients are sufficiently high (0.89 and 0.92 for Opsis and Luminox data respectively) to suggest a strong correlation between the data sets. The slopes of the regression lines however, indicate a relatively poor level of agreement between the analysers, with the Luminox and Opsis measurements being on average, 22% and 38% lower than the Teco 42 measurements. In both cases this proportional bias is significant at the 5% level.

When using the slope of the regression line to determine the level of agreement between analysers it must be remembered that the slope can be biased by a few points that do not reflect the majority of the data. The values for σ_m and σ_y given in Table 5.6 show that the error in the slope and the scatter of data points around the regression line are larger for the Opsis data than for the Luminox data, by 90% and 10% respectively. In fact, Figure 5.14(b) shows that the Opsis measurements appear to be divided into two data sub-sets, one of which depicts a fairly good level of agreement at high NO_2 concentrations and the other showing lower values than the Teco 42.

The intercept of the regression graphs indicate an offset of 0.8 ppb for the Luminox data and –0.7 ppb for the Opsis data. As discussed earlier in Section 5.3.2.3, the Luminox analyser offset may be due to the positive interference effects from ozone. Background ozone concentrations of approximately 30 ppb, with minimal seasonal variation, are typical for

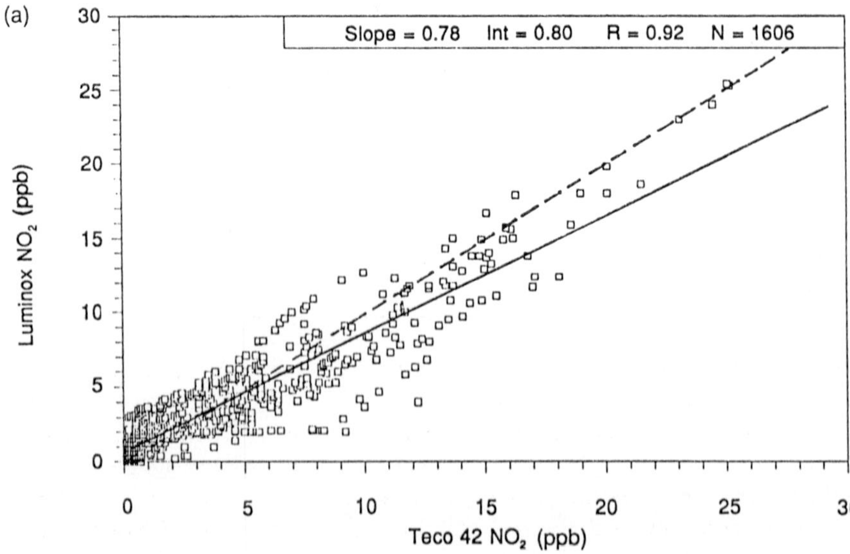

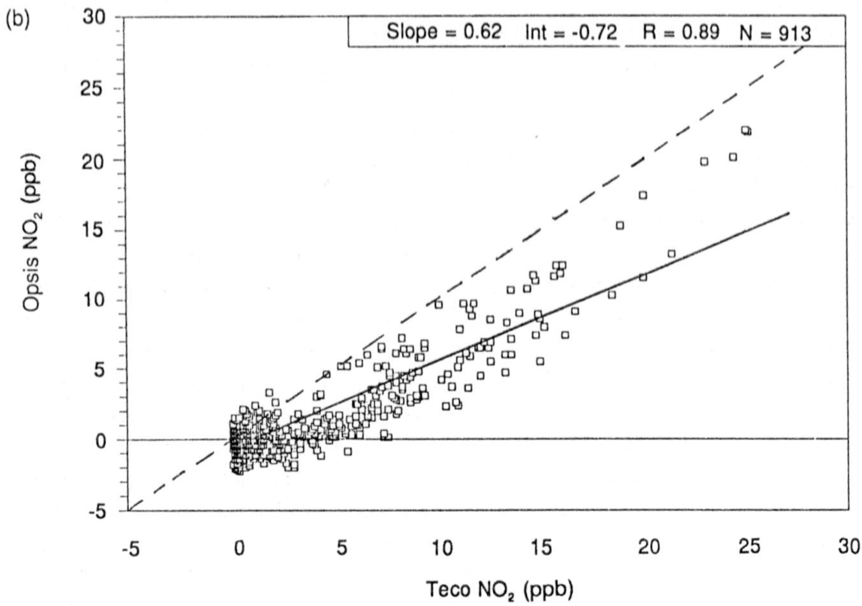

Figure 5.14 *Linear regression graphs comparing (a) Luminox and (b) Opsis NO_2 measurements against the Teco 42 data. Strath Vaich, November 1990–March 1991*

Table 5.7 Strath Vaich hourly average NO_2 data statistics: November 1990–March 1991

	Teco 42 *NO_2 (ppb)*	*Luminox* *NO_2 (ppb)*	*Opsis* *NO_2 (ppb)*
Average	3.6	3.5	1.3
Minimum	0.1	0.4	–2.0
Maximum	25.2	25.4	22.0
Std dev.	5.1	4.4	3.7
No. hours	404	404	404

Strath Vaich (Bower *et al.*, 1990). The negative offset for the Opsis data may be due to the zero calibration used.

The average NO_2 concentrations reported by each analyser calculated from 404 hours of coincident measurement data are shown in Table 5.7, together with the maximum and minimum concentration values. The average concentration as determined by the Luminox analyser (3.5 ppb) is in excellent agreement with the average value recorded by the Teco 42 (3.6 ppb), however, the Opsis reports an average NO_2 concentration ~65% lower at 1.3 ppb. Maximum concentrations recorded by the Teco 42, Luminox and Opsis are in good agreement (25.4, 25.2. and 22.0 ppb), however, the minimum concentrations recorded appear to differ for reasons mentioned above. The minimum hourly average NO_2 concentration determined by Opsis of –2 ppb gives an approximate estimate of the magnitude of the zero offset. If a correction of 2 ppb was applied to the Opsis data, then the average and maximum NO_2 concentrations would be in good agreement with the corresponding Teco 42 and Luminox values. However, this estimate of the minimum hourly average NO_2 value excludes very negative data, which have been automatically rejected by the Opsis software (see Section 5.2.4), therefore, it cannot be used to correctly determine the zero offset.

Figure 5.15 shows the differences between the Teco 42 data and the hourly average NO_2 concentrations recorded by the other analysers, plotted as a function of time. Periods when the differences are large, >5 ppb, correspond to the episodes of elevated ambient NO_2 data which are shown in Figure 5.11. The increase in differences as a function of increasing concentration is more clearly illustrated in Figure 5.16. This relationship is expected as all the analysers employ light-sensitive detectors and the noise associated with each measurement is proportional to the square root of the light intensity reaching the detector. An alternative approach to examining the differences between the measurements is to essentially normalize the differences, with respect to the ambient NO_2 levels at which the measurements were made, i.e. divide the difference between the analysers by the concentrations recorded by the Teco 42. The normalized difference is plotted against the Teco 42 NO_2 concentration in Figure 5.17. The relative level of agreement between the Luminox and Teco 42 at NO_2 concentrations above 5 ppb is excellent. Below 3 ppb the difference between the Luminox and Teco 42 is up to five times the measured concentration, and below 1 ppb by up to 15 times, with the

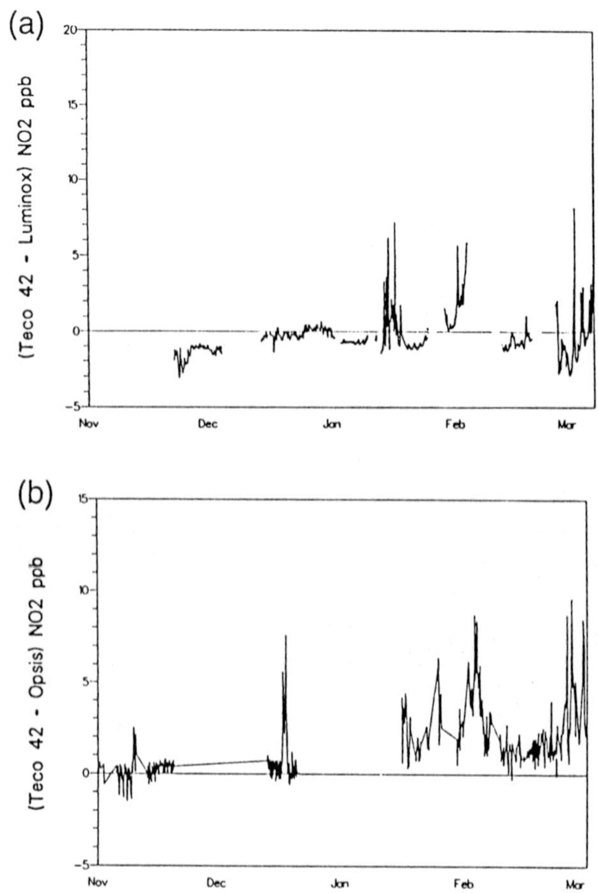

Figure 5.15 *Differences in the hourly average NO_2 concentrations against time for Strath Vaich, November 1990 – March 1991. (a) Teco 42 and Luminox data and between (b) Teco 42 and Opsis data*

Luminox consistently reading higher than the Teco 42. Similarly, the Opsis measurements show large levels of disagreement at NO_2 concentrations below 1 ppb, with differences of up to ten times the measured concentration, however, in this case the Opsis recorded concentrations both above and below the Teco 42 NO_2 levels. These large differences highlight the problems of accurately determining low levels of NO_2 when using different techniques which are subject to interferences.

5.4 Summary and conclusion

Three fundamentally different methods for determining low levels of NO_2 were compared during two intercomparison studies, one of which was conducted in a suburban site at Stevenage, and the other under more realistic

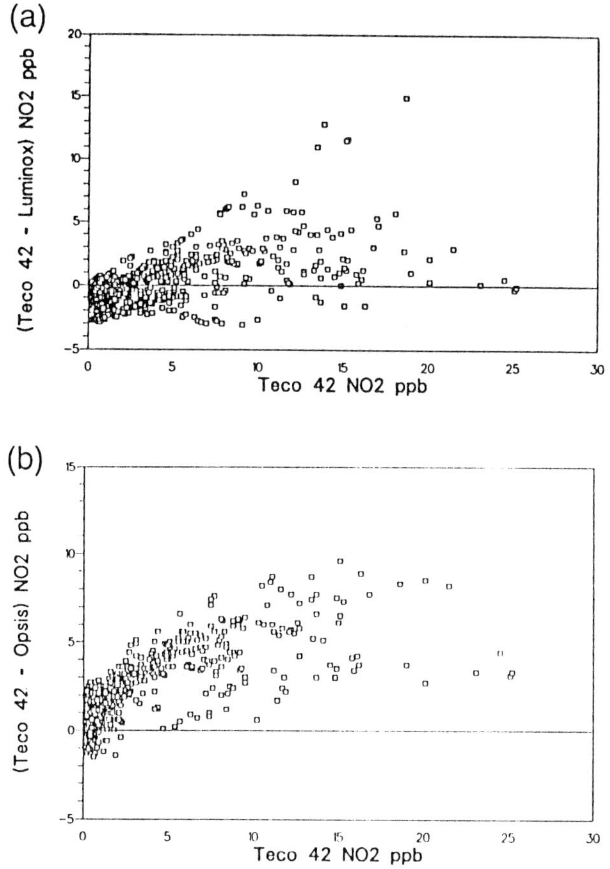

Figure 5.16 *The variation with NO_2 concentration of the differences in NO_2 concentration for (a) Teco 42 and Luminox data; (b) Teco 42 and Opsis data. Strath Vaich, November 1990–March 1991*

field conditions at the remote Strath Vaich site. The instruments under evaluation were a gas phase chemiluminescence NO_x analyser (Teco 42), a luminol chemiluminescent NO_2 analyser (Luminox) and a differential optical absorption spectrometer (Opsis). The aim of this project was to evaluate each analyser's performance capability and reliability for determining low levels of NO_2, with a view to selecting the most appropriate method for use in remote unattended locations.

In terms of operational reliability, the Teco 42 chemiluminescent analyser proved to be the most suited to continuous unattended operation in a remote location. The Teco 42 does have a distinct advantage over the other analysers, in that it uses an established detection technique, hence improvements in instrument sensitivity and reliability have been developed over many years. This technique does however, still suffer from interference effects from other nitrogen containing species.

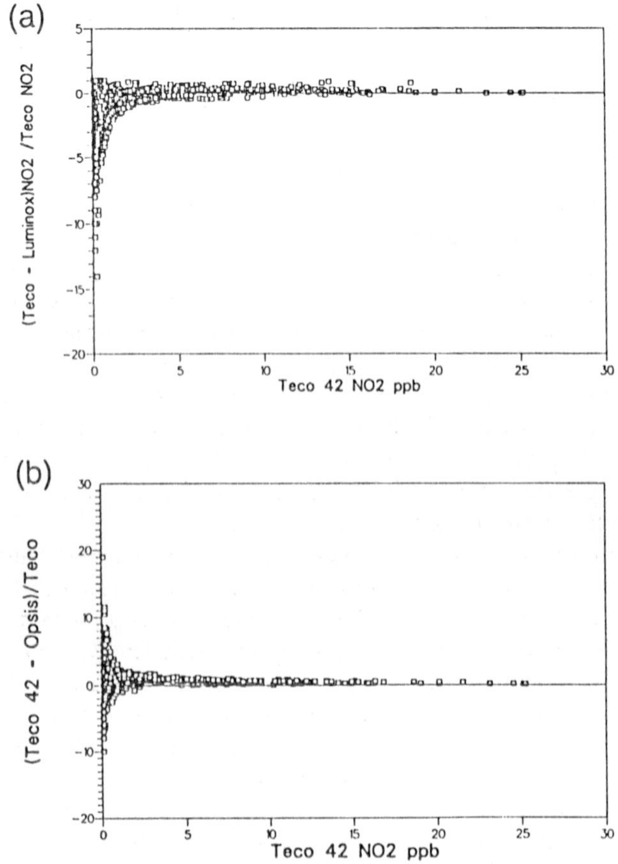

Figure 5.17 *Normalized differences in NO_2 concentration against the Teco NO_2 concentration for (a) the Luminox and (b) Opsis. Strath Vaich, November 1990–March 1991*

The Luminox analyser demonstrated a potential to operate reliably, since no major operational problems were encountered. However, replacement of the peristaltic pump tubing was required on a weekly basis. The Luminox analyser is therefore only suitable for use at attended sites unless a more sophisticated pump or positive pressure luminol feed mechanism, as used in the Luminox PAN analyser, is implemented.

The non-linear response of the Luminox analyser has a significant disadvantage in terms of data processing and calibration. Multi-point calibrations at levels of NO_2 below ~3 ppb are both difficult and time-consuming to perform on site. The effects of interference from ozone and PAN must also be corrected if this technique is to be used for accurate low level determinations of NO_2.

Long term operation of the Opsis analyser showed that its performance in the field was unreliable, mainly as a result of the unstable optical alignment of the beam. In addition, repair and maintenance of the spectrometer and computer by specialist Opsis engineers was frequently required. Problems associated with the use of elaborate electronics and delicate optical components made the Opsis analyser particularly unsuitable for use at sites in remote locations, where mains power supplies are often unstable and site visits infrequent.

When comparing the different measurement methods it should be remembered that the Opsis instrument utilizes sampling and analysis techniques that are distinctly different from the currently accepted conventional ambient monitoring methods. Open path sampling, involves the use of sophisticated optical devices such as retroreflectors, fibre optic cables and the spectrometer itself. Pollutants concentrations are derived from complex mathematical manipulation of the absorption spectrum, which includes information on not only the pollutant species under investigation, but also other species which have absorption characteristics in the same wavelength region. Attempts to calibrate the Opsis analyser, and hence verify the measurement results, have been unsuccessful, and further work to develop calibration methodologies is required.

Intercomparison of the NO_2 measurements focused on determining the correlation, and level of agreement observed between the Teco 42 chemiluminescent analyser and the measurements reported by the other instruments. Results from the Stevenage intercomparison showed that measurements from all the analysers were well correlated ($R > 0.9$). The agreement between the Teco 42 and ML 8840 was excellent, as would be expected from techniques employing the same detection principle. The Luminox and Opsis measurements agreed with the Teco 42 to within 14% and 6% respectively. In all cases the level of agreement was within the limits of uncertainty associated with each technique ($\sim$15%).

Some degree of discrepancy between the measurements was expected since the techniques are susceptible to different sources of measurement error arising from interferences, calibration and sampling errors. For example, Opsis measurements were performed over the entire absorption path as opposed to at a fixed point. In addition, there was no physical contact with the sample gas, therefore possible effects of wall losses within sample lines or within the analyser itself are excluded. Differences in the sampling and integrations times utilized by each instrument may also lead to reduced temporal overlap of the data reported.

The results from the Strath Vaich intercomparison study showed a significant decline in the level of agreement observed between the Teco 42 NO_2 data and the other analysers' measurements; regression slopes demonstrated levels of agreement to within 22% for the Luminox analyser and only 38% for the Opsis analyser. These results reflect the difficulty in accurately determining NO_2 at very low ambient concentrations, even when using high sensitivity analysers.

The largest differences between the Luminox and Teco 42 measurements were observed at low levels of NO_2 below 3 ppb, where effects of interferences and errors associated with linearizing the Luminox response were most likely to be significant. Figure 5.14(a) shows however, that at higher concentrations, e.g. 5 to 25 ppb NO_2, there was relatively good agreement between the Luminox and Teco 42 measurements.

Agreement between the Teco 42 and Opsis analyser was significantly worse for the Strath Vaich NO_2 data set (38%), than for the corresponding Stevenage data set (6%). This reflects the overall poorer performance of Opsis at Strath Vaich, which led to reduced data capture and quality. Errors (or deviations) associated with the ambient measurements by Opsis were, on average, higher for the Strath Vaich NO_2 data (~3 ppb) than for the Stevenage data (~ 1 ppb). The high deviations observed were either a result of increased signal noise due to the low and fluctuating light levels and/or from increased interference from other ambient species. Paradoxically, the Opsis analyser appeared less capable of accurately determining low levels of NO_2 whilst installed at Strath Vaich than it had been in Stevenage.

All Opsis data had to be carefully screened using appropriately selected data quality criteria to remove data with unacceptably low light levels or high deviations. Less stringent data quality criteria were used to validate the Strath Vaich NO_2 data in order to avoid rejecting the majority of the data. Problems with performing a satisfactory zero calibration at Strath Vaich, gave rise to a significant negative offset in the Opsis data set, and this, together with the large deviations associated with the ambient measurements contributed to the overall poor agreement between the Opsis and Teco 42 data sets. The Strath Vaich intercomparison results therefore showed that the single-ended Opsis analyser at this stage of development was neither reliable nor sensitive enough to be used in a remote field location for measuring low levels of NO_2. Operation of the Opsis analyser at Stevenage did however show that under more favourable operating conditions, and with sufficient technical support, satisfactory NO_2 measurements could be obtained for short periods of time.

In conclusion, this intercomparison project has shown that the analysers, each employing a different detection principle, also differed considerably in terms of reliability, sensitivity, interference effects and ease of operation in the field. All these factors must be taken into consideration when selecting the most appropriate instrument to suit the monitoring requirements and location.

Although none of the methods evaluated are yet ideal, the well-proven and tested chemiluminescence method demonstrated significant advantages over the other methods, in that it proved to be reliable, simple to operate and calibrate and sufficiently sensitive for determining low levels of NO_2 at remote unattended rural sites.

5.5 References

Kelly T J, Spicer C W and Ward G F (1990). An assessment of the luminol chemiluminescence technique for measurements of NO_2 in ambient air. *Atmos. Environ.* **24A**, 2397-2403.

Eaton S and Stevenson K J (1993). Primary standard calibration methods for ambient

air quality analysers. In this publication.
Bower J S, Broughton G F, Dando M T, Lees A J, Stevenson K J, Lampert J E, Sweeney B P, Parker V J, Driver G S, Waddon C J and Wood A J (1989). Monitoring for the European Community Nitrogen Dioxide Directive in the UK : A Review of Data for 1987. Stevenage: Warren Spring Laboratory, (1989), Report LR 734 (AP).
Shiff H I, Mackay G I, Castledine C, Harris G W and Tran Q (1986). Atmospheric measurements of nitrogen dioxide with a sensitive luminol instrument. *Water, Air, Soil Pollution,* **30**, 105-114.
Bower J S, Stevenson K J, Broughton G F, Lampert J E, Sweeney B P and Wilken J (1990). Ozone in the UK: A Review of 1989/90 Data. Stevenage: Warren Spring Laboratory, Report LR 793 (AP).

— 6 —

Primary standard calibration techniques for ambient air quality analysers

S Eaton and K Stevenson

6.1 Abstract

The Air Pollution Division of Warren Spring Laboratory has developed a primary standard calibration laboratory for the calibration of ambient air quality analysers compressed gas cylinders containing mixtures at near-ambient concentrations. The facility includes a number of independent primary standards for each pollutant species measured. Currently, standards exist for nitric oxide (NO), nitrogen dioxide (NO_2), carbon monoxide (CO), sulphur dioxide (SO_2) and ozone (O_3). These can be reliably generated over a range of 0-1 ppm for ozone, NO_x and SO_2, and 0-50 ppm for CO. The calibration techniques used include static (volumetric) dilution, dynamic dilution, permeation tubes, gas phase titration and gravimetric standard cylinders.

These primary standards for NO, NO_2, SO_2 and CO are used to calibrate automatic analysers, or compressed gas cylinders obtained from commercial suppliers. These are used as transfer standards for on-site calibration of analysers. Special techniques have also been developed for calibrating O_3 analysers in both laboratory and field locations.

6.2 Introduction

The Air Pollution Division of Warren Spring Laboratory has been involved in operating or managing UK national automatic air quality monitoring networks on behalf of the Department of the Environment since 1972. To support this work, gas calibration standards have had to be developed for the ambient air quality analysers used, since usable and traceable standards were not available elsewhere. In recent years, EC Directives, as well as advisory guidelines from, for instance, the World Health Organization, have increased the need for reliable and traceable calibration standards for ambient air quality monitoring. It was against this background that a national air quality calibration laboratory has been developed.

This facility, in addition to supporting existing national monitoring networks, is available at cost to any organization requiring calibration of air quality analysers or compressed gas cylinders. Currently, standards have been developed at WSL for nitric oxide (NO), nitrogen dioxide (NO_2), carbon monoxide (CO), sulphur dioxide (SO_2) and ozone (O_3). Standards can reliably be generated over a range of 0-1 ppm for ozone, NO, NO_2 and SO_2, and 0-50 ppm for CO. To ensure reliability, at least two independent calibration methods are provided for each pollutant species. Each method has been developed such that all measurements are fully traceable within the National Measurement System. This requires the documentation of all procedures involved, as well as traceable calibrations of all equipment used, such as pressure, flow and mass. Regular quality audits are held to ensure that the documented procedures are rigorously adhered to and that comprehensive records are kept of all measurements made. It is our intention to apply for accreditation for each pollutant species individually, starting with carbon monoxide.

Regular participation in intercomparison exercises ensures harmonization with other monitoring organizations. These exercises are discussed in detail in Section 6.6.

6.3 Calibration techniques

Table 6.1 summarizes the several different methods which are employed for each pollutant species. These methods are based on ISO procedures, but have been developed into fully traceable and documented measurement

Table 6.1 Methods used for primary calibration

Pollutant Species	*Method*
Nitric oxide	Static dilution Dynamic dilution Gas phase titration
Nitrogen dioxide	Static dilution Dynamic dilution Permeation tubes Gas phase titration
Sulphur dioxide	Static dilution Dynamic dilution Permeation tubes
Carbon monoxide	Dynamic dilution Gravimetric cylinder
Ozone	UV photometry Gas phase titration

systems appropriate for the gases and concentrations of interest. Much of the specialist equipment required has been constructed in-house to meet specific performance criteria.

6.3.2 Calibration methods

6.3.2.1 Permeation tubes

Permeation tubes are small stainless steel vessels capped with a porous PTFE plug. The required pollutant species is placed in the tube as a pure liquid, which establishes an equilibrium with the vapour phase within the tube. The tubes are commercially available, although these need to be recalibrated by regular weighing. The vapour diffuses through the PTFE at a constant rate, provided the tube is kept at constant temperature and pressure. A commercially available oven is used to provide a low flow of scrubbed air which sweeps across the tube constantly. The flow is then further diluted with scrubbed air to obtain a useful, near-ambient, concentration. A schematic diagram is shown in Figure 6.1. The permeation oven is electronically controlled to maintain the temperature within 0.1°C. The permeation rate of the tube is determined by regular weighings. The concentration of the span gas can be calculated by measuring the diluent air flowrate, using the formula:

$$\text{Concentration(ppb)} = \frac{W \times R \times T}{F \times a \times P}$$

where W is the measured weight loss of tube (ng min^{-1}), R is the gas constant (8.2054×10^{-2} l atm K^{-1} mol^{-1}), T is the ambient temperature (K), F is the measured flowrate (l min^{-1}), a is the molecular weight of species (g mol^{-1}) and P is the pressure (atm).

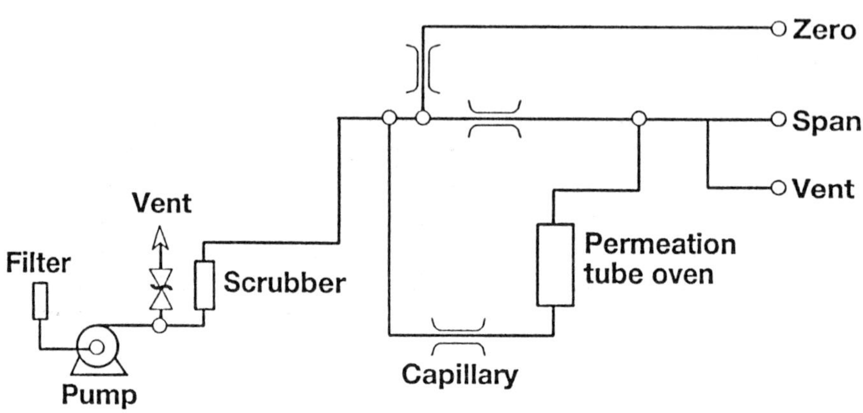

Figure 6.1 *A schematic of the permeation oven*

A careful check is maintained on the permeation system, including routine weekly weighing of permeation tubes, and determination of diluent flows using traceable flow standards. All gas scrubbers and driers are regularly checked and replaced if necessary. An estimate of the uncertainty associated with this technique can be calculated thus:

$$\frac{\Delta C}{C} = \left[\left(\frac{\Delta W}{W}\right)^2 + \left(\frac{\Delta T}{T}\right)^2 + \left(\frac{\Delta F}{F}\right)^2 + \left(\frac{\Delta P}{P}\right)^2\right]^{1/2}$$

From the calibration certificates of the measuring devices involved, we can calculate the relative uncertainties in these measurements for typical values. These give an uncertainty of calibration of 2.6%.

There is a further uncertainty which may be associated with instabilities in the permeation rate, which cannot be independently determined. Thus, an estimate of the overall uncertainty is best made by calculation of the standard deviation (σ) from weekly determinations of span gas concentrations over several months. The interval $\pm 2\sigma$ then defines a 95% confidence interval for the span gas concentration. Examples of such permeation data are shown in Table 6.2. The overall uncertainty in the concentration of span gas from a permeation tube system can be considered to be typically $\pm 10\%$.

Table 6.2 Typical permeation tube span concentrations determined over several months

Tube no.	*Time period*	*Mean Conc (ppb)*	σ *(ppb)*	2σ as %age of Conc.
SO_2-1166	Jul 90–Oct 91	108.0	4.9	9.1
NO_2-624	Jul 90–Feb 91	123.7	6.1	9.8
NO_2-631	Jun 90–Feb 91	193.1	6.5	6.7
SO_2-1173	Dec 90–Oct 91	104.3	4.9	9.4

6.3.2.2 Static dilution

In principle, the static dilution system consists of a container of fixed volume in which a known small volume of pure pollutant species is mixed with a known volume of diluent gas, to obtain directly a known gas volume ratio concentration. This technique is very flexible, as it can be used for many of the pollutant species and any make-up gas. This is important as nitric oxide must be used with nitrogen as balance gas to prevent oxidation. Additionally, by selecting the appropriate volume of pure gas, the nominal concentration of the calibration gas can be selected by the operator. Currently, this calibration method can be used for concentrations in the range 0.1–1 ppm, although this could be extended to higher concentrations by using a larger volume of injected pollutant species.

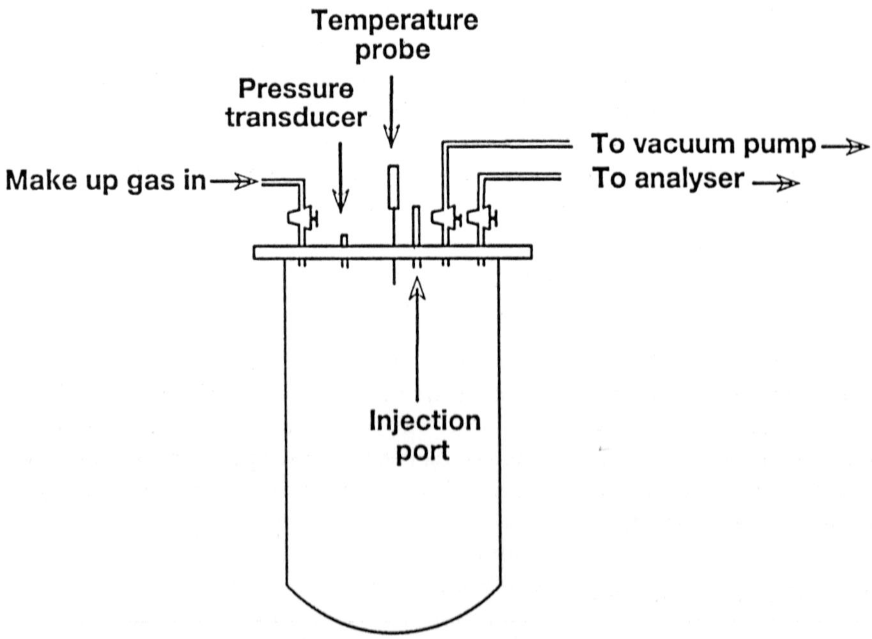

Figure 6.2 *A schematic of the static dilution apparatus*

The static dilution equipment constructed by WSL consists of a cylindrical glass vessel of accurately known volume (77 l ±0.5%), fitted with two inlet and two outlet ports, a pressure gauge and an internal temperature gauge (see Figure 6.2). A vacuum pump is connected to one of the outlet ports, and an inlet port is fitted with an injection septum. The vessel is evacuated and then filled with either dried scrubbed air or zero grade (99.999%) nitrogen to just above atmospheric pressure. This is vented through the analyser to establish a zero for the calibration, which eliminates error from any small carry-over from a previous calibration. The vessel is vented until it is at equilibrium with atmospheric pressure, and the measured atmospheric pressure is recorded. A known volume of pure pollutant species is introduced into the vessel using a 100 μl syringe, and the vessel is then filled to about 1.9 bar with air or nitrogen. Filling in this order ensures adequate mixing of the gases. While the vessel is being filled, the gas temperature rises; the gas must be allowed to cool to ambient temperature before the final pressure is recorded. The cooled gas is put through the analyser and the response recorded. The concentration of the calibration gas can be calculated from:

$$\text{Concentration(ppb)} = \frac{10^9 \times V_\text{p} \times P_\text{a}}{V_\text{v} \times P_\text{f}}$$

where V_p is the volume of pure pollutant species added (litres), V_v is the volume of static dilution vessel (litres), P_a is the atmospheric pressure (bar) and P_f is the final pressure of filled vessel (bar).

When the system is used for NO_2 the calculated concentration must be multiplied by an additional factor (dependent on temperature and pressure, but typically 1.7) to account for the NO_2/N_2O_4 equilibrium in the pure species. The exact value of this factor can be obtained from tables. It can be assumed that all the N_2O_4 dissociates on dilution.

An estimate of the uncertainty of the gas concentration can be made from the uncertainties of the individual measurements thus:

$$\frac{\Delta C}{C} = \left[\left(\frac{\Delta V_p}{V_p}\right)^2 + \left(\frac{\Delta P_a}{P_a}\right)^2 + \left(\frac{\Delta V_v}{V_v}\right)^2 + \left(\frac{\Delta P_f}{P_f}\right)^2\right]^{1/2}$$

The uncertainty of the measurements are shown in Table 6.3 for a typical calibration gas mixture of approximately 350 ppb SO_2 in air. The uncertainties quoted are obtained from calibration certificates where possible. These are then expressed as percentages of typical values used in the calibration, and are then combined to obtain an overall uncertainty for the calibration method.

Table 6.3 Uncertainties for static dilution method

Source of Uncertainty	*Uncertainty*	*% Uncertainty*
Purity of pollutant gas	99.7 ± 0.1%	0.1
Volume of dilution vessel	77.0 ± 0.4 l	0.5
Volume of pollutant	50 ± 2 μl	4.0
Atmospheric pressure	1.013 ± 0.001 bar	0.1
Vessel fill pressure	1.9 ± 0.001 bar	0.05
Uncertainty of calibration		± 4%

6.3.2.3 Dynamic dilution

This involves the dilution of a high concentration cylinder with air or nitrogen by mixing gas flows to produce near-ambient concentrations of calibration gas. A diagram of the apparatus used is shown in Figure 6.3. The flows are controlled by thermal mass flow controllers, which can be continuously varied across their range, thus allowing the production of any concentration of calibration gas. The mass flow controllers are carefully calibrated before use against a traceable soap-film flowmeter.

All connectors used for the dilution must be PTFE, and kept as short as possible to minimize wall reactions; even so, a considerable conditioning period (up to 2 hours) must be allowed between setting the flows and taking the analyser reading. This technique allows multipoint calibrations to be carried out easily by simply altering one or both flows. This provides a measure of the linearity of the analyser response.

The high concentration cylinders used for dynamic dilution, typically 100 ppm, can be produced gravimetrically, that is, a known mass of pure gas is

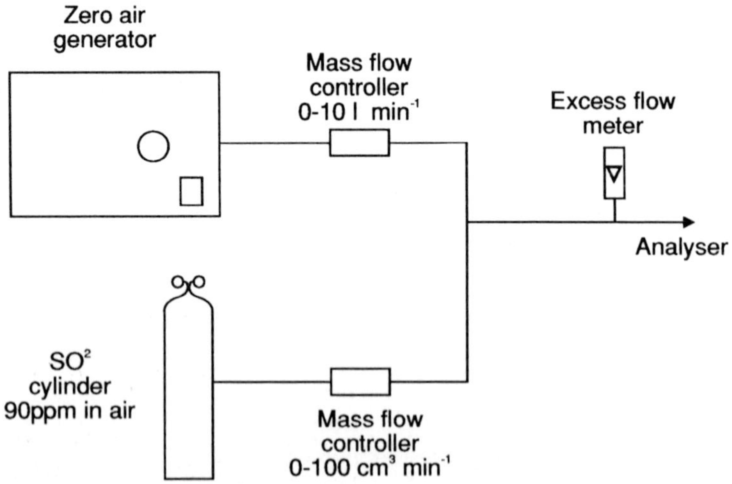

Figure 6.3 *A schematic of the dynamic dilution apparatus*

diluted with a known mass of air or nitrogen. This type of cylinder is suitable for use as a primary standard as its concentration can be accurately determined. Such cylinders of NO and CO are available from the National Physical Laboratory, and typically have accuracies within 1%.

The concentration of calibration gas produced by dynamic dilution is calculated from:

$$\text{Concentration(ppb)} = 10^3 \times C_{\text{std}} \times \frac{F_{\text{std}}}{F_{\text{std}} + F_{\text{dil}}}$$

where C_{std} is the concentration of standard cylinder (ppm), F_{std} is the flowrate of standard gas (l min^{-1}) and F_{dil} is the flowrate of diluent gas (l min^{-1}). An estimate of the uncertainty of the concentration of the calibration gas produced is given thus:

$$\frac{\Delta C}{C} = \left[\left(\frac{\Delta C_{\text{std}}}{C_{\text{std}}}\right)^2 + \left(\frac{\Delta F_{\text{std}}}{F_{\text{std}}}\right)^2 + \left(\frac{\Delta F_{\text{dil}}}{F_{\text{dil}}}\right)^2\right]^{1/2}.$$

The uncertainty in these measurements is shown in Table 6.4 for a typical dynamic dilution calibration.

Table 6.4 Uncertainties for dynamic dilution method

Source of uncertainty	*Uncertainty*	*% Uncertainty*
Std cylinder conc.	100 ± 1 ppm	1.0
Std gas flow	$30 \pm .6\ \text{cm}^3\text{min}^{-1}$	2.0
Diluent gas flow	$3 \pm 0.05\ \text{l min}^{-1}$	1.7
Uncertainty of calibration		$\pm$ 4 %

6.3.2.4 Gas phase titration

Gas phase titration (GPT) uses the stoichiometric reaction between NO and ozone producing NO_2:

$$NO + O_3 \rightarrow NO_2 + O_2$$

The technique can be used for either NO_x or O_3 calibration, with the species to be determined titrated against an excess of the other reactant. At near-ambient concentrations, the reaction cannot be considered infinitely fast, and a mixing vessel of sufficient volume must be incorporated into the system to allow the reaction to exceed 99% completion. For NO calibrations, an ozone analyser can be connected to the mixing vessel to check for any unreacted ozone.

Variable NO concentrations are produced in a similar manner as for dynamic dilution, using a high concentration cylinder diluted with air via mass flow controllers. Ozone is generated using a stabilized ultraviolet light source, which can produce a variable O_3 concentration by moving a shutter across the lamp. A diagram of the GPT apparatus is shown in Figure 6.4.

GPT can be used to calibrate either the NO or NO_2 channels of a NO_x analyser as NO is quantitatively converted to NO_2 by reaction with O_3. Excess NO is reacted with O_3, the concentration of which is measured using a photometer. This measurement directly corresponds with the drop in measured NO concentration, and also to the observed increase in NO_2 concentration.

The system can also be used to calibrate an ozone analyser, and thus can provide traceability for the O_3 standard by the use of a primary NO standard prepared, for example, by dynamic dilution. In this case, excess O_3 is titrated

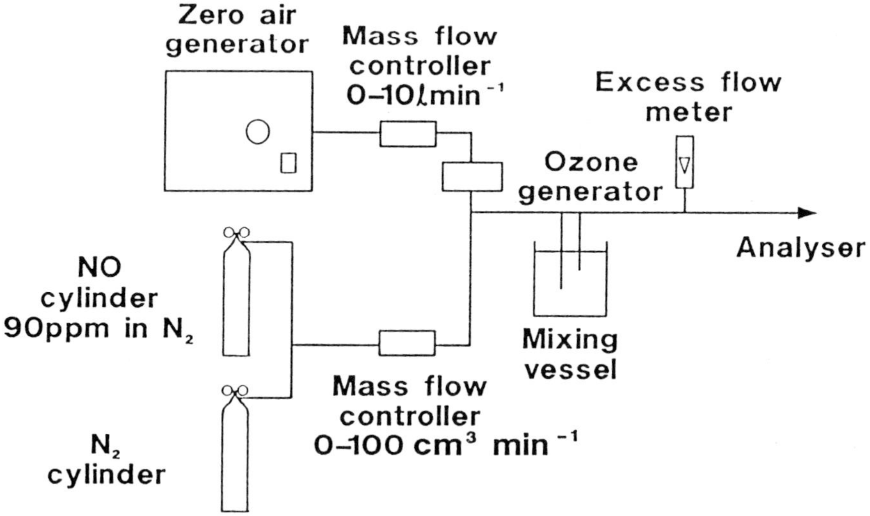

Figure 6.4 *A schematic of the gas phase titration apparatus*

against a known NO concentration, since the drop in O_3 concentration is equal that of the NO added. Thus, a calibration standard for O_3 is obtained.

Care must be exercised when carrying out GPT to ensure that the reactants are adequately mixed, and that the reaction vessel is sufficiently large to allow the reaction to proceed to completion. The vessel and tubing require considerable conditioning before a reliable reading is obtained, as any moisture present will result in unstable readings. Additionally, the efficiency of the NO_2 to NO converter in the NO_x analyser should be checked prior to use.

The estimate of the uncertainty of the concentration in the calibration gas is made as for that of dynamic dilution above, with the addition of a factor to account for the completeness of reaction between NO and O_3. An estimate of this uncertainty is given in Table 6.5, but this can only be considered as an approximation because the reaction efficiency is variable, and insufficient data exist at present to allow an estimate to be made from the standard deviation of individual calibrations.

Table 6.5 Uncertainties for gas phase titration method

Source of uncertainty	*Uncertainty*	*% Uncertainty*
Standard gas conc.	0.5 ± 0.02 ppm	4.7
Reaction efficiency	99 ± 1%	1.0
Uncertainty of calibration		± 4.8%

6.3.2.5 UV Photometry

This method of ozone determination uses an ultraviolet photometer as the primary standard for O_3 concentration. The photometer uses the Beer-Lambert equation:

$$I = I_0 e^{-aLc}$$

where I is the light intensity through ozone, I_0 is the light intensity through scrubbed air, a is absorption coefficient at 254 nm (308 $atm^{-1}\,cm^{-1}$), L is the path length and c is the concentration ratio by volume.

Since intensity measurements need only be determined as a ratio, this method can be considered as a primary standard where the cell path length is accurately known and the photometer is demonstrated to be functioning correctly by, for instance, comparison with a second instrument or a GPT experiment. It is important that the photometer used corrects for temperature and pressure variations.

To calibrate an analyser, an O_3 generator is used to produce a suitable concentration, and the analyser and photometer are set up in parallel to measure the calibration gas. The output signal from the two instruments are compared for several different O_3 concentrations. The UV photometry method is particularly useful for field work, as the equipment is reasonably portable.

The estimation of the uncertainty associated with UV photometry is difficult to quantify, as individual uncertainties in the terms in the Beer-Lambert equation above are not all available. An overall estimate will be made from comparisons with calibration data from gas phase titration, but at present, insufficient data are available.

6.3.2.6 Gravimetric cylinders

Gravimetric standard cylinders are available for CO in nitrogen at ambient levels, typically 10 ppm. These are obtained from the National Physical Laboratory with a typical uncertainty of $\pm 1\%$. These cylinders provide a convenient and reliable primary standard for calibration of gas filter correlation analysers. Many older analysers use the non-dispersive IR technique, for which standards made with air as the balance gas are required. Such standard cylinders are also available for NO in nitrogen mixtures, but these require (dynamic) dilution to ambient levels as described previously.

6.3.2 Calibration methodology

The calibration methodology for air quality analysers, which has been developed in the calibration laboratory involves three separate operations:

1. checking the linearity of the analyser response;
2. determination of the instrument zero response;
3. determination of the instrument calibration factor, i.e. concentration units measured per volt of signal output.

In the case of NO_x analysers, the NO_2 converter efficiency is also determined. The linearity of the analyser is checked by performing a multi-point calibration using dynamic dilution as described previously, using a minimum of five points. Note that cylinders of accurately determined concentration are not required for this stage, only accurately determined dilution ratios.

The instrument zero response is determined by introducing zero air to the analyser. Zero air is produced in the calibration laboratory by use of chemical scrubbers for SO_2, NO_2 and water vapour, and catalytic scrubbers for NO and CO.

The instrument calibration factor is then determined by use of a standard calibration gas prepared by one of the primary standard techniques described in Section 6.2. In some cases, the calibration procedure is repeated after 24 hours to check the stability of the analyser.

6.4 Data quality control

It is important that the uncertainties involved in the measurements are known and are traceable. In addition, efforts must be made to ensure that these uncertainties are minimized wherever possible.

In order to maintain traceability, all physical measurements such as flow, pressure and temperature are regularly calibrated against primary standards. Such standards are not necessarily held by WSL, and some of this work is contracted out to suitably equipped and accredited external organizations. Systematic errors in calibrations are minimized by the use of more than one primary standard technique for each pollutant. Random errors are controlled by the use of quality control charts.

Within the calibration laboratory there are four reference cylinders, one each of NO, NO_2, CO, and SO_2. These cylinders are of near-ambient concentration and are of proven stability. Each time a primary standard calibration is performed by any of the methods detailed, the relevant reference cylinder is also calibrated. The concentration of each cylinder is recorded on a quality control chart in the laboratory; an example is shown in Figure 6.5. This provides an ongoing check of calibration data over time, and allows poorly performed calibrations to be identified. As several independent calibration techniques are used for each cylinder, any systematic drift in the reference cylinder concentration over a period of time can be clearly

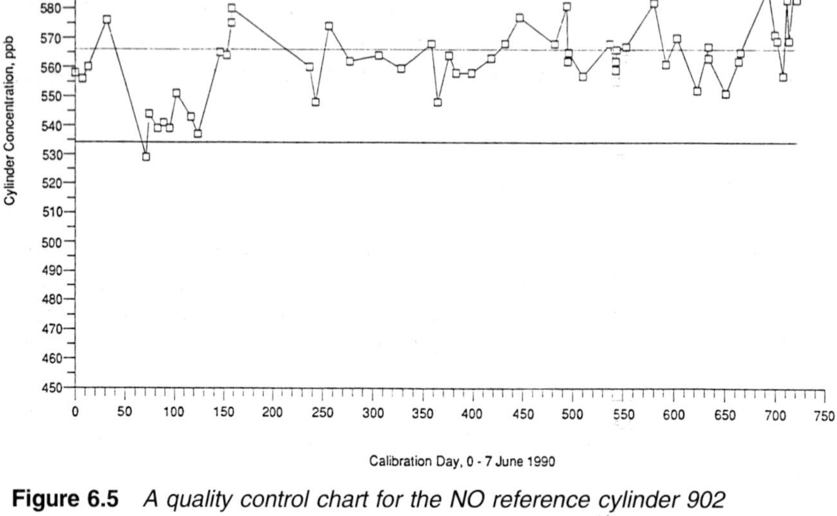

Figure 6.5 *A quality control chart for the NO reference cylinder 902*

identified. The results for each reference cylinder taken over an 18-month period are shown in Table 6.6.

The table shows that taking 2σ as an estimate for the overall uncertainty for the calibration methods gives a relative uncertainty of about 5% for NO, CO and SO_2, and about 7% for NO_2. This higher uncertainty for NO_2 reflects the reactive nature of this species. The estimation of the uncertainties of these cylinder concentrations includes the precision of the analyser used for the determination.

Table 6.6 Mean concentrations obtained for reference cylinders

Cylinder	*Species*	*Mean conc.*	*Std dev.*	*Number of measurements*
902	NO	559 ppb	13 ppb	34
903	CO	20.46 ppm	0.46 ppm	7
904	NO_2	679 ppb	22 ppb	28
905	SO_2	512 ppb	12 ppb	20

6.5 NAMAS accreditation of gas standards laboratory

NAMAS accreditation is currently being sought for the gas standards calibration laboratory. This will cover ambient air quality analysers for the pollutants mentioned above, as well as compressed gas cylinders. NAMAS accreditation involves establishing a comprehensive traceability base for all measurements, and necessitates all measurement equipment used in the calibration laboratory being calibrated by organizations accredited by NAMAS. All such equipment must be recalibrated at regular intervals, usually annually.

In order for NAMAS accreditation to be granted, all procedures involved in the calibration must be comprehensively documented, and detailed records kept of all calibrations, instrument service and repairs. Regular audits by NAMAS recognized staff will ensure that such procedures are rigorously adhered to.

6.6 Intercomparison exercises

WSL has undertaken a number of standards intercomparisons in recent years, on both national and international levels. The most recent intercomparison exercise was carried out in 1990 at RIVM, the Dutch National Institute of Public Health and Environmental Protection. In this exercise, a set of cylinders (one each of NO, NO_2, SO_2 and CO) plus the ozone photometer were calibrated using the WSL calibration facilities. These were then compared against the RIVM standards. The results are shown in Table 6.7.

Table 6.7 Results of 1990 intercomparison with RIVM Holland

Species	*WSL*	*RIVM*	*% Difference*
NO	260 ppb	260 ppb	0
NO_2	254 ppb	256 ppb	1
SO_2	507 ppb	512 ppb	1
CO	20.7 ppm	21.5 ppm	4
O_3	145 ppb	148 ppb	2

The WSL O_3 standard photometer has recently been compared to the long-path ozone photometer at AEA Technology, Harwell. This showed agreement to within 4%. In addition, a comparison held at the Mace Head monitoring site in Eire demonstrated good agreement with the United States NOAA ozone standards.

6.7 Future developments

The gas calibration laboratory, as presently constructed, provides scope for useful future expansion in the scope and coverage of its calibration services. Wider concentration ranges could be covered for existing pollutants, together with additional species such as ammonia and hydrogen sulphide. These additional capabilities could be provided by using existing calibration equipment, requiring only the purchase of the additional gases and analysers.

The ranges of current calibration standards for NO, NO_2, SO_2 and CO could be extended to include both higher and lower concentrations. Standards at lower ranges, for example 5 to 50 ppb, will be useful for calibration of high sensitivity rural analysers. Higher ranges, up to $\sim$1000 ppm, will be appropriate for emission, occupational and vehicle monitoring standards.

A mobile calibration facility could also be developed to allow on-site calibration of air quality analysers. This will involve construction of portable versions of the facilities already available in the fixed laboratory. On-site calibration of analysers is desirable as disconnecting the power supply for transport is likely to change the calibration factor of the analyser. The calibration equipment in the mobile calibration laboratory would need to be easily demounted from the vehicle, as many monitoring sites do not allow close approach of a vehicle. This would probably require a modular design, where only the equipment necessary is transported to the site. Equipment such as permeation ovens could be connected to a mains power supply at a fixed location when not in use, thus removing the need for high capacity portable power supplies.

— 7 —

Checking precipitation gauge performance

B Sevruk

7.1 Abstract

The necessity to evaluate precipitation gauge performance and the procedures which can be used to achieve this goal are discussed. Standard precipitation measurement techniques and reference instruments, such as the pit gauge for rain and the double-fence gauge for snow, are presented and their physical characteristics discussed. The focus is on the systematic error in precipitation measurement, in particular wind-induced losses and wetting and evaporation losses. Methods to evaluate and correct these losses are described and examples of their application presented. The results of international intercomparison exercises using reference instruments are analysed and global aspects of precipitation measurement discussed.

7.2 Introduction

The problem with precipitation measurement techniques is that, despite the simplicity of the common precipitation gauge, the process of precipitation measurement is loaded with a number of systematic errors. These cause considerable losses in precipitation catch, generally amounting to 5 to 10% for rain and up to 50% for snow. Published precipitation data do not include these losses; therefore, the results of many hydrological, meteorological and climatological computations should be treated with caution. Better results can be expected when precipitation data are corrected for systematic measurement errors.

The following sections introduce the basic problems in estimating systematic errors in precipitation measurement, particularly the possibilities for checking precipitation gauges in the field, the estimation and correction of both wind-induced losses and wetting and evaporation losses. Examples of their application are discussed. More detailed information has been published elsewhere by Sevruk (1987a, 1989a).

Within this volume, this paper is also intended to introduce and complement the papers dealing with wind-tunnel studies of precipitation collector and dust gauge design.

7.3 Systematic errors in measurement

When considering systematic errors in precipitation measurement, it is important to note the following basic features of current gauges and measurement techniques.

- The precipitation gauge is a universal instrument to measure rain and snow at the same time.
- It essentially consists of an upright cylinder where precipitation is collected and stored until it is measured.
- The upper end of the precipitation gauge is reinforced to prevent deformation of the orifice and, hence, change of the orifice area.
- The gauge is raised above the ground to provide protection against in-splash and drifting snow.
- Precipitation is usually measured by volume once a day at a fixed observing time.
- The stored water is either poured out from the container into a calibrated vessel, or its level in the container is read directly or measured with a graduated stick.
- Precipitation gauges are installed outdoors in enclosed or open areas under various environmental conditions (e.g. cities, airfields, gardens, parks, woods, mountains, etc.).

A common type of precipitation gauge and its component parts are shown in Figure 7.1.

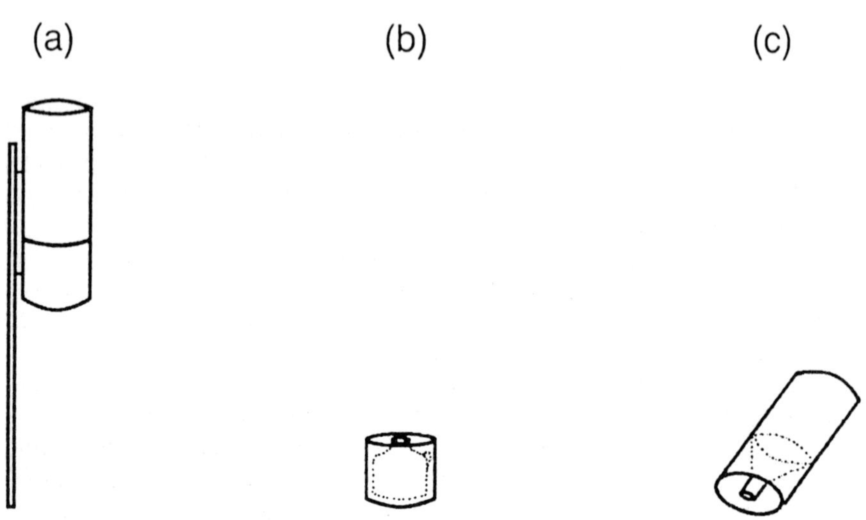

Figure 7.1 *Precipitation gauge: (a) elevated above the ground and fixed on a pile. The component parts: (b) the container with protecting can; and (c) the collector with the funnel*

The characteristics mentioned above give rise to a series of potential errors. The most important of which are summarized below:

- An elevated precipitation gauge distorts the wind field above its orifice. As a result some of the lighter precipitation particles may be deflected from their fall trajectories and fail to enter the collector (wind-induced losses).
- Internal walls of the precipitation gauge are wetted during precipitation events and adhering drops are lost through evaporation (collector wetting losses).
- When emptying a gauge, part of the stored water remains in the container and evaporates (container wetting losses).
- The water or snow stored in the precipitation gauge during the time between the end of a precipitation event and the observing time may evaporate (evaporation losses).

Further sources of error include the splash in and out of raindrops and the blowing in and out of snow.

As noted in the introduction, systematic errors in precipitation measurement can be large. This is not surprising when we consider that:

- precipitation is usually accompanied by wind;
- precipitation is distributed over the year in small amounts occurring with varying frequency, in some regions many hundreds of times per year;
- there are breaks of a few minutes to some hours between events on days with precipitation;
- the duration of evaporation of accumulated precipitation in precipitation gauges is on average between 6 and 12 hours.

7.3.1 Influencing factors

The magnitude of systematic error in precipitation measurement depends on three groups of variables; instrumental, observational and environmental. The latter includes meteorological variables such as wind speed, temperature and humidity, and site characteristics such as the degree of protection of the precipitation gauge against the wind (exposure). Observational variables include the height of the precipitation gauge orifice above the ground and the number and time of precipitation measurements per day. Of the instrumental variables the following are believed to be the most important: the use of a wind shield; the shape of the precipitation gauge body and orifice rim, the thickness of the orifice rim, the material and colour.

A variety of precipitation gauge types having different instrumental parameters are in use throughout the world. The elevation heights differ according to country from 0.2 to 2.0 m. Relevant meteorological variables change with latitude, and in the middle latitudes according to the season. Moreover, there can be significant local variations depending on topography and altitude. For example, precipitation gauge sites in forest areas or in valleys are better protected against wind than those in open areas or in the mountains. Similarly, the fraction of snow increases with increasing altitude and degree of latitude; systematic errors in precipitation measurements of snow are considerably greater than for rain because of the increased sensitivity to windspeed. In addition, the variability of the degree of exposure of precipitation gauge sites in a network contributes to the special variability of systematic errors in measurement.

7.3.2 Consequences

As a result of the sensitivity of errors in precipitation measurement to these factors it is normal for different results to be obtained when (i) different types of precipitation gauges are installed at the same site, and (ii) when gauges of the same type are installed in the same vicinity but under different exposure conditions.

Consequently, changing precipitation gauge types or the degree of exposure of a site can cause an increase or decrease in measured precipitation figures. This, of course, is unrelated to actual changes in precipitation conditions, arising for example from climate change.

Considering (i) the variety of types of precipitation gauge in use worldwide, (ii) the frequent changes in measurement techniques, (iii) the differing exposures of precipitation gauge sites, (iv) the different latitudes and altitudes of sites and (v) the different landscape zones of precipitation gauge sites, the magnitude of the systematic error in precipitation measurements can vary considerably not only between countries using different types of precipitation gauge but also within a national network. Similarly, differences can arise at the same site after changes of instruments and exposure have taken place. As a result, precipitation data sets are not in general compatible and precipitation time-series show multiple inhomogeneities. Thus they are hardly suitable for use to compute 'climatic norms' or to study precipitation trends, nor to draw reliable conclusions concerning global precipitation and climate changes, unless corrected for systematic errors.

The correction procedures are based mainly on results of field intercomparisons and hence have an empirical character (Sevruk 1982). They are helpful for eliminating the effect of different precipitation measurement standards as used in different countries and degrees of exposure within one country. Using this approach the steps in isohyets at national borders disappear and the global compatibility of precipitation time-series can be established. Thus the application of corrections opens new avenues for dealing with the spatial and temporal inhomogeneities in precipitation data sets and makes conclusions regarding global precipitation more certain. In addition, the corrections are also important for hydrological computations such as the global hydrological cycle, the world water balance, local and regional water balances, precipitation-runoff modelling and many other environmental analyses.

7.3.3 The role of international organizations

The existence of temporal and spatial inhomogeneities in precipitation time-series is well known and disturbing. So far attempts have been made to eliminate it on a national level only in the countries of the former USSR. Recognizing that it will not be possible in the near future to obtain the aim of global standardization of precipitation measurement by (i) replacing the 60 types of national, standard gauges as used world-wide with one universal precipitation gauge and (ii) installing all gauges under the same exposure conditions, the World Meteorological Organization decided to introduce two reference gauges. One design is for rain (pit gauge) and one for snow (double fence gauge). These have been used to carry out international intercomparison measurements for liquid and solid precipitation.

The main purpose of these World Meteorological Organization intercomparisons was (i) to check precipitation measurements in the field and to estimate the regional magnitude of systematic errors and, (ii) to develop operative correction procedures for different, national standard gauges.

In this way it will at least be possible to move towards the standardization of global precipitation data sets in the near future, if not the standardization of measurement techniques. The latter is a more complex problem including not only technical and scientific questions but also some political aspects. In addition, the World Meteorological Organization have organized or co-sponsored workshops and published reports and proceedings which provide information on the subject (Sevruk 1982, 1986a, 1989a). The importance of the problem of systematic error in precipitation measurement is underlined by the fact that the Intergovernmental Panel on Climate Change, IPCC, included it in its deliberations as did the Second World Climate Conference.

7.4 Basic models of standard precipitation gauge

According to Sevruk and Klemm (1989b), there are three basic models of standard precipitation gauge. Figure 7.1 shows the most common type, elevated above ground (Figure 7.1(a)) and with its parts disassembled: a removable container situated in the protecting can (Figure 7.1(b)) and a removable collector with a fixed funnel (Figure 7.1(c)). The collected water is stored in the container and is either poured out for measurement or the container itself is graduated. Examples include the Swiss and German standard gauges of the Hellmann type and the UK standard Mk2 precipitation gauge of the Snowdon type. Besides these gauges there are two other models as shown in Figure 7.2, where the cross-sections of all three precipitation gauge models are presented. The second model consists of a single body having a removable funnel. In winter this funnel is removed to enlarge the volume of the container for snow (Figure 7.2, centre). Examples include the Russian Tretyakov gauge, the Swedish and Chinese precipitation

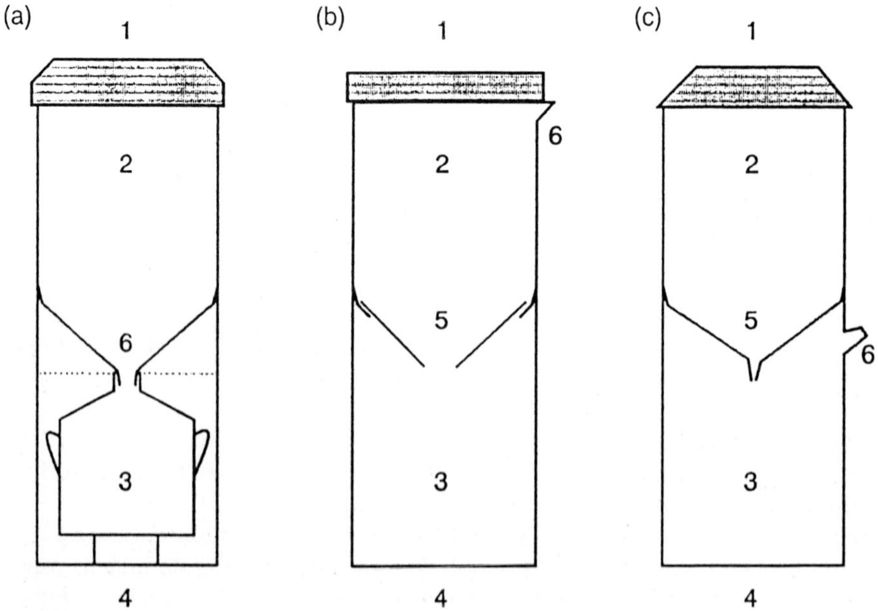

Figure 7.2 *Cross-section of three basic models of precipitation gauges, from the left: a precipitation gauge with removable collector and container (see Figure 7.1); a precipitation gauge with removable funnel and finally a compact precipitation gauge without removable parts and with outflow. Key: (1) orifice rim (these have different shapes and thickness); (2) collector; (3) container; (4) protecting can; (5) removable funnel; (6) outflow*

gauges. The third model of precipitation gauge consists of a compact body, without removable parts, thus joining the collector and container via the fixed funnel (Figure 7.2, right). The water is poured out into a calibrated jar through a tap fixed at the side of the container (e.g. Dutch gauge). The upper edge of the collector is reinforced with a thicker orifice rim to prevent its deformation and, hence, any change in the strictly defined size of the orifice area.

The different constructions of the three basic models of standard precipitation gauge described above considerably affect the magnitude of evaporative losses. Further relevant construction parameters, separated according to the type of losses, are shown in Figure 7.3. For example in the case of wind-induced losses, critical roles are played by the use of a wind shield (see Figure 7.11) and the thickness and shape of the orifice rim, which can vary considerably depending on the type of precipitation gauge as shown in Figures 7.2 and 7.4. In contrast, the area of the internal walls rather than the orifice area is the main parameter influencing wetting losses; additional parameters to be considered are the material and colour of the gauge.

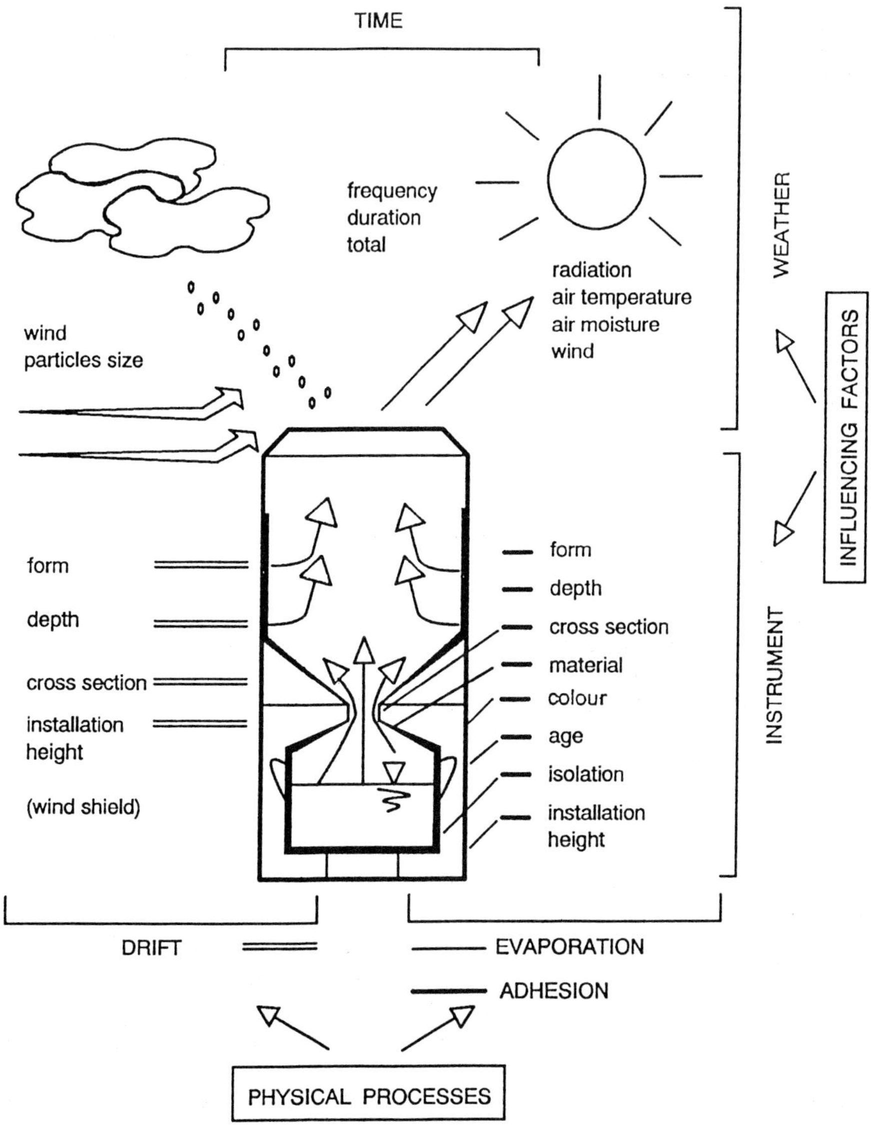

Figure 7.3 *Schematic representation of relevant physical processes affecting the systematic error of precipitation measurement, according to Sevruk (1981). In the upper part are the most important weather variables, in the centre are instrument variables and at the bottom are the relevant physical processes, such as, on the left, wind field deformation (drift) and, on the right, evaporation and adhesion. Each process is represented by a special legend as follows: empty double solid lines indicates wind field deformation, small line indicates evaporation and the filled double solid line indicates adhesion (wetting)*

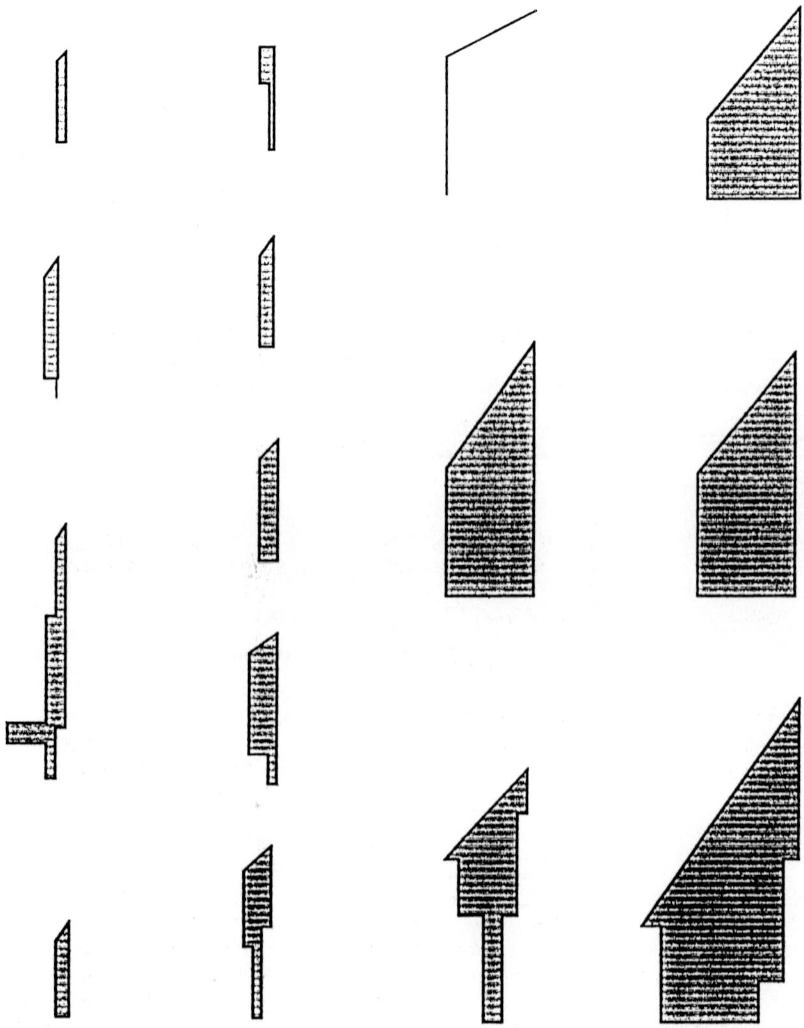

Figure 7.4 *Cross-section of orifice rims of selected precipitation gauges*

Apart from the standard precipitation gauges, there are also recording and storage gauges designed to measure precipitation amounts over shorter or longer time intervals respectively.

7.5 Wetting losses

In Figure 7.5, the factors influencing wetting losses and the measurement techniques used to assess them are summarized. Different procedures may be applied for the collector and container. The ratio of the area of the internal surfaces of the collector to its orifice area provides an index for the magnitude of wetting losses; it expresses the relative size of wetted area for a

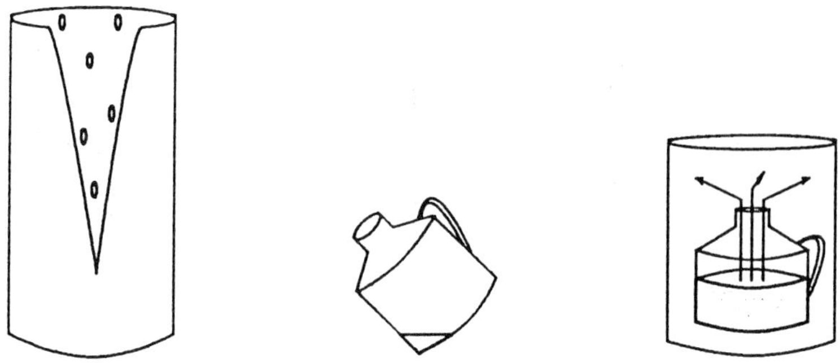

Wetting of collector: Drops hanging on internal walls of collector	Wetting of container: water rest after emptying the container	Evaporation of water accumulated in the container

Influencing factors

Weather

Frequency, type and amount of precipitation, drying time of collector internal walls	Amount of precipitation, frequency of emptying of container, drying time of container	Type of precipitation, saturation deficit, wind speed

Instrument

Orifice area, depth, material, color and age of collector	Material, color and age of container	Type of guage, isolation of container, color and age of collector and size of opening of container

Measurement techniques

<u>Wetting test</u>
Internal walls of collector are wetted using a syringe, and the weight difference of dry and wet collector gives wetting losses per precipitation event

<u>Emptying tests</u>
Water is poured in and out of the container and the weight difference of wetted and dry container gives wetting losses per one emptying the of container

Precipitation is measured at the same site immediately after the end of precipitation event using two gauges: One gauge is measured using a balance and the other one by volumetric method. The difference between precipitation values shows wetting losses.

<u>Active and passive experiment</u>
The difference between a given amount of water as poured into the gauge in the morning and measured in the evening on a precipitation-free day. Difference between the precipitation as measured by two gauges; one measured immediately after the end of storm and the other as usual at the observation time.

Figure 7.5 *Wetting and evaporation losses and the influencing factors*

particular precipitation gauge type. The greater the ratio, the greater the relative wetting loss. This relationship is shown in Figure 7.6 for three construction materials. Roughly one half of standard precipitation gauges world-wide have ratios (S_w/S_o) between two and six. Most gauges are made from galvanized iron sheet and copper but aluminium, brass and plastic are

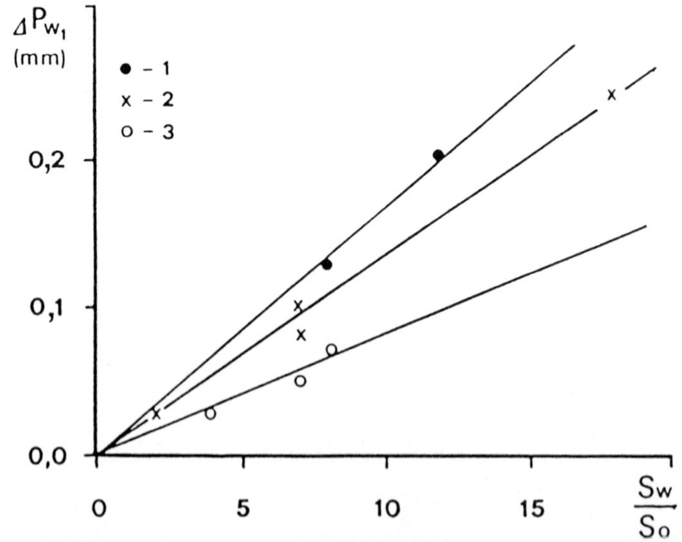

Figure 7.6 *Wetting loss, P_{w1}, as a function of the total area of inner walls, S_w, of the precipitation gauge collector and the orifice area, S_0, for various materials, according to Golubev (1986): (1) galvanized steel sheet; (2) lacquered steel sheet; (3) polished metal or glass*

also used. This, of course, also affects the magnitude of wetting losses, as polished metals have considerably smaller wetting losses than galvanized ones. Wetting losses can amount to up to 0.2 mm per rain event, as shown in Figure 7.6. In the case of snow, however, losses are roughly one half of those for rain as, in contrast to raindrops, snowflakes do not adhere to the internal collector surfaces. When there are several precipitation events in a day, the total wetting loss for the collector increases.

7.6 Evaporation losses

Figure 7.5 also lists factors influencing evaporation losses and summarizes how they may be measured. In Figure 7.7, the relationship between evaporation losses and meteorological variables is shown for the three basic models of precipitation gauge shown in Figure 7.2.

Evaporation is generally small for precipitation gauges with a protected container such as the Hellmann and Snowdon gauges (Figure 7.2(a)), but can be three to four times larger for gauges with a removable funnel, such as the Tretyakov, Swedish and Chinese models (Figure 7.2, centre). This is particularly true during wintertime, when the funnel is taken out of the gauge and the accumulated water or snow in the container is exposed to sunshine and wind. The smallest evaporation losses, particularly in the summer, occur with the compact gauges with a fixed container and restricted ventilation, such as the Nipher precipitation gauge and the Finnish

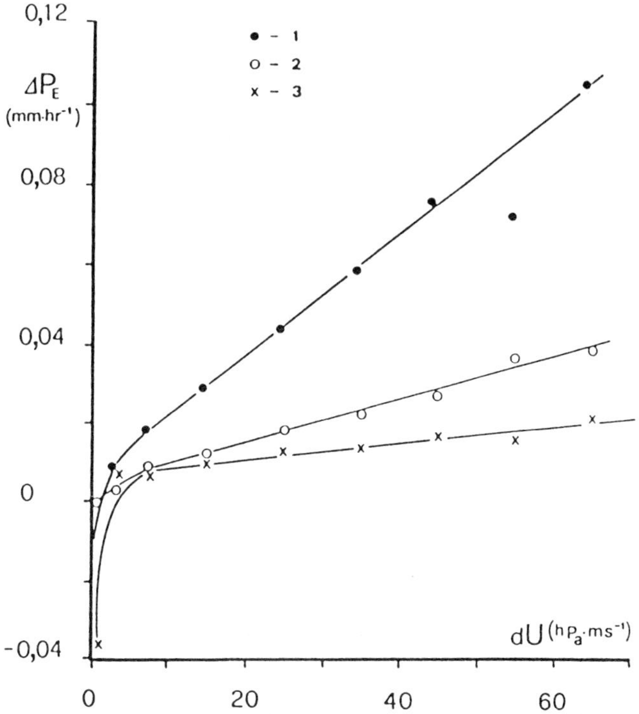

Figure 7.7 *Intensity of evaporation,* P_E*, from the precipitation gauge container as a function of the product of the saturation deficit and wind speed,* dU*, for various gauges, according to Golubev (1986): (1) Tretyakov gauge with removable funnel; (2) Hellmann gauge with fixed funnel and removable container; (3) Nipher compact gauge with fixed funnel and container*

Wild precipitation gauge (Figure 7.2, right). Most types of precipitation gauge experience evaporation losses similar to those for the Hellmann gauge (curve 2 in Figure 7.7). Apart from the type of gauge, the evaporation loss depends on the saturation deficit and wind speed as indicated in Figure 7.7 and, of course, on the duration of evaporation and the time of day (day/night). However, the model of precipitation gauge seems to play the most important role; differences between various models under the same climatic conditions can be as much as a factor of five.

7.7 Wind-induced losses

The physical background to wind-induced loss is shown in Figure 7.8. An important factor is whether or not a wind shield is used. This helps to reduce wind speed over the precipitation gauge orifice and, hence, increases the collection efficiency. In addition, the shape of the precipitation gauge body and the orifice rim shape and thickness are also important. Precipitation gauges fitted with a wind shield or with a thin orifice rim have smaller wind-

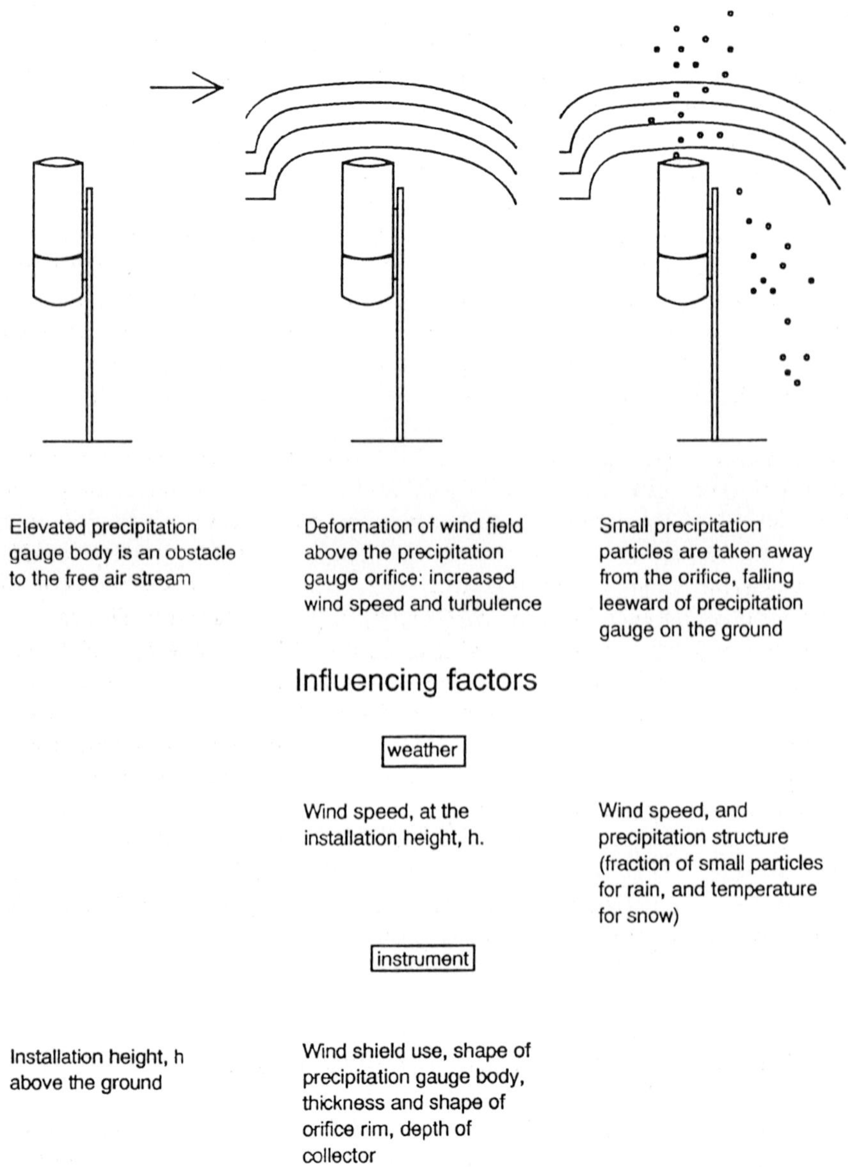

Figure 7.8 *Physical background and the influencing factors of wind-induced loss*

induced losses, particularly during snowfall as discussed in the following section. Precipitation gauges may be checked for wind-induced losses in the field using reference instruments, the pit precipitation gauge for rain and the double-fence gauge for snow (Goodison *et al.*, 1989). In addition, the Tretyakov gauge is used as a working reference and the 'bush precipitation gauge' as the absolute reference. Good results can also be obtained using water equivalent of new snow (Sevruk, 1983, 1987b). Wind-tunnel studies allow detailed investigation of wind-induced losses under controlled conditions and are extensively discussed elsewhere in this volume. By measuring wind speed acceleration over the precipitation gauge orifice it is possible to evaluate and to compare gauge performance. An example of this approach can also be found in Sevruk *et al.* (1989c). The numerical simulation of the wind field above the gauge orifice also offers great potential for evaluating and improving design features (Sevruk *et al.*, 1992a).

The four field reference precipitation gauges discussed above are shown in Figure 7.9. The double-fence gauge consists of two circular, lath fences, the outer one having a diameter of 12 m and a height of 3 m and the inner one having a diameter of 4 m and a height of 2.5 m; in the centre is a Tretyakov gauge. The 'bush precipitation gauge' consists of a bush area of 100 x 100 m^2 cut to the level of the gauge orifice, which is placed near the centre of the area.

The results of field measurements can be used as follows: the difference in precipitation values between an elevated and a reference precipitation gauge of the same type is related to the average wind speed during a precipitation event and to a parameter of precipitation structure as shown in Figure 7.10. On the basis of these known variables, the correction for a specific type of precipitation gauge can be estimated. For example, as shown in Figure 7.10, the Tretyakov gauge with a wind shield catches at least twice the volume of snow compared to the unshielded Hellmann gauge. The difference increases quadratically with increasing wind speed and decreasing air temperature defining the snow structure. Consequently, field intercomparisons aiming to develop correction procedures for wind-induced error have to be repeated for each different precipitation gauge type. Unfortunately, as yet there are only a few types of gauge for which similar relationships to those in Figure 7.10 have been developed. It is intended to improve this situation with data from the current international intercomparisons organized by the World Meteorological Organization (Goodison *et al.*, 1989).

7.8 International intercomparison measurements

During the period 1972 to 1976 the World Meteorological Organization organized intercomparison measurements of national precipitation gauges with a reference pit precipitation gauge (Figure 7.9) at 60 sites in 22 countries. The difference in precipitation values which was used to express the wind-induced loss for rain amounted on average to 3% but varied from 1 to 20%. For example, in Europe the difference at sites at altitudes below 100 m above sea level was 2% in Belgium, 4% in the Netherlands and 12% in Denmark. This clearly illustrates the fact that measured precipitation values from different countries are not necessarily compatible and explains why isohyets show steps at national borders. The three countries mentioned above use not only different types of national standard gauge, but the elevation height varies as well. In Belgium, the national gauge is equipped

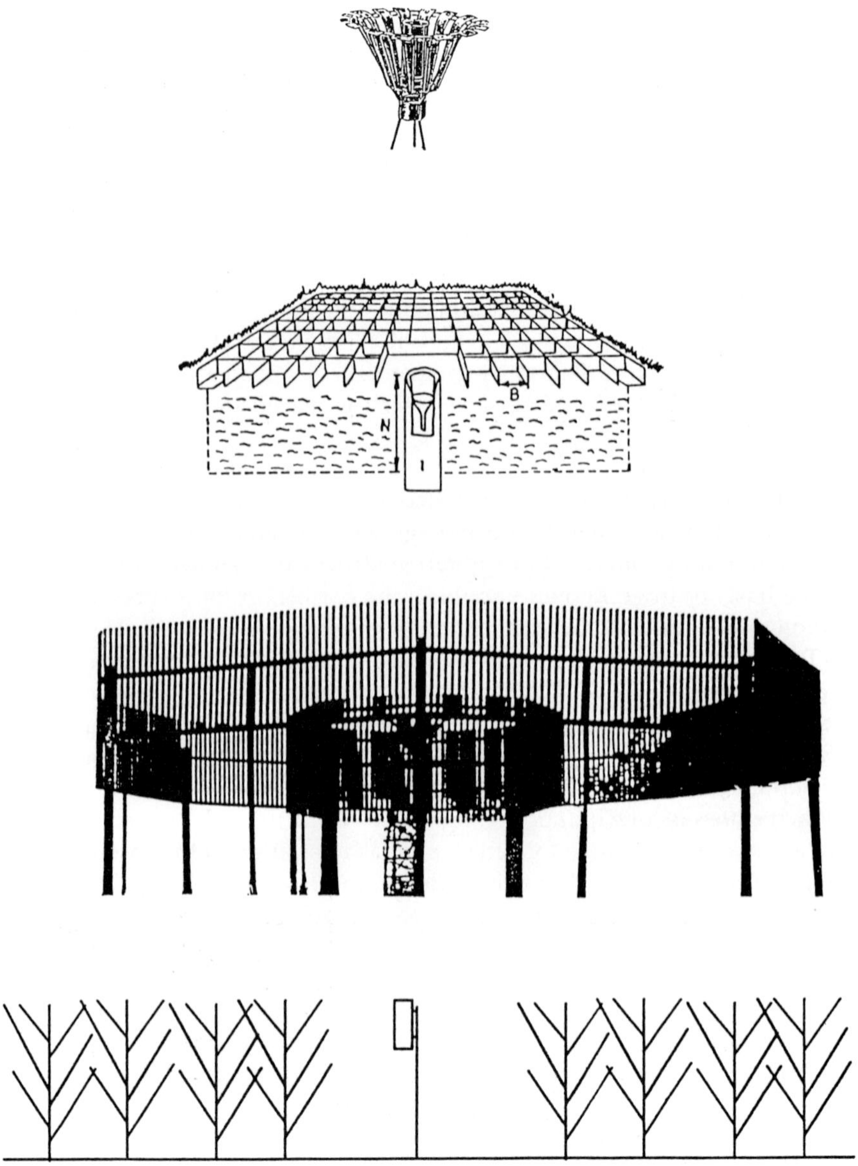

Figure 7.9 *Reference instruments to check wind-induced losses, from the top: the working reference (Tretyakov precipitation gauge with wind shield), the pit gauge with anti-splash protection grid, the double-fenced gauge and the bush gauge*

with a wind shield and the height is 0.6 m. This gauge catches the most precipitation and is in best agreement with the reference gauge. In the Netherlands, the national gauge has no shield and is installed 0.4 m above ground; the difference was larger than that in Belgium. In Denmark, the precipitation gauge is elevated 1.5 m above ground and is not shielded. This

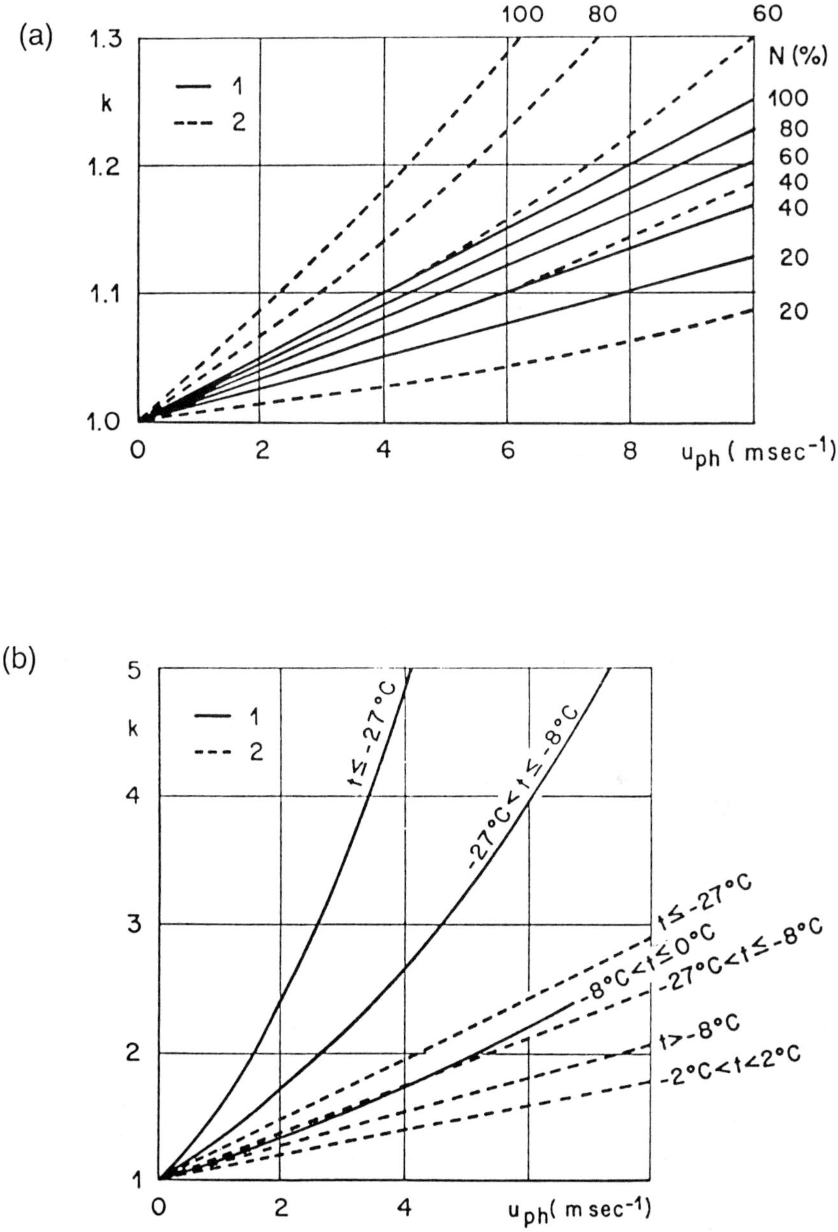

Figure 7.10 *Wind-induced losses expressed by the conversion factor, k, of 'true' values to measured precipitation values as a function of the wind velocity, u, during precipitation periods: (a) at the installation height of the precipitation gauge and the structure parameter, N, for rain; and (b) as a function of air temperature for snow. Adapted by Sevruk (1982) from UNESCO (1978). Gauge types: (1) Hellmann precipitation gauge; (2) Tretyakov precipitation gauge*

explains the 12% difference compared to the reference gauge. In addition, different climatic conditions in the three countries have some effect. Therefore, when making comparisons of precipitation gauge performance, wind speed and the structure of precipitation have also to be considered, as noted by Sevruk and Hamon (1984) in evaluating the results of these intercomparisons. The effect of wind speed was more evident at sites with low rainfall intensities. Introducing the percentage of rainfall with low intensity ($i_p \leq 0.03$ mm min^{-1}) as a further parameter, *N*, (see Figure 7.10), a non-linear relationship was found for the estimation of the above difference at six sites.

The differences for snow are larger than for rain. To complement the pit precipitation gauge results, the World Meteorological Organization organized between 1986 and 1992, intercomparison measurements for solid precipitation using the double-fence precipitation gauge as a reference (Goodison *et al.*, 1989). Eight countries and 13 sites are taking part in this

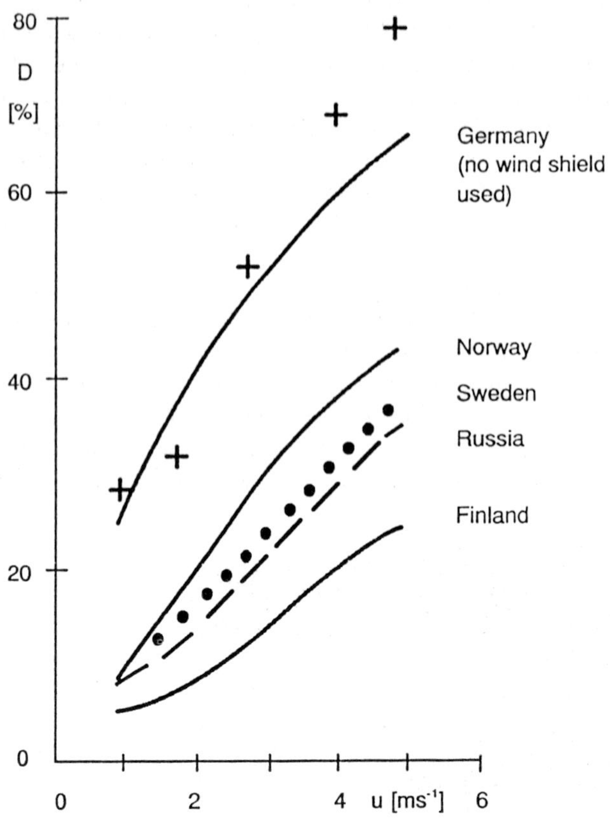

Figure 7.11 *Wind-induced losses,* D, *for snow, as a function of wind speed,* u. D *is expressed by differences between precipitation values as measured by the double-fenced gauge and various national standard gauges, with and without wind shield, according to WMO (1991). Crosses indicate control measurement from other places for the Hellmann gauge*

intercomparison. Preliminary results indicate that the wind-induced losses for snow depend on the type of precipitation gauge used and on the wind speed, as shown in Figure 7.11. For example, for a wind speed of 4 m s^{-1} losses range from 20% for the Finnish gauge with a wind shield to 55% for the unshielded Hellmann gauge (WMO, 1991). The results also show agreement between the undercatch in precipitation as related to wind speed for the same type of gauge at different sites (shown by the comparison of the crosses and the upper curve in Figure 7.11).

7.9 Applications

Important areas of application of systematic error in precipitation measurement include the mapping of precipitation and the elimination of inhomogeneities in precipitation time-series. The latter is particularly relevant to studies of climate change. Because of the systematic wind-induced losses discussed earlier, even small changes in the construction parameters or installation height of a gauge and/or site exposure can cause serious inhomogeneities which can be mistaken for evidence of climate change. Knowledge of so-called metadata (data about the collection of data) for a particular precipitation gauge site or a study of the site's history can help to detect relevant changes. To this end the degree of exposure and changes during its period of observation have to be assessed for each site by drawing on national meteorological archives. Four classes of exposure can be defined, based on the average vertical angle, α of obstacles around the gauge as measured in eight 45° sectors: open site, α = 0-5°; partly open, α = 6-12°; partly protected, α = 13-20° and protected site, α = 21-27°. Figure 7.12 shows a partly protected precipitation gauge site.

When mapping precipitation, corrections should be introduced into the precipitation data set. In this case, besides the development of correction procedures for a particular type of gauge, there are additional problems connected with the implementation of corrections for a particular site. These

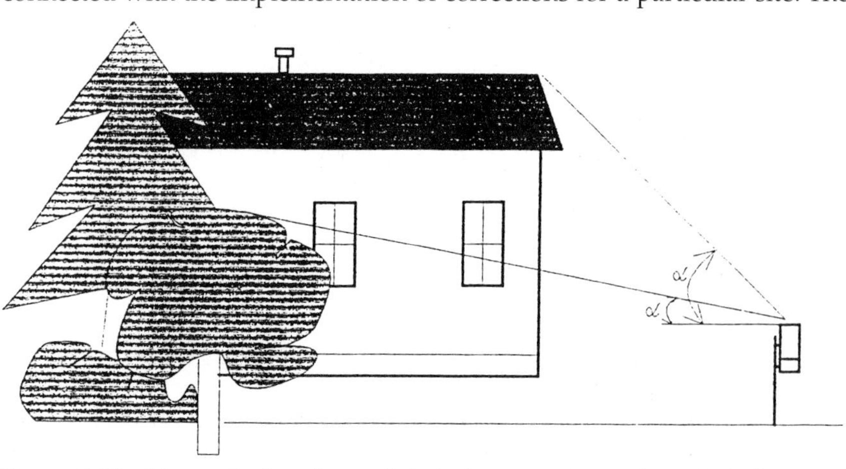

Figure 7.12 *The vertical angle, α, of obstacles around a partly protected gauge site*

include, for example, the assessment of wind speed, temperature and precipitation structure for sites where these variables have not been measured directly. Further, the degree of exposure of the gauge should be estimated and the wind speed reduced to that at the level of the orifice. Examples are given in Sevruk (1986b) and Sevruk and Kirchhofer (1992b).

7.10 Conclusions

Precipitation measurement is subject to considerable systematic errors. By checking precipitation measurements using laboratory tests and field intercomparisons with reference instruments such as the pit gauge and double-fence gauge, the magnitude of these systematic errors can be assessed. This will depend on three types of variable; instrumental, observational and environmental. The most important instrumental variables are the use of a wind shield and the shape and thickness of the orifice rim. An important observational variable is the height of the precipitation gauge orifice above the ground. Of the environmental variables, besides wind speed, the type and structure of precipitation and the exposure of the gauge site must be considered. The introduction of corrections to account for these systematic errors in precipitation measurements can considerably improve the quality of precipitation data and the accuracy of climatological and hydrological computations.

7.11 References

Golubev V S (1986). Random and systematic errors of precipitation measurements by network instruments. In: B Sevruk (ed), Proc. International Workshop on Correction of Precipitation Measurements, Swiss Federal Institute of Technology, ETH Zurich, *Zurcher Geographische Schriften*, **23**, 131-140.

Goodison B E, Sevruk B and Klemm S (1989). WMO solid precipitation measurement intercomparison: objectives, methodology, analysis. In: J W Delleur (ed), Atmospheric deposition. Internat. Assoc. Hydrol. Sci., Publ. No. 179, 57-64.

Sevruk B (1981). Methodical investigation of systematic errors of Hellmann rain gauges in the summer season in Switzerland (in German). Trans. Versuchsanstalt für Wasserbau, Hydrologie und Glaziologie, Zürich, Mitt No 52, 297.

Sevruk B (1982). Methods of correction for systematic error in point precipitation measurement for operational use. Operational Hydrol. Rep., 21, WMO-No. 589, World Meteorol. Org., Geneva.

Sevruk B (1983). Correction of measured precipitation in the Alps using the water equivalent of new snow. *Nordic Hydrology*, **14**, 49-58.

Sevruk B and Hamon W R (1984). International comparison of national precipitation gauges with a reference pit gauge. Instruments and Observing Methods Rep., No. 17, World Meteorol. Org., Geneva.

Sevruk B (ed) (1986a). Correction of precipitation measurements. Proc. International Workshop on the Correction of Precipitation Measurements, Swiss Federal Institute of Technology, ETH Zurich, *Zurcher Georgraphische Schriften*, **23**.

Sevruk B (1986b). Correction of precipitation measurements: Swiss experience. In: B Sevruk (ed), Correction of Precipitation Measurements, Swiss Federal Institute of Technology, ETH Zurich, *Zurcher Georgraphische Schriften*, **23**, 187-196.

Sevruk B (1987a). Point precipitation measurements: why are they not corrected? In: Rodda J C and Matalas N C (eds). Proc. Water for the Future: Hydrology in Perspective. Internat. Assoc. Hydrol. Sci., Pub. No. 164, 477-486.

Sevruk B (1987b). On difficulties of checking storage gauges in winter. In: Proc. 13th Conference of Carpathian Meteorology. Institute of Meteorology and Hydrology, Bucharest, 169-174.

Sevruk B (ed) (1989a). Precipitation Measurement. Proc. International Workshop on Precipitation Measurement, St Moritz, Switzerland, 3-7 December, 1989. Institute of Geography, Swiss Federal Institute of Technology, ETH Zurich.

Sevruk B and Klemm S (1989b). Types of standard precipitation gauges. In: B. Sevruk (ed) Precipitation Measurement. Proc. International Workshop on Precipitation Measurement, St Moritz, Switzerland, 3-7. December, 1989. Institute of Geography, Swiss Federal Institute of Technology, ETH Zurich, 227-232.

Sevruk B, Hertig J-A and Spiess R (1989c). Wind field deformation above precipitation gauge orifices. In: J W Delleur (ed) Atmospheric deposition. Internat. Assoc. Hydrol. Sci., Pub. No. 179, 65-70.

Sevruk B, Nespor V and Spiess R (1992a). The effect of bird protecting ring on wind field deformation above a precipitation gauge. WMO, Instruments and Observing Methods Rep., No.49, WMO/TD-No. 462, World Meteorol. Org., Geneva.

Sevruk B and Kirchhofer W (1992b). The mean annual corrections of measured precipitation amounts, 1951-1980. In: Hydrological Atlas of Switzerland, Part 2.3, Bern.

UNESCO (1978). World water balance and water resources of the earth. UNESCO series: Studies and Reports in Hydrology, No 25, 663, Paris.

WMO (1991). International organizing committee for the WMO solid precipitation measurement intercomparison, 5th Session, 100 pp, World Meteorol. Org., Geneva.

— 8 —

Aerodynamic considerations in precipitation collector design

D J Hall, J G Irwin, B H Stone and S L Upton

8.1 Abstract

The chapter describes investigations aimed at improving the performance of precipitation samplers for chemistry studies. The two features of these samplers which most affect their performance are: firstly, loss of sample due to aerodynamic effects on the entry and retention of precipitation; and, secondly, the accuracy of precipitation sensors that control the opening of wet-only collectors. The aerodynamic behaviour of a variety of existing collector shapes has been examined, including the collection of some reference data on basic cylindrical shapes. The aerodynamic performance of a wet-only collector has been examined in detail and a number of deficiencies identified. Improved collector shapes have been investigated, using streamlined gauge shapes and flow deflectors. A combination of these two features has produced a collector with aerodynamic performance approaching the ideal, though it presents some problems for precipitation chemistry studies. A number of precipitation sensors have been investigated. Sensors detecting changes in electrical resistance across intermeshed electrodes can provide a cheap and effective method for detecting precipitation state but electrode separation, heating and wettability all significantly influence performance.

8.2 Introduction

Precipitation is an important pathway for the removal of pollutant species from the atmosphere and networks of collectors have been established in many parts of the world to collect samples for subsequent analysis. Collecting these samples imposes a number of requirements beyond those normally associated with conventional precipitation measurement, where only the collected volume is of interest. Firstly, contamination of the sample must be avoided and, secondly, a chemically representative sample must be collected. To achieve the first, precipitation collectors are set well above the ground, typically about 2 m, in order to avoid contamination by wind-raised material (whether dust, snow or water) from the ground. This is unlike most conventional gauges which are normally set on the ground, at heights well below 1 m or, preferably, with their openings flush with the ground (Rodda and Smith, 1986).

One of the main reasons for this low-level siting of conventional gauges is to avoid wind effects. The great majority of precipitation collectors use either funnels or cylindrical containers. In a wind, their substantial bulk causes a vertical displacement and acceleration of the windflow over the gauge opening. This distorts the trajectories of the falling precipitation and carries some of it over the opening, resulting in a reduced catch. There is also a wind-driven circulating flow inside the collector, which is capable of ejecting precipitation which has already entered the collector. The problem increases in severity as windspeed increases and as the falling speed of the precipitation decreases. Thus undercollection is greatest at high windspeeds, and for snow and fine rain as against larger sized raindrops.

This directly affects the second requirement of precipitation chemistry studies noted above, that a representative sample should be collected. Being set above the usual height for rain gauges in order to avoid contamination, the collectors are well exposed to the wind. Over a grass surface (of aerodynamic roughness height 2 cm) the windspeed at 2 m height is about 60% higher than at 0.4 m (the height of most conventional gauges). As a result undercollection can be significantly greater than with conventional precipitation gauges. Furthermore, since the deficiencies are strongly dependent upon windspeed and the falling speed of the precipitation, the samples become unrepresentative, being weighted towards that material precipitating at low windspeeds and with the highest falling speed.

From comparative studies in the field, it is known that these collection deficiencies are highly variable. For example, Stone (1991) reports efficiencies for the Warren Spring Laboratory (WSL) bulk collector of between 77% and 99% when compared with a collocated Meteorological Office standard gauge; lower collection efficiencies being observed at the more exposed, windier sites. At an international intercomparison between collectors designed for precipitation chemistry studies at Aspvreten in Sweden, the WSL wet-only collector collected less than 70% in winter (when there was considerable snowfall) and about 95% in summer of the rainfall collected by a Swedish Meteorological Office gauge. It was not unusual in exhibiting these deficiencies or this variation, considering its bulk and height above the ground; comparisons between other collectors show similar variation.

Although the basic causes of wind effects on precipitation collectors have been understood since the last century, the subject has received only limited and intermittent attention, most relatively recently. Rodda (1967) measured the changes in windspeed over rain gauges and, more recently, Sevruk has carried out more comprehensive measurements (Sevruk *et al.*, 1989a, Sevruk, 1990). Attention has also been given to the shapes of precipitation collectors (Folland, 1988). The most common method of trying to reduce wind effects has been to use some sort of shield or fence around the collector, in order to reduce effective windspeeds (Goodison *et al.*, 1989). Designs of this sort have proved popular and of some effect with snow collection. However, their use in precipitation chemistry studies is unwelcome as they are regarded as a possible source of contamination of the collected sample. It is considered preferable that the collector opening should be well clear of any surface upon which material might accumulate and then find its way into the collector.

An attempt to address this problem in a novel way, with regard to snow collection in Scotland, was made a few years ago (Barrett *et al.*, 1985; Hall *et al.*, 1988a, 1989). Over a 4-year programme for the Scottish Development Department and the European Community, a collector was developed which used improved aerodynamic properties rather than shields to control the flow over the opening thereby greatly reducing the displacement and acceleration. Field measurements revealed some initial problems, mainly related to sample retention, particularly of small amounts of precipitation. Snow, for which this collector was designed, is significantly more difficult to collect than rain. Though some problems remained, preliminary measurements suggested it performed quite well as a rain collector. This paper describes further investigations of the fundamental features of wind effects on precipitation collectors and the application of the findings to the production of practical designs. The work is continuing and this paper is, therefore, of an interim nature.

In order to minimize the contribution of dry deposition, wet-only collectors (that is, those which are only open when it is raining) are preferred for monitoring acidity and other species in precipitation. The sensors employed to detect the onset of precipitation are potentially a significant source of undercatch. For this reason a number of sensors have been tested and their performance related to precipitation characteristics typical of Europe.

8.3 Experimental

8.3.1 Wind-tunnel

Measurements of the aerodynamic properties of the various designs of precipitation collector were made in the Warren Spring Laboratory (WSL) number 1 wind-tunnel. This is an open-circuit fan-driven wind tunnel with a working section 22 m long, 1.5 m high and 4.3 m wide (Figure 8.1). There is a faired inlet containing four screens and two honeycombs, for flow conditioning and straightening, and a wide-angle diffuser at the exit. This is also fitted with a honeycomb to control any fan-generated swirl which would otherwise recirculate back into the tunnel inlet. The maximum operating windspeed of the tunnel is about 10 $m\,s^{-1}$. Most of the experiments reported here were conducted at windspeeds of 3 $m\,s^{-1}$.

It is desirable to mimic as well as is practicable the atmospheric turbulence to which samplers are normally exposed as well as the mean windspeed. This is done by setting up a rough-walled channel flow, which produces a flow with the highest scales and intensities of turbulence that are readily manageable in the wind-tunnel. In order to do this, the wind-tunnel roof and floor were roughened using battens of 45 mm height and 22 mm depth

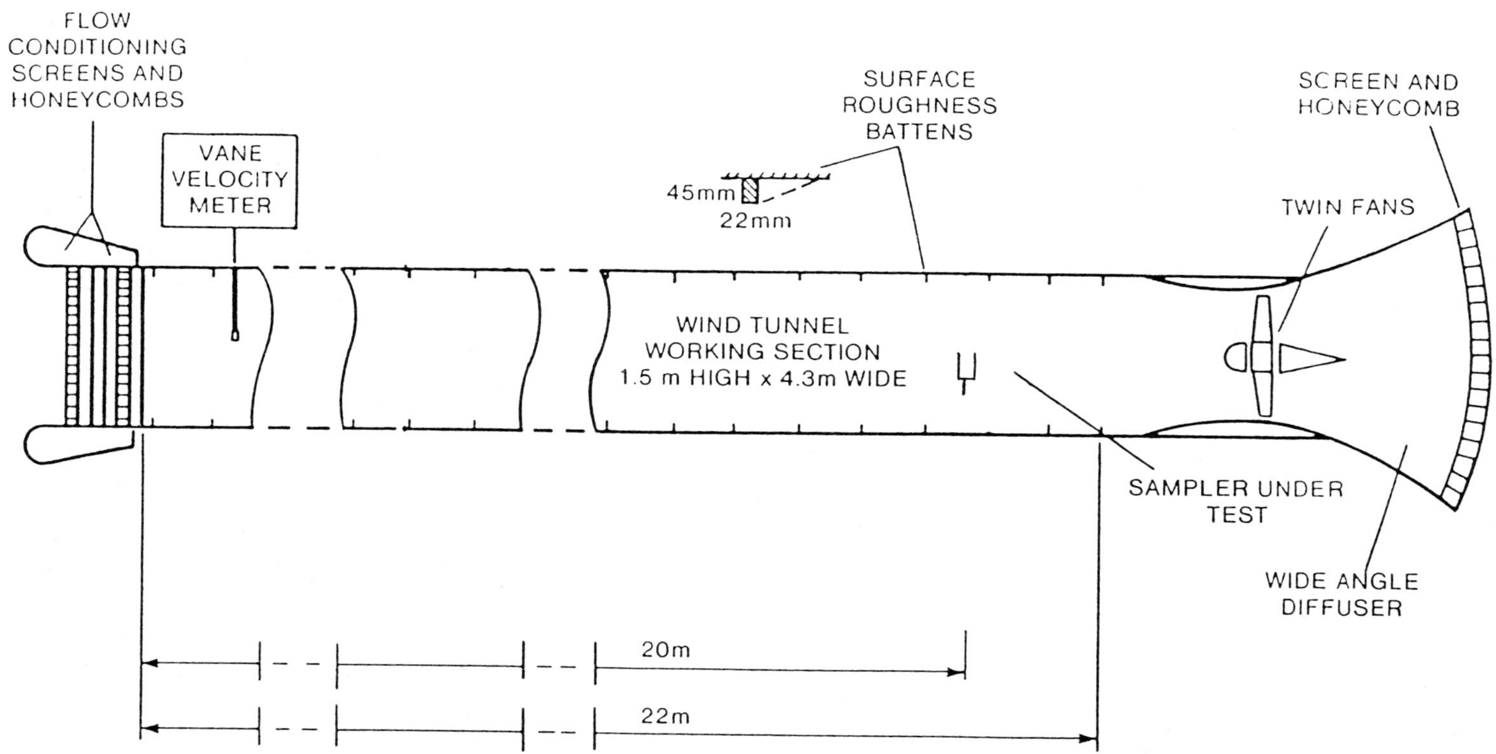

Figure 8.1 *Schematic layout of wind-tunnel experiment*

uniformly spaced 600 mm apart, fixed along the whole length of the tunnel. The same system has been used with work on dust and aerosol samplers. Typical velocity and turbulence profiles are shown in Figure 8.2. The longitudinal turbulence intensity in the central region of the tunnel was around 6%. The measured longitudinal turbulence spectrum is shown in Figure 8.3. The velocity profile has a region of fairly uniform windspeed across the central region of the tunnel where the majority of the measurements over the gauges were made. The velocity profile is consistent with an aerodynamic roughness height of about 3 cm, similar to that of a rough grass surface with a few shrubs or trees present. Some of the gauges investigated, the Institute of Hydrology (IoH) tipping bucket rain gauge and the Meteorological Office 5" rain gauge, are normally set on the ground. For the tipping bucket gauge, measurements were made at two positions in the tunnel: at the tunnel centre (where all of the other measurements were made, and where there would be a flow all around the gauge); and on the floor of the tunnel so that it would be at the same height above the surface as in field usage. The Meteorological Office 5" gauge was also tested on the wind-tunnel floor, at approximately the same height above the surface as is used in the field.

All the collectors tested had a drain hole to a collecting vessel. In some cases, the large funnels for example, this hole was of significant size. The hole was always plugged in order to avoid any possible effects on flow patterns around the collector.

8.3.2 Velocity measurements

Measurements of the windflow over the openings of the various precipitation collectors were made using a pulsed-wire anemometer (Bradbury and Castro, 1971). The pulsed wire is essentially a 'time of flight' device for detecting a pulse of heated air travelling between a horizontal pulsing wire, and two vertical sensor wires positioned either side of it. Its main advantage is that the use of the two opposed sensor wires allows the direction of the wind to be identified as well as its speed. It was thus possible to traverse the vertical windspeed profile down through the shear layer above the collector opening into the region of recirculating flow inside the collector, where negative velocities occur, and to obtain an accurate velocity profile over the whole region. This is a valuable asset for this particular type of work.

Velocity profiles were measured above the centres of the openings of the collectors. Where they were of sufficient depth, measurements were made down into the collectors themselves. The pulsed-wire sampling probe was mounted in the wind-tunnel on a traverse, set up just behind the gauge. Once set up, the traverse and the pulsed-wire anemometer were automatically controlled by the wind-tunnel's main computer system.

Figures 8.4 and 8.5 show typical velocity profiles over deep and shallow collectors respectively. From the velocity profile, the maximum velocity over

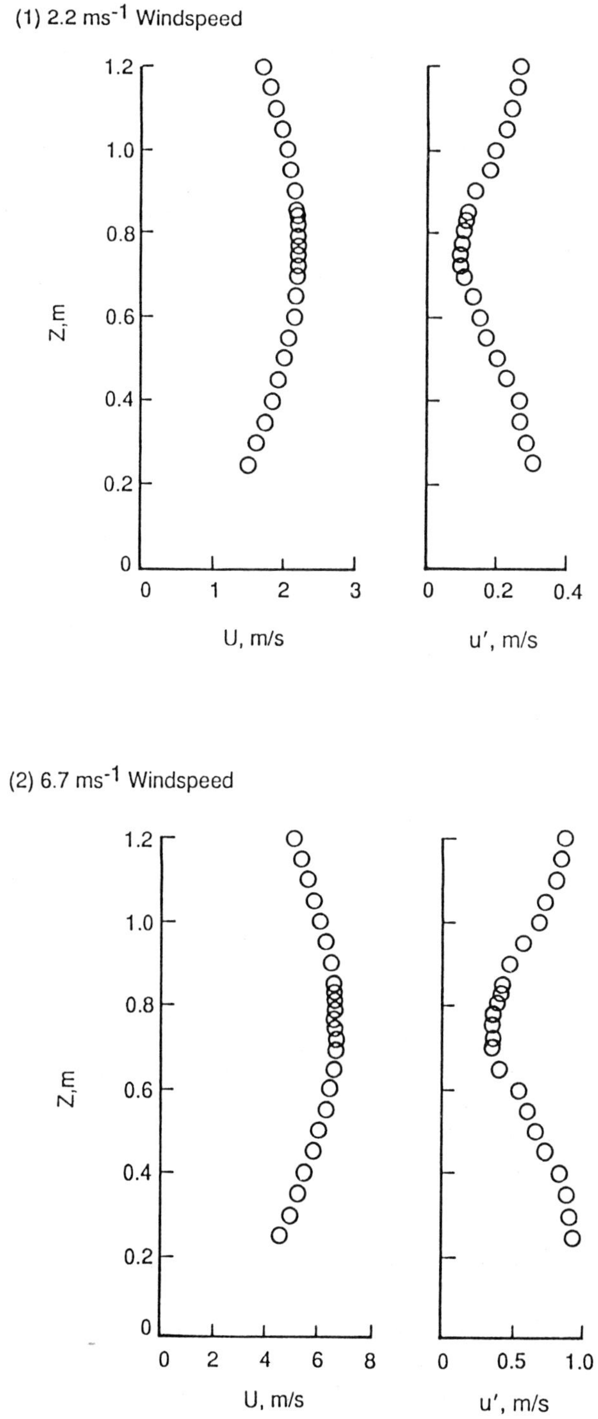

Figure 8.2 *Typical windspeeds and turbulence profiles in wind-tunnel at collector test point*

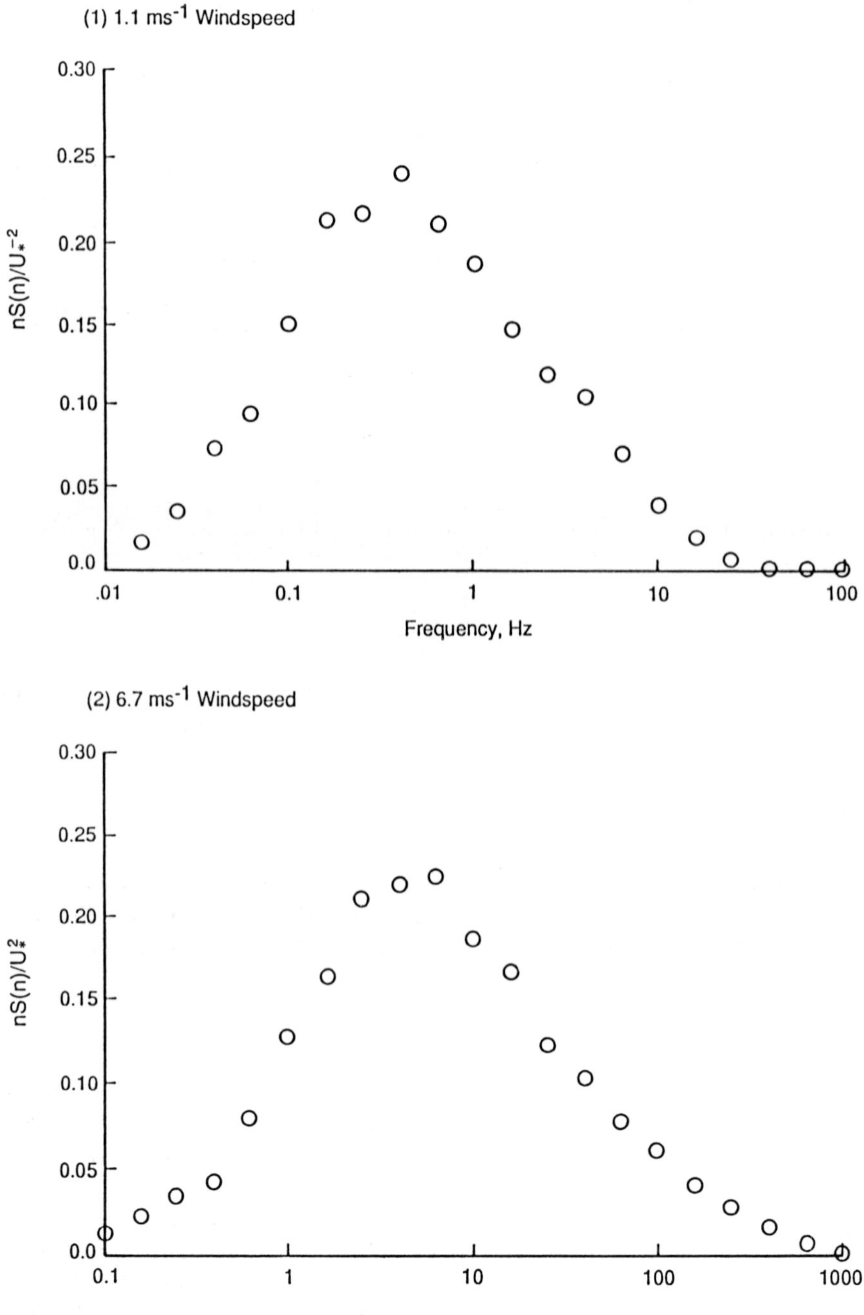

Figure 8.3 *Longitudinal turbulence spectrum at collector test point*

the collector opening and the height above the opening at which it occurred were determined. The method of doing so was by extrapolating curves through the two parts of the profile (above and below the shear layer) and taking the values at their point of intersection, as shown in the figures. This is a somewhat arbitrary definition, but it is a fairly easy point to determine and convenient for rapid assessment of collector performance. It is normally close to the upper edge of the mixing layer at the boundary between the main windflow and the separated flow region within the collector, and so tends to overestimate the velocity increment and displacement a little. The 'acceleration' and streamline 'displacement' were then defined as:

$$\text{acceleration} = \frac{\text{increase in windspeed measured above collector inlet}}{\text{tunnel free stream windspeed at inlet height}}$$

$$= \frac{\Delta U}{U}$$

$$\text{displacement} = \frac{\text{height of maximum windspeed above inlet opening}}{\text{diameter of inlet opening}}$$

$$= \frac{\Delta h}{d}$$

The accuracy of determination of these parameters appeared to be within ± 0.02, limited mainly by the accuracy of velocity measurement.

8.3.3 Precipitation sensors

As noted earlier, wet-only collectors require a sensor to detect the onset of precipitation. Five sensors employing different principles to determine precipitation state were tested and compared under field conditions.

8.3.3.1 Optical sensors

The standard sensor for comparison purposes was an ORG-700 (Scientific Technology Incorporated (STI)). This detects the intensity of scintillations characteristic of precipitation induced in a 0.5 m long beam of partially coherent infra-red light by falling rain droplets. The quoted range for measured intensity is from 0.1 to 3000 mm hr^{-1}; however it will detect rainfall down to an intensity of 0.07 mm hr^{-1}. The lenses are heated to reduce the effect of misting and the unit will operate at temperatures between –50° and +50°C. Comparing rainfall amounts recorded by a collocated tipping bucket rain gauge with the integrated results from the ORG-700 showed good agreement, to within 0.5%.

The ORG-100 (STI) is a smaller and simpler version of the 700 model using the same operating principle. The intensity range covered is 0.5 to 1600

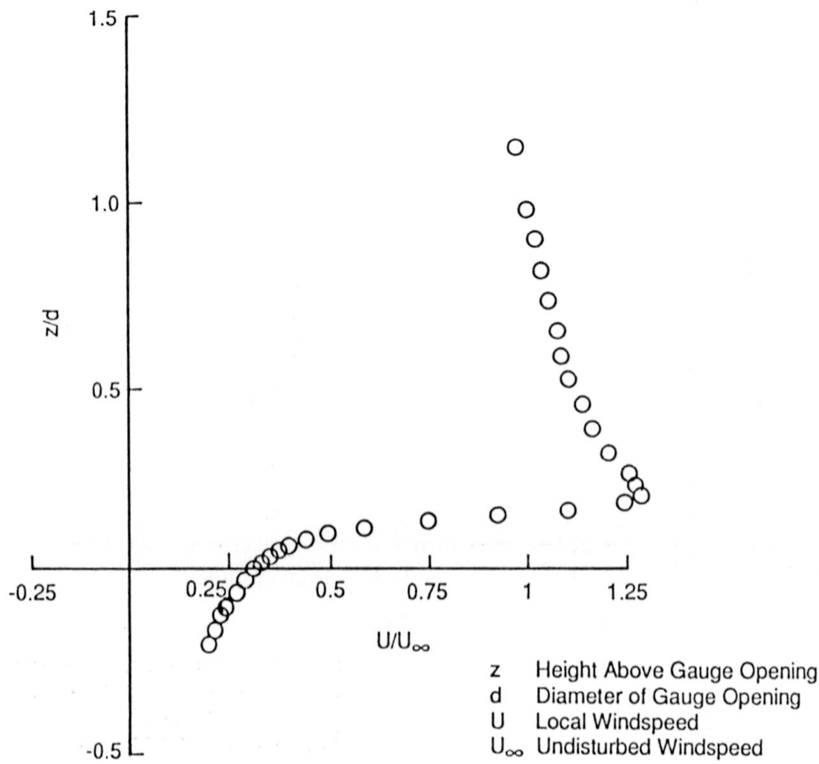

Figure 8.4 *Profile of windspeed over a typical deep cylindrical collector (British Standard deposit gauge)*

mm hr^{-1}. This model operates over the temperature range +1° to +40°C; the lenses are not heated.

8.3.3.2 Electrical resistance sensors

Detecting changes in an electrical circuit caused by rain droplets is the method most widely used to detect precipitation state. Two sensors of this type were used in these tests.

The OMC 270 (Obsermet Ltd) has two intermeshed electrodes each with 23 strips of silver-palladium, 1.2 mm wide, silk screened onto a flat enamelled steel insulating surface giving a total sensing area of 48 cm^2. The electrodes are separated by 0.6 mm with a potential of 10 V at 50 Hz AC applied across them. When the resistance between the electrodes falls to below 25 kΩ a relay operates the switch closure. This switching resistance is adjustable but was preset at the factory to the mid value on the model used. The surface is continually heated at a variable rate between 0.08 and 0.96 W which maintains the sensor surface temperature above ambient temperature

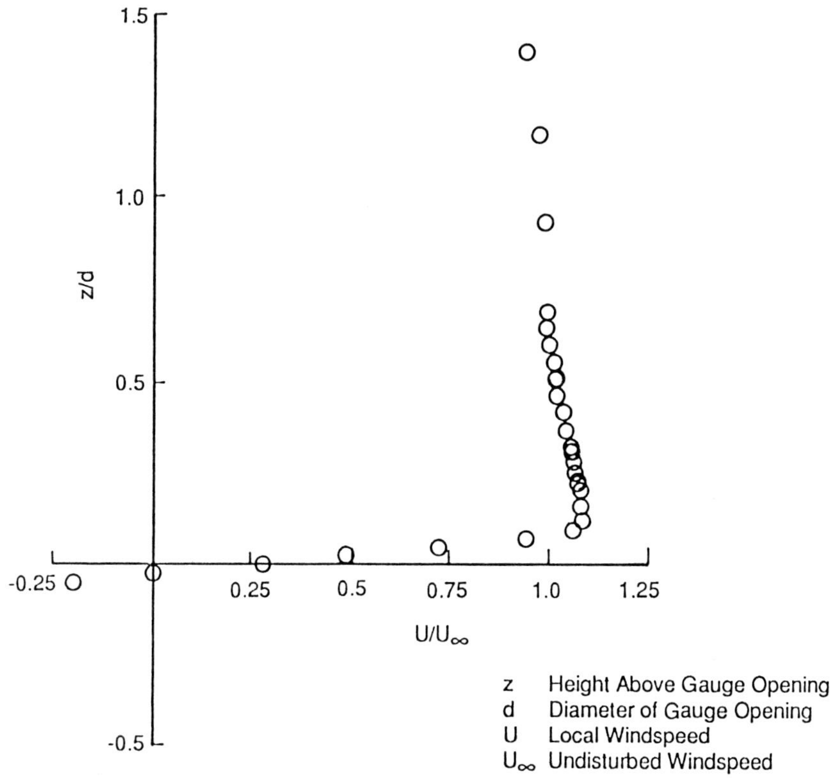

Figure 8.5 *Profiles of windspeed over a typical shallow collector (inverted frisbee)*

by between 1 and 4°C. On detecting precipitation this is increased to 7.2 W. A switch-off delay, adjustable between 1 and 10 minutes is provided.

The WOC sensor is the sensor currently used on wet-only collectors in the UK. The two electrodes of tinned copper are attached to a flexible plastic insulating plate. This is curved gently along one dimension. A potential of 1.5 V at 6 kHz AC is applied across the electrodes which are separated by 1.0 mm. The relay closes the switch contacts when the resistance between the electrodes falls to 50 kΩ. The sensor surface heating is regulated to switch on below about 14°C. Heating power is approximately 13 W which maintains a surface temperature of about 7°C at an ambient temperature of –5°C. Adjustable switch-off and next-event delays were set to 25 s and 20 s respectively.

8.3.3.3 Vibrational sensor

One sensor using an entirely different principle to detect precipitation was the automated dry/wet deposition separator of aerial precipitates (ADDSOAP) sensor developed by the Institute of Terrestrial Ecology. This detects the amplified sound of precipitation as it impacts on a 900 cm^2

stainless steel sheet angled at $\sim 30°$ to the horizontal. It can be set to respond to different numbers and intensities of impacts. For the experiments reported here, settings were selected to respond to one or two impacts of comparatively large droplets. Response can also be set to take account of humidity but this feature was disabled during these experiments. The switch-off delay was set to the minimum of 50 s throughout.

8.3.3.4 Field experiments

During the first 5 months of 1991, these precipitation sensors were installed over a grassy area, surrounded by trees and low buildings. As some trees were up to 10 m high and within 10 m of the sensors, the site was not suitable for representative sampling of precipitation but conditions were the same for each sensor. The ADDSOAP sensor was at a height of 48 cm and all the others at approximately 150 cm. The surface of the OMC 270 instrument was inclined at 15° to the horizontal, and in the same direction as the inclined surface of the ADDSOAP sensor.

Temperature, also at a height of 150 cm, and windspeed and direction (at an approximate height of 2 m) were recorded along with the outputs from the sensors. All inputs were sampled every 0.5 s. Sensor responses were averaged and stored every 15 s, windspeed and direction every 5 minutes, and temperature every 15 minutes. Data were collected for a total of 104 days during the period. After removing periods affected by snow, which is not recorded accurately by the ORG-700 sensor, 101 days of data remained. In the case of the ORG-100, 5% of the data were lost due to the temperature falling below its environmental limit of +1°C.

8.4 Results of experimental programme and discussion

8.4.1 Effect of aspect ratio on the performance of cylindrical collectors

8.4.1.1 Background

The aerodynamic performance of exposed precipitation collectors depends both upon their relative depth and the ratio of depth to diameter, as well as other details of their shape. Most precipitation collectors are of cylindrical form (the other most common form is the conical funnel or a combination of funnel and cylinder), though the ratio of depth to diameter is variable. Reducing the collector depth reduces the displacement and acceleration of the flow over its opening, by reducing the aerodynamic blockage caused by the collector. Also variable are many details of collector shapes, particularly the detailed form of the edge of the collector opening. This is of importance as it also affects the flow over the opening (Sevruk *et al.*, 1989). In our experiments on snow collectors, described above, we took advantage of both reduced collector depth as well as careful shaping in order to improve aerodynamic performance.

One of the problems in assessing the aerodynamic performance of different collectors is in determining to what extent this is a result of the relative depth of the collector or of the details of its shape. In order to clarify this problem we carried out a preliminary exercise to determine the basic flow characteristics of precipitation collectors, using a series of standard cylinders of varying depth to diameter. The intention was to provide a reference data set against which different collectors could be compared, so that it would be clear to what extent changes in collector shape resulted in improvements or deterioration over the simple form. The cylinders used in the experiments were all straight sided and of 200 mm diameter; the heights varied between 10 mm and 600 mm. Up to 300 mm depth, the cylinders were manufactured of aluminium sheet, beyond this the tallest metal cylinder was extended by wrapping an extension of thin card tightly around it. The depth to diameter, or aspect ratio of the collector was defined as:

$$\text{aspect ratio} = \frac{\text{height of cylinder}}{\text{diameter of cylinder}}$$

With the heights of cylinder used, aspect ratios varied between 0.05 and 3. The walls of the cylinders were about 1 mm thick and the edges of the openings were simply machined off square. With such a small wall thickness the shape of the edge is not important.

Two aspects of the effect of aspect ratio on flow behaviour were investigated, firstly the variation in acceleration and displacement and, secondly, the ejection of material from inside the collector due to the internal wind-driven circulation.

8.4.1.2 Effect of aspect ratio on the flow over the collector opening

Figures 8.6 and 8.7 show respectively the variation in the acceleration and displacement of the wind flow over the cylinders. Two sets of data are plotted, firstly, the measurements made in the relatively high turbulence flow in which all the other measurements discussed in this paper were made. These are the triangular symbols on the figure. Secondly, there are some earlier measurements made in a low turbulence flow, of about 1-2% intensity compared with the 6% intensity of the more recent work. Both show the same general trends and the same values of displacement to within the level of accuracy of the measurements. There is a difference of about 10-15% between the two sets of acceleration measurements, the lower turbulence measurements showing the higher acceleration. There is initially a rapid increase in both acceleration and displacement as the aspect ratio of the collector increases. Beyond an aspect ratio of about 0.5, there is an abrupt change and though both the displacement and the acceleration continue to increase it is at a slow rate. The highest value of the displacement for the high turbulence measurements is about 0.22 and of the acceleration 0.25; it

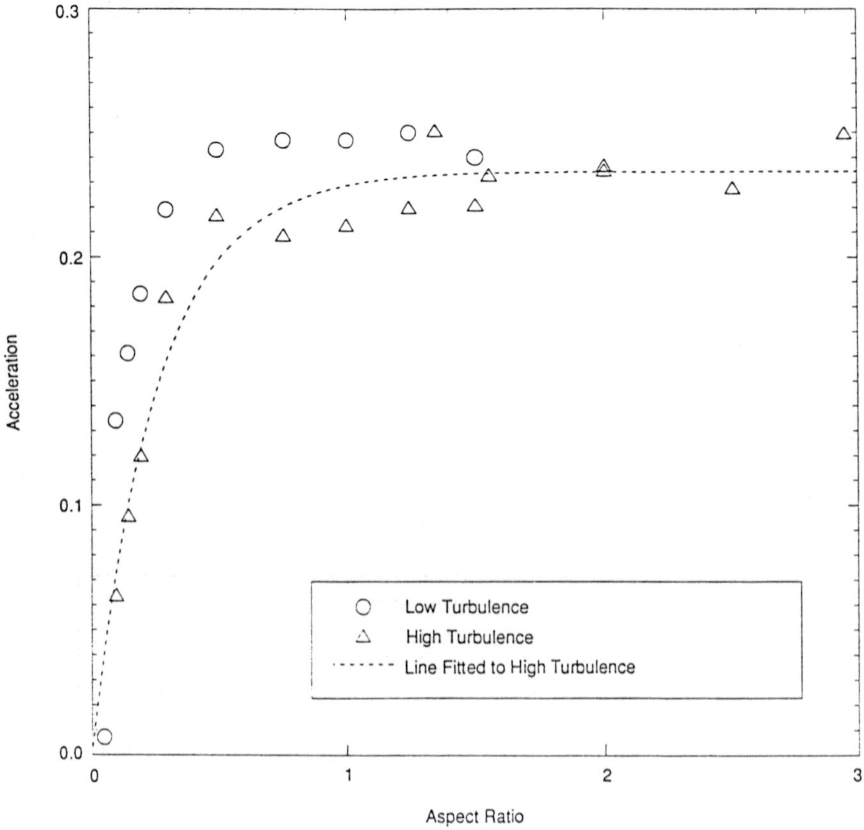

Figure 8.6 *Effect of aspect ratio on acceleration over cylindrical collectors*

seems likely that these values are close to the maximum that will occur with a cylinder of any aspect ratio.

It is not surprising that some limit to the displacement and acceleration should be reached. There clearly must be some depth of cylinder beyond which the flow at its end becomes a purely 'local' phenomenon independent of its depth. What is perhaps surprising is the relatively small depth of cylinder at which this limit is approached, with an aspect ratio well below unity. It follows from this that if a precipitation collector is to make use of reduced depth in order to improve the flow over the opening, it must be quite shallow. The aspect ratio would need to be below 0.5 and preferably below 0.2. There are practical difficulties associated with such shallow collectors, for example with effective drainage of collected samples and with problems due to splashing by large raindrops, particularly during heavy rainfall. These matters are not beyond solution, the subject is considered in more detail in our earlier work on snow collectors and also that on dust gauges (described in another part of this volume). However, it also follows that, if deeper collectors are preferred for practical reasons, then improve-

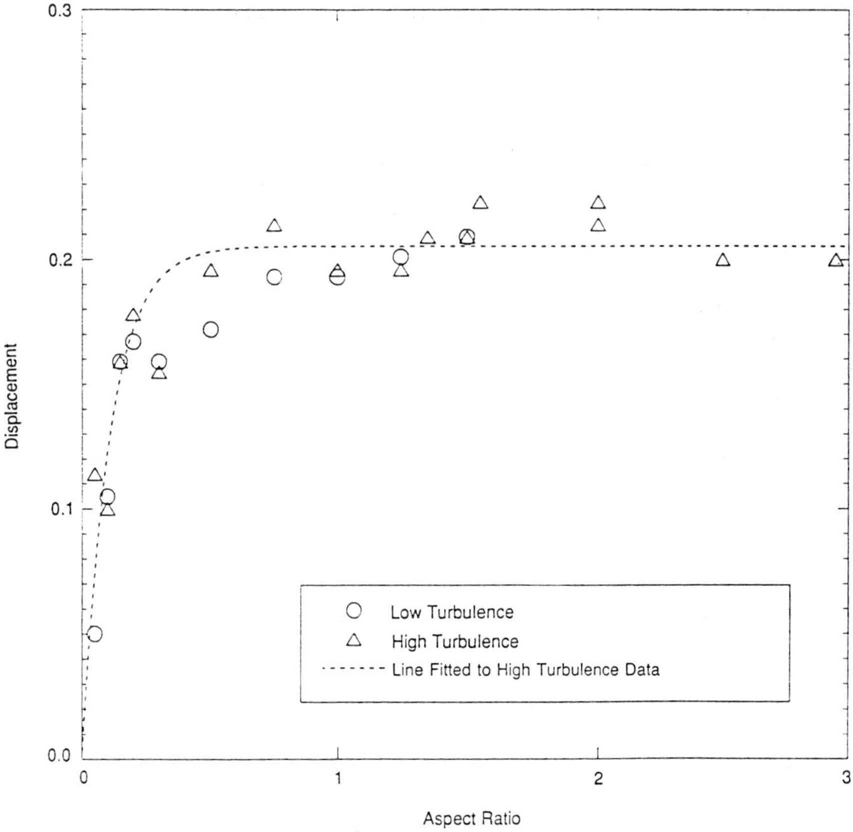

Figure 8.7 *Effect of aspect ratio on displacement over cylindrical collectors*

ments to the flow over the collector opening must come from improved aerodynamic styling.

The curves drawn through the data in Figures 8.6 and 8.7 are best fits through the high turbulence data. They are used in the subsequent figures as references for comparison.

8.4.1.3 Effect of aspect ratio on the ejection of precipitation from the collector

The retention of precipitation is just as important as its initial collection. The ejection of already collected precipitation by the wind-driven circulation inside the collector can occur quite readily. The problem is well known with dust gauges where the dust lies on the gauge surface. It might be thought that this problem was not important for precipitation, since rainfall is drained away after contacting the surface. However, solid precipitation (mainly snow) can be removed from a collector surface. Also fine precipitation with low falling speeds may circulate within the collector and be ejected without contacting the surface. These effects will clearly affect mainly the collection of fine rain and snow, but deficiencies in the collection of these types of precipitation is a significant problem. Besides the ejection of collected precipitation, the internal air circulation within the collector encourages the evaporation of precipitate lying on the collector's surfaces.

The wind-driven circulation inside the collector and its effects on particle ejection is known to vary with collector shape. Hall and Waters (1986a) noted that the windspeed at which particles were blown out of a shallow dust gauge (an inverted frisbee) with an aspect ratio of about 0.15 was about 50% higher than that required to blow the same particles out of a conventional dust gauge with an aspect ratio of 0.6. They also remarked that gauges with aspect ratios around unity seemed to have the strongest internal circulation, but there are no systematic data on this effect. Measurements on the behaviour of cylindrical collectors were therefore extended to address this question.

The measurements of particle ejection were made with two different materials. Firstly, with glass spheres, which have very good sample uniformity and whose characteristics are closely definable. They are quite mobile when placed on a surface, so are moved and blown out of a collector more readily than most materials; they have a specific gravity of 2.9; secondly, with expanded polystyrene, either spheres or irregular pieces. These are also quite mobile and their specific gravity is about 0.025, near to the lowest values which usually occur for snow. The material was the same as that used in earlier snow collector experiments (Hall *et al.*, 1989). Thus, the two types of particle bracketed the range of specific gravities that might be expected of natural precipitation. This density variation is significant for ejection measurements, which are affected by the particle size as well as by its falling speed (normally the dominant parameter). A range of particle sizes were used corresponding to aerodynamic diameters between about 50 μm and 500 μm, whose falling speeds varied between about 7 and 200 $cm\,s^{-1}$; this covers a range consistent with fine rain and snow. The sizes of the two different material types overlapped, the polystyrene particles having the lowest falling speeds. The technique is a subjective one. Some of the material was placed in the bottom of the collector and the windspeed increased until it was ejected from the collector. There was not a sharp dividing windspeed at which this occurred, but beyond a minimum value there was a gradual increase in the rate of ejection of material. In the present experiments the same operator judged the lowest windspeed at which a significant rate of particle ejection occurred, so that the measurements, though subjective, should be self-consistent.

Figure 8.8 shows the results of the experiments. All the material sizes and types showed the same basic pattern. With increasing aspect ratio there was a sharp rise in the windspeed at which ejection occurred up to an aspect ratio of about 0.2. Beyond this there was initially a fall in the critical windspeed to a minimum at aspect ratios around 0.5–0.8, followed by a further rise. The highest aspect ratio in the present experiments was 1.5 as this was the deepest collector that could be used in the small wind-tunnel used for the experiments, but some earlier work using a deep collector of aspect ratio 2

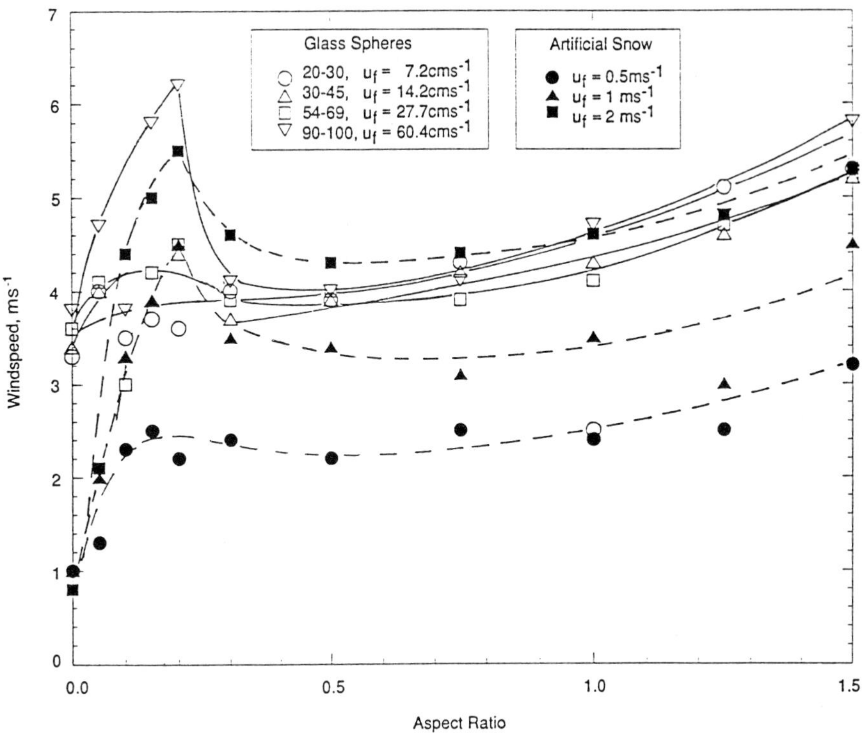

Figure 8.8 *Windspeeds for the ejection of materials from cylindrical collectors*

had shown that it was very difficult to persuade particles to eject from it, windspeeds of 9 $m\,s^{-1}$ and abpve being required. Thus, with increasing aspect ratios beyond 1.5, the critical windspeed can be expected to continue rising.

There was some variation within the basic pattern of the results depending upon the material size and type. The smallest polystyrene particles showed more of a plateau beyond an aspect ratio of 0.2 rather than a minimum. The smaller of the glass spheres also showed a more limited peak in the critical velocity at the aspect ratio of 0.2. Also the critical windspeeds did not vary in a simple way with the material size or its type. For the polystyrene, the critical windspeed increased markedly with increases in the material falling speed. For the glass spheres the variation of windspeed with the falling speed was less. Except for the largest spheres it was about the same, irrespective of the falling speed, within the accuracy of the experiment. The overlap between the critical windspeeds for the two types of material was not consistent with their falling speeds, the critical windspeeds for the glass spheres were closest to that of the largest polystyrene particles. These differences are probably related to the way in which the particles were saltated from the inner surface of the collector. The mean windspeed to which a particle on a surface is exposed depends upon its size. Lower density particles are larger for a given falling speed, protrude further from the surface and are thus exposed to higher mean windspeeds. In the present experiments the density difference between the two types of particle was very large, so that the saltating forces were also very different.

It is of practical interest that many existing precipitation collectors have internal aspect ratios around unity or a little less, close to the region of the minimum in the windspeed for particle ejection and the highest internal airspeeds.

8.4.2 Windflow over existing and novel designs of precipitation collectors

8.4.2.1 Measurements over existing designs

For comparison with the base data on rectangular cylinders, described above, measurements of flow displacement and acceleration were made over the openings of a variety of collectors. The measurements were not intended to be comprehensive; Sevruk and Klemm (1989b) have identified over 30 different precipitation collectors in official use, world-wide, besides the great variety of collectors used for special purposes. The collectors chosen covered those forms which had been the subject of previous investigations, together with others which could be obtained fairly readily over the period of the measurements. However this was sufficient to cover a wide variety of different types, including both conventional collector shapes and some novel forms, and made an interesting comparison with the base data. Not all of them are formally precipitation collectors. Some are dust gauges, but have shapes or properties of interest for precipitation collection.

A list of the different collectors tested, together with a brief description of each, is given below. Diagrams of the outlines of the collector shapes are shown in Figures 8.9 and 8.10. Some modifications to these designs have also been tested, these are referred to through the text of the paper. The approximate aspect ratios of the collectors are noted and a list of these values, both internal and external, is given in Table 8.1.

Meteorological Office 5″ rain gauge (Figure 8.9(a)). This is the standard UK precipitation collector. It is of plain cylindrical form with a sharpened edge to the opening. The interior is cylindrical for a depth of about one diameter, below which it is funnel shaped. It is normally set on or in the ground (sometimes partly, sometimes wholly), but was tested here set on the ground with the opening about 340 mm above the ground. The aspect ratio is 2.7 externally (when set at this height above the ground) and 1.2 internally.

IoH tipping bucket rain gauge (Figure 8.9(b)). This collector sits on the ground with the opening at a height of about 380 mm. It is nominally cylindrical, having a tapering convex external form with a cusped rebate at the top forming a sharp vertical edge to the opening. The interior is a concave funnel, to encourage efficient drainage of the sample. The tipping

Table 8.1 Measurement of accelerations and displacements for various collectors

Type of collector	*Diameter (mm)*	*Aspect ratio*	*Windspeed ($m\,s^{-1}$)*	*Acceleration*	*Displacement*
Met. Office 5" rain gauge	127	2.68	3	0.2	0.27
NILU dust gauge	200	2	2.8	0.175	0.16
Proposed ISO dust gauge	200	2	2.8	0.235	0.223
BS dust gauge	307	0.62	3	0.27	0.18
Tipping bucket gauge (elevated)	200	1.9	3	0.37	0.23
Tipping bucket gauge (ground)	200	1.9	3	0.29	0.18
UK wet–only (0°)	297	1.82	3	0.11	0.55
UK wet–only (other°)	297	1.82	3	0.39	0.22
ASTM dust gauge (45°shield)	150	2	2.8	0.07	0.28
UK rain funnel (large)	150	0.73	3	0.28	0.17
UK rain funnel (small)	115	0.74	3	0.25	0.1
UK rain funnel (small + bottle)	115	0.74	3	0.26	0.13
Large funnel 1	198	0.75	2.8	0.2	0.17
Large funnel 2	245	0.6	2.8	0.2	0.17
Metal frisbee	227	0.15	3	0.17	0.15
Snow collector (Mk II)	202	0.32	2.8	0.117	0.1038
Snow collector (Mk III)	310	0.29	2.8	0.126	0.085
Deep bowl	204	0.44	3	0.22	0.12
Deep bowl (+ cone insert)	204	0.44	3	0.23	0.13

bucket is just below the drain point. The aspect ratio is 1.9 externally and 0.55 internally.

60° funnel, WSL bulk collector (Figure 8.9(c)). This is used in the UK precipitation chemistry monitoring network. Full details can be found in Hall (1986b). It is a standard polypropylene funnel of 60° included angle in two inlet diameters, 115 mm and 152 mm (the size used depends upon the local precipitation). The opening is 1.75 m above the ground. The aspect ratio is 0.74, based on the funnel depth and opening diameter.

Two large funnels (Figure 8.9(d)). These are used in Ireland as dust deposit gauges. They were included because funnels are a commonly used collector

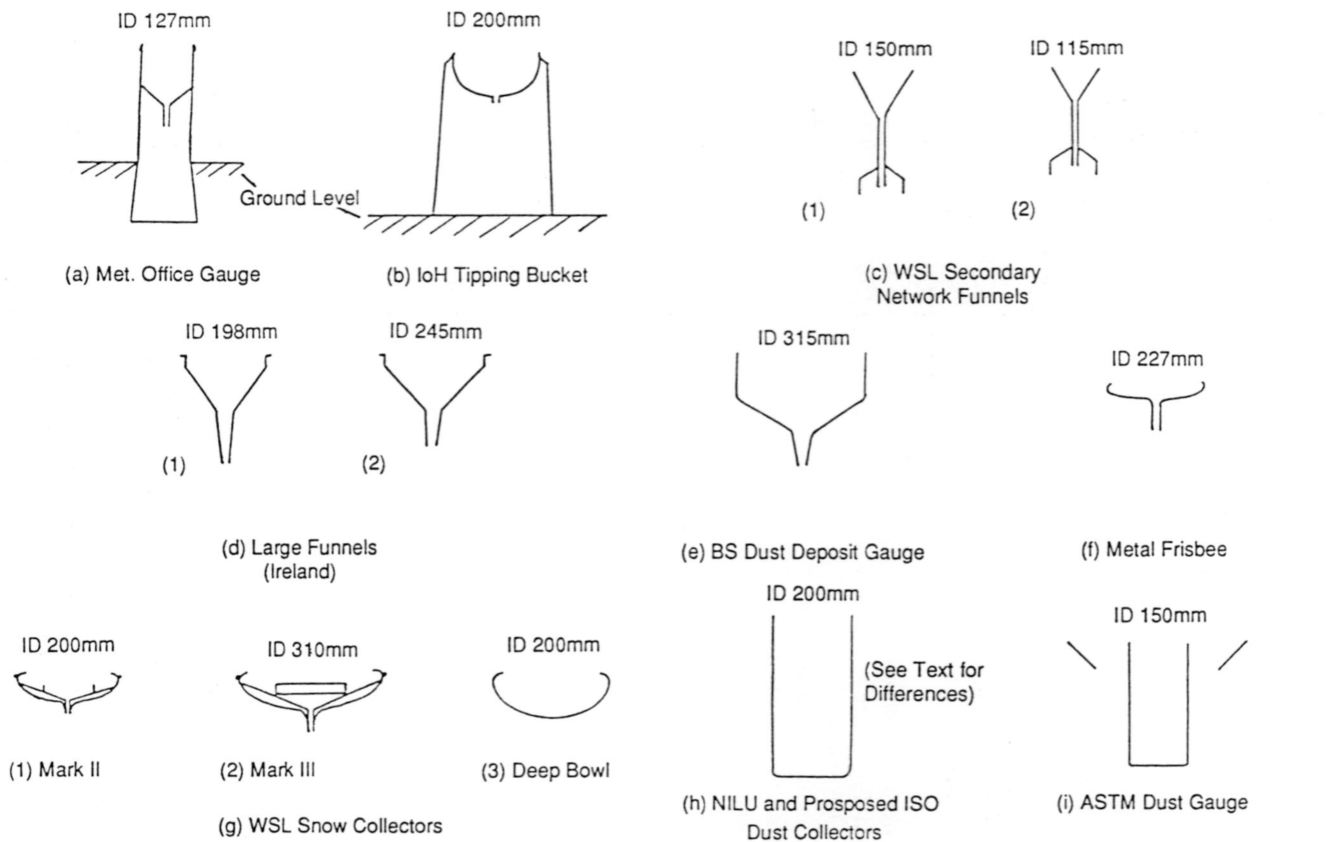

Figure 8.9 *Outline diagrams of existing collector shapes tested*

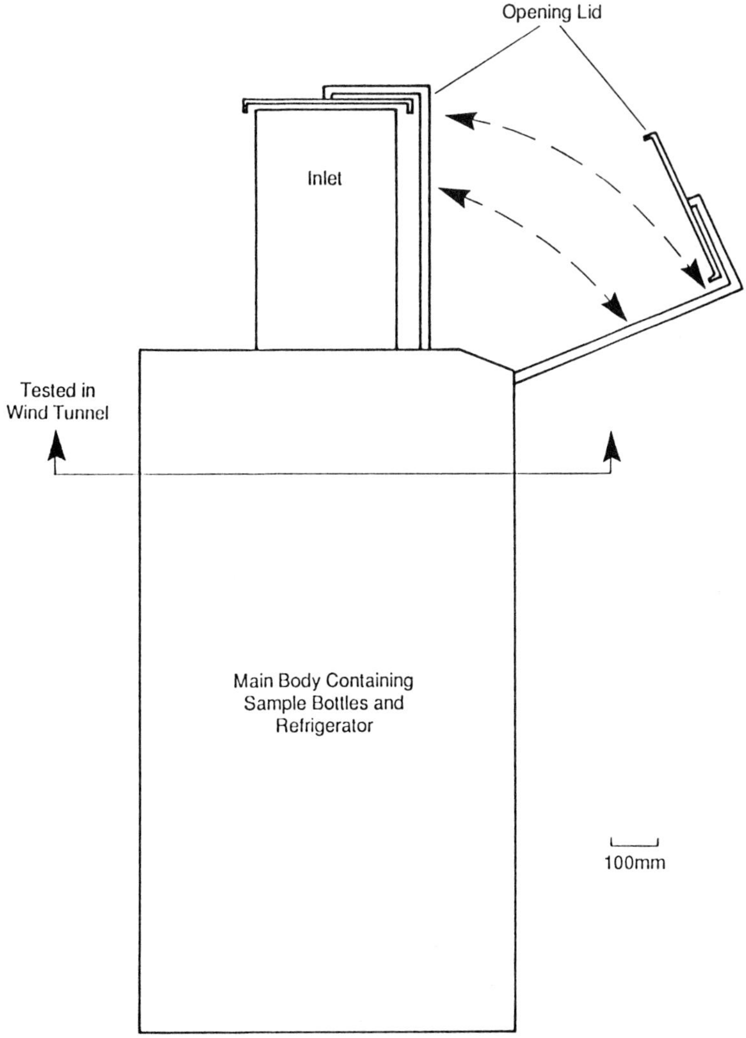

Figure 8.10 *Wet-only collector*

shape and we wished to see if there was much variation in the flows over different types. These two were immediately available. They were both large polypropylene funnels with a short parallel portion at the top and included angles in the cone of 72° and 85° respectively. Their aspect ratios are 0.75 and 0.6 respectively.

British Standard dust deposit gauge bowl (Figure 8.9(e)). This has been used as a bulk collector in UK precipitation chemistry studies. Also, its collection performance for small precipitation (up to 1000 μm) has been investigated in detail (Ralph and Barrett, 1984) in relation to its use as a dust collector. It is a cylinder of 307 mm diameter with a shallow funnel shaped bottom and is normally set about 1 m above the ground (BS, 1969). The aspect ratio is 0.62.

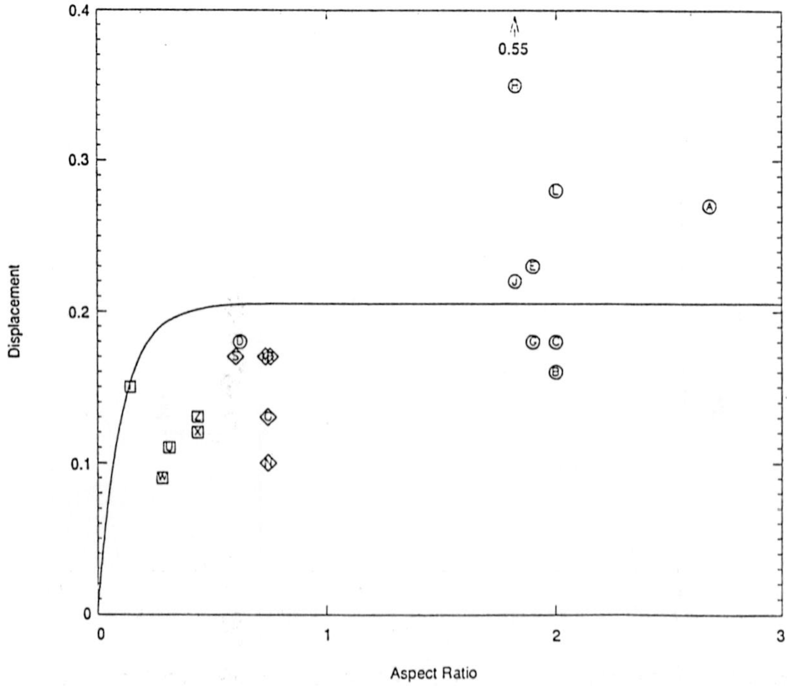

Figure 8.11 *Displacement as a function of aspect ratio for various collector shapes*

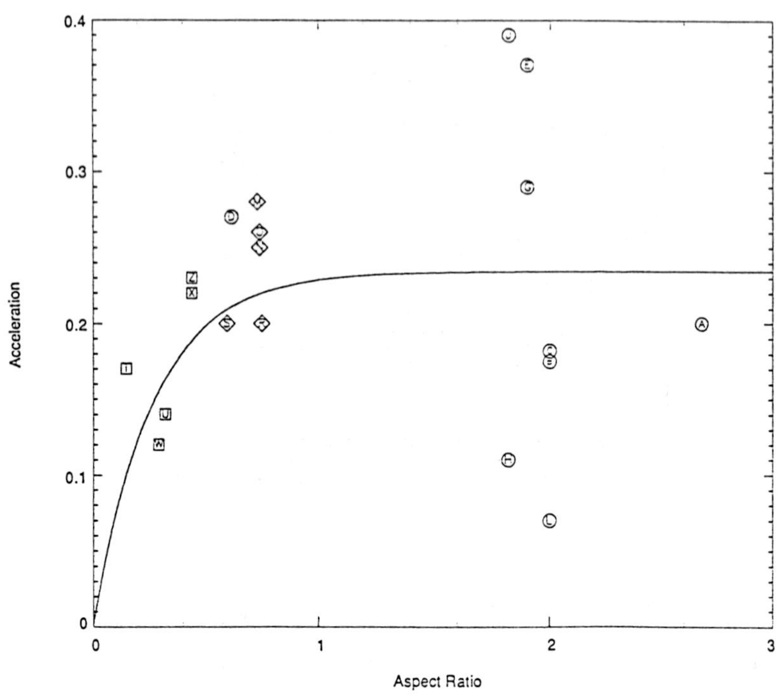

Figure 8.12 *Acceleration as a function of aspect ratio for various collector shapes*

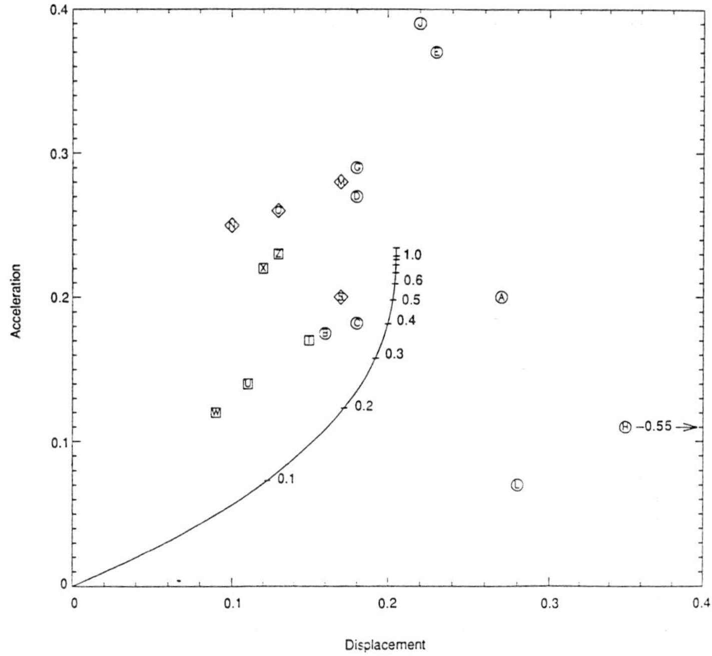

Figure 8.13 *Acceleration against displacement for various collector shapes*

○ Cylinders
◇ Funnels
□ Bowls
—— Metal Cylinders

Key to Figures 8.11, 8.12 and 8.13

Cylinders

A Met. Office 5" Rain Gauge
B NILU Dust Gauge
C Proposed ISO Dust Gauge
D BS Dust Gauge
E IoH Tipping Bucket Rain Gauge (elevated)
G IoH Tipping Bucket Rain Gauge (on ground)
H UK Acid Deposition Wet-Only Collector (0°)
J UK Acid Deposition Wet-Only Collector (other directions)
L ASTM dust gauge for method D 1739

Funnels

M UK Acid Deposition Large Funnel (funnel only)
N UK Acid Deposition Small Funnel (funnel only)
O UK Acid Deposition Small Funnel (in normal mounting, with bottle)
R Large Funnel type 1
S Large Funnel type 2

Streamlined Shapes

T Metal Frisbee-type Dust Gauge
U Snow Collector Mk2
W Snow Collector Mk3
X Deep Bowl
Z Deep Bowl with Inverted Cone Insert

An inverted frisbee type of dust collector (Figure 8.9(f)). This is described in Hall and Waters (1986a) and Hall and Upton (1988b). There is some streamlining in the shape, but it also has a low aspect ratio, of about 0.15.

Two types of aerodynamically styled snow collector (Figure 8.9(g)). These are described in Hall *et al.* (1988a, 1989). Both have all curved surfaces intended to produce a downforce and turned-in lips to the collector opening which are fitted with flow control devices in order to encourage a horizontal flow over the opening. One has an aspect ratio of 0.32, the other of 0.29. Also tested was a deeper shape of this sort, referred to as a 'deep bowl', which is described in Hall *et al.* (1988a). It has an aspect ratio of 0.44.

The NILU and proposed ISO dust collectors (Figure 8.9(h)). The NILU design is also used as a precipitation collector. Both are rectangular cylinders of aspect ratio 2. The proposed ISO dust gauge has a sharp vertical edge to the opening, while the NILU has a small turned-in lip at a 45° angle. Both have been the subject of detailed investigation as to their performance as small particle collectors by Ralph and Barrett (1984).

The ASTM dust gauge (Figure 8.9(i)). This is described in detail in the relevant standard (ASTM, 1989). Though it is intended as a dust gauge, it is typical of the design of many precipitation collectors in being a rectangular cylinder of aspect ratio about 2. It is normally set about 2 m above the ground. The main point of interest in this gauge is that it is fitted with a flow deflector, a separated ring around the opening. The ring is in the form of a truncated cone, set at an angle of 45° to the horizontal. The intention of the deflector ring is to displace the approaching airflow downwards in opposition to the normal upwards displacement caused by the bulk of the gauge. This design is typical of the sort of modification which has been attempted on several precipitation collectors, so it is thus of more general interest. In its original form, the deflector ring was of only limited effectiveness, but formed the basis for some further experiments, which are described later in this chapter.

The WSL wet-only collector (Figure 8.10). In this collector the opening is fitted with a cover which is removed automatically when precipitation occurs, in order to minimize dry deposition into the collector. Also samples are separated, by period or event, and refrigerated in order to maintain integrity. Events, and the associated precipitation amount, are recorded in order to assist data analysis. It is inevitable that such a complex instrument will have a substantial bulk. The figure shows the body of the instrument, which is about 1 m square and 1.4 m high, on top of which sits the collector head. This is a cylinder of 0.3 m diameter and 0.54 m height, so that the height of the opening is 2 m. The aspect ratio of this cylinder is 1.8 externally and 0.6 internally. It is clear that this collector must suffer from the aerodynamic difficulties discussed above and that a fair proportion of the deficiency in collection discussed by Stone (1991) must be attributable to this cause. There are additional difficulties caused by the cover for the opening when the collector is open. These are discussed later.

Figures 8.11, 8.12 and 8.13 show plots of the displacement and acceleration over the collector opening. Figures 8.11 and 8.12 show respectively the displacement and acceleration as a function of aspect ratio, and Figure 8.13 shows a plot of displacement against acceleration. Tabulated values of the data can be found in Table 8.1. Because of the large variety of collector types tested, the individual plotted points are identified by letter and there is a separate key. The generic types are indicated by the shape of symbol, circles are for cylindrical (or closely so) collectors, diamonds are for funnels and squares for 'bowls' and streamlined shapes. In Figures 8.11 and 8.12, the data from the rectangular cylinders have also been reproduced, for reference, as the lines drawn through the data in Figures 8.6 and 8.7. Similarly in Figure 8.13, there is a line for the rectangular cylinders derived from these data. The aspect ratio of the collectors is not directly evident on this line, but values are appended at the appropriate points. It can be seen that the line in Figure 8.13 terminates at values of the displacement and acceleration corresponding to the respective limits approached in Figures 8.6 and 8.7.

Figures 8.11 and 8.12 show that there was a wide variation in the acceleration and displacement over the different collectors that was not merely a function of their aspect ratio. Values of the acceleration ranged between 0.07 and 0.4, and of the displacement between 0.09 and 0.55. When compared with the cylinder of equivalent aspect ratio, overall the streamlined shapes showed similar levels of acceleration but significantly reduced displacement, the funnels showed greater acceleration but reduced displacement, while the cylindrical shapes showed a very wide spread in both acceleration and displacement.

Of the cylindrical gauges, those closest to simple cylindrical shapes were the Met Office gauge and the NILU and proposed ISO gauges, A, B and C on the plots. Compared with the basic cylinder data, the Met Office gauge showed a higher displacement and slightly lower acceleration. These differences were presumably a result of its being set on the ground and in a shear layer, rather than elevated and in a region of approximately constant velocity as were the basic cylinder measurements. The NILU and proposed ISO gauges produced similar results to one another, with a little less displacement and acceleration than the basic cylinder data. The differences are only just outside the order of accuracy of the experiment. The IoH tipping bucket rain gauge was tested both elevated, E, and on the ground, G (its normal operating position). Setting the gauge on the ground appeared to reduce both the displacement and the acceleration. The displacement was then similar to the standard cylinder measurements, but the acceleration was much greater. Two results are shown for the wet-only collector, at H with the opened lid facing into the wind, and at J for the lid facing in other directions. It can be seen that with the lid facing into the wind the displacement was unusually high and the acceleration unusually low. With the lid facing in other directions, the acceleration was the highest of those measured, but the displacement was similar to that of the equivalent standard cylinder. The behaviour of this gauge is discussed later in more detail. The remaining cylindrical gauge tested was the ASTM dust gauge, L, which had a flow deflector around it set at an incidence of 45°, similar to some precipitation collector designs. This exhibited one of the lowest values of the acceleration (0.07), but a relatively high value of the displacement, well beyond that of the equivalent cylinder. The behaviour of this sort of collector was the subject of further investigation, discussed later. The main feature of the measurements over the different cylindrical gauges is the confusing picture that they present. The measurement of acceleration and displacement is not a very precise one (partly due to the highly turbulent flow in which the measurements were made) and is of about 10% accuracy. However, this is not sufficient to explain the differences between designs that have been observed.

The funnels tested were all of about the same aspect ratio, around 0.8. The two larger funnels, R and S, showed similar levels of acceleration and displacement, not far from the values of the equivalent cylinders. The two smaller funnels, M and N, with a steeper side of 60° included angle, showed larger values of the acceleration but similar or lower levels of displacement, the smaller funnel showing the smallest value of 0.1, distinctly lower than for the other three. Though of different sizes, the two smaller funnels were identical in shape and this variation between them is a little surprising. However, an earlier measurement (Barrett *et al.*, 1985) on a larger funnel, of 200 mm diameter as against 115 and 152 mm used here, also gave a value of the displacement of 0.11. These variations may simply be due to experimental error, but there could be deeper reasons for the differences. It is possible for Reynolds number effects to occur readily in flows at these scales, but in the present case this seems a little unlikely. The funnels tested also had a variety of shapes of lip to the opening (the larger funnels had vertical edges to the openings, while the smaller funnels had sharp cut edges at the 60° funnel angle) and this is known to affect the resultant flow. The smaller funnels are normally mounted over a shrouded collecting bottle, of about 210 mm diameter and 250 mm below the funnel opening (Hall, 1986b). This condition is plotted as point O; there was no significant change compared with the isolated funnel.

The results for the streamlined shapes show the clear advantage that can be obtained from carefully shaped collectors, though this seems to be in reduced displacement more than in reduced acceleration. One of the collectors plotted under this heading (labelled 'T') was the inverted frisbee dust gauge, which has some limited streamlining. It had the same displacement but worse acceleration than for its equivalent basic cylinder, though the slope of the graph in this region is so steep that the accuracy of the measurement may not be very good. Since this gauge had a distinctly improved collection performance over the conventional deeper cylinders, it appears that this was due to its reduced depth rather than its shape. The performance of the streamlined collectors, and improvements to them, will be discussed later in the paper.

Figure 8.13, which shows acceleration plotted against displacement for the collectors tested, largely reinforces the impression left by Figures 8.11 and 8.12, of the large range of values of both parameters that can be obtained

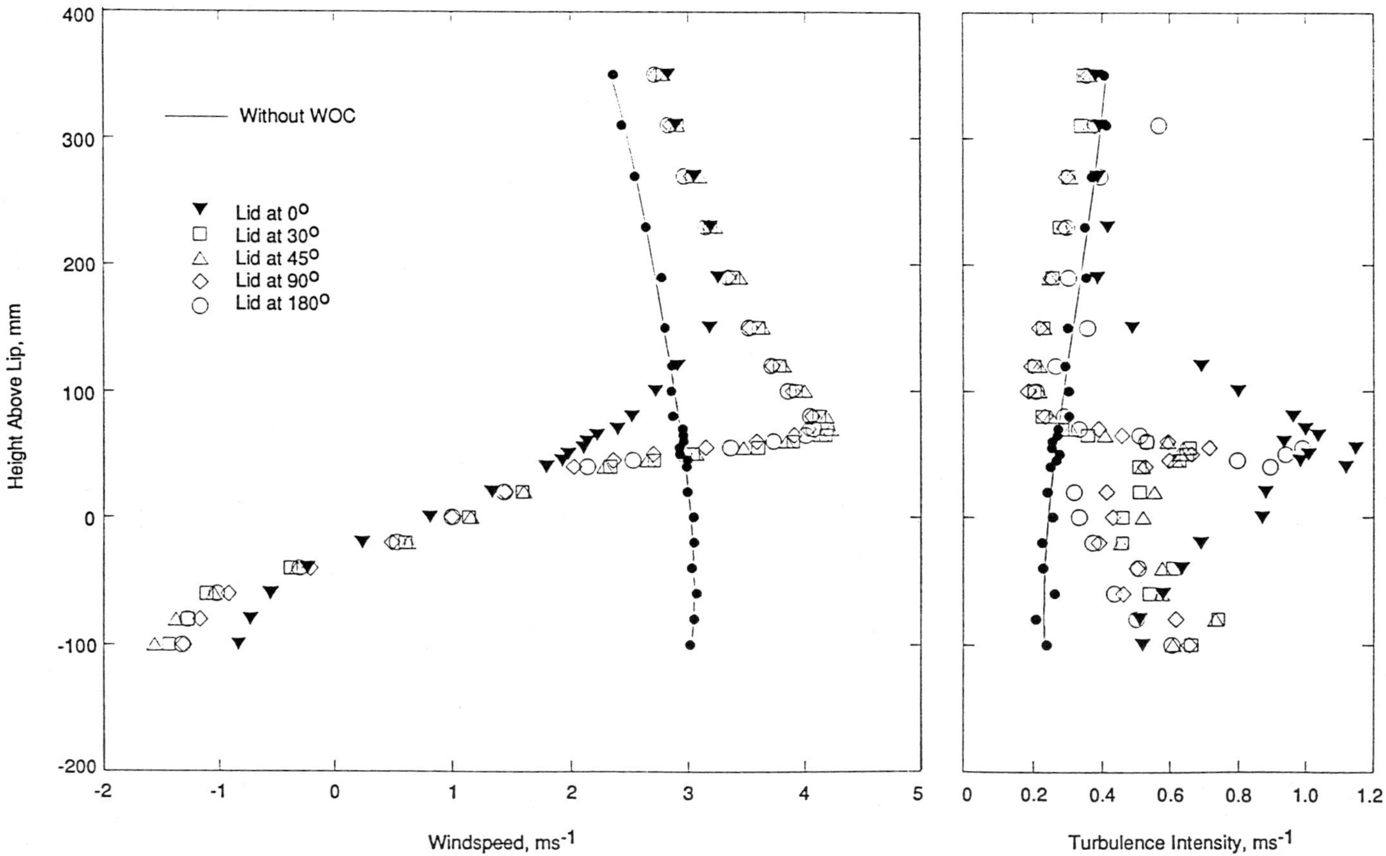

Figure 8.14 *Windspeed profiles over wet-only precipitation collectors in various wind directions*

with different collectors. It also shows that some designs of collector perform distinctly less well than a plain cylinder.

For the remaining investigations described here only this last form of presentation, a plot of acceleration against displacement, will be used.

8.4.2.2 Performance of the WSL wet-only collector

Because a desire to improve the performance of the Warren Spring Laboratory wet-only collector was one of the main reasons for the present programme, its performance was examined in more detail than the other collectors. Vertical profiles of windspeed and longitudinal turbulence intensity over the centre of the collector opening are shown in Figure 8.14 for five wind directions, shown in Figure 8.15; the direction 0° is that for the opened lid of the collector facing into the wind. It can be seen that the windspeed profile for the direction 0° was markedly different from those of the other directions, which showed only small differences. The same variation was shown by the longitudinal turbulence intensity, the measurements for the wind direction 0° being markedly different from the others. The differences in the acceleration and displacement from these wind profiles were noted in the previous section.

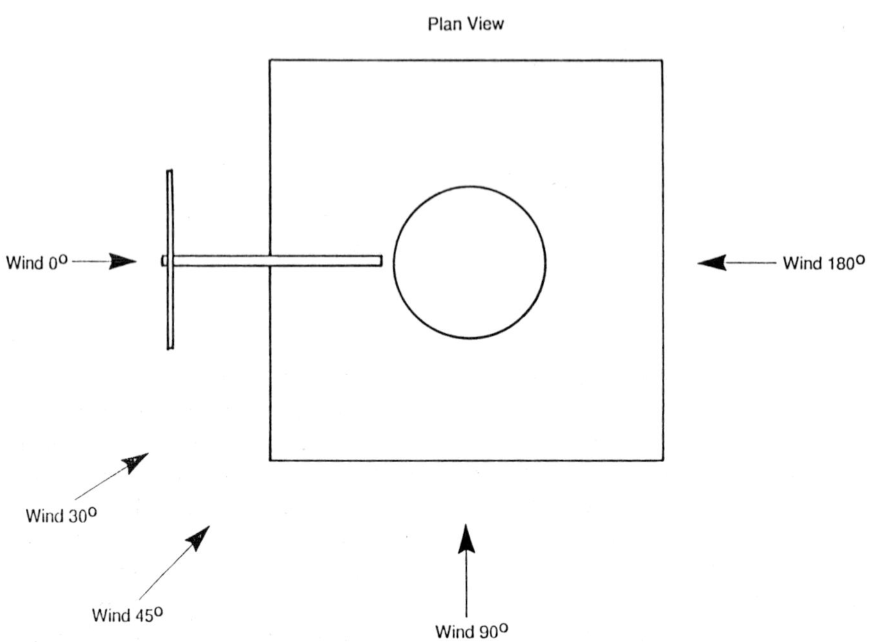

Figure 8.15 *Wet-only collector orientations for velocity profile measurements*

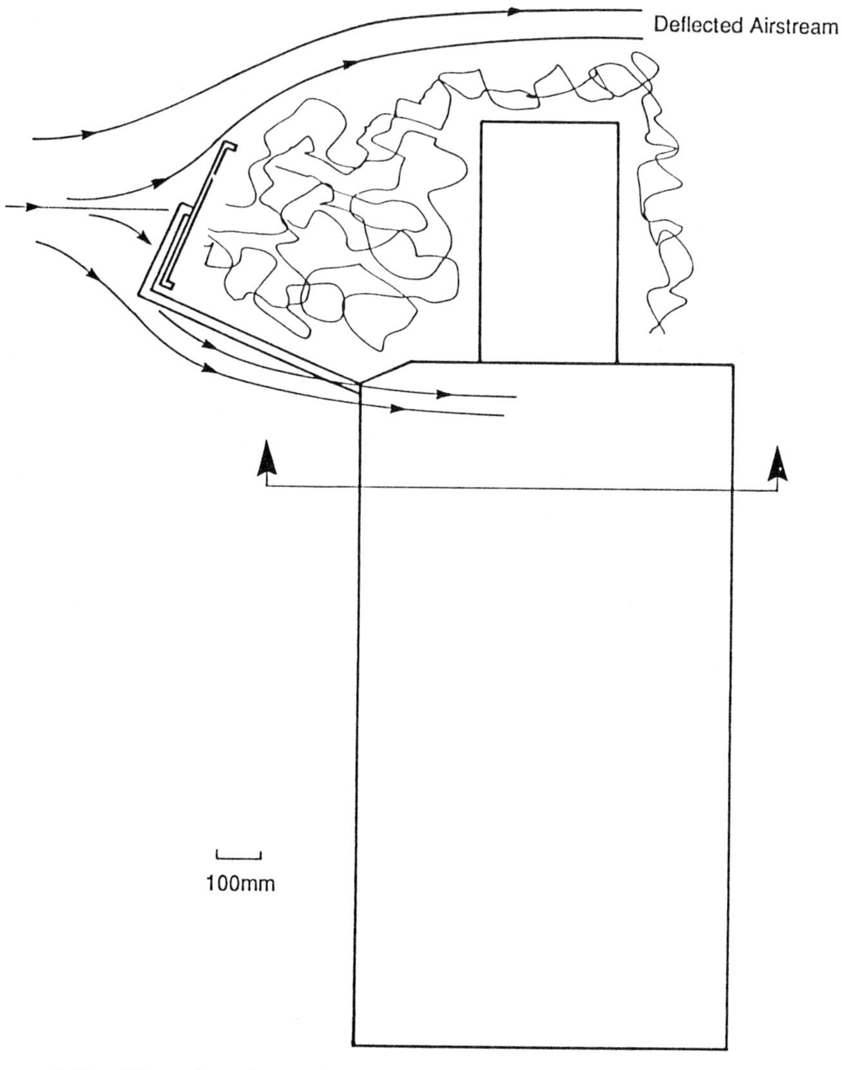

Figure 8.16 *Wet-only collector flow pattern over lid*

It was clear from the measurements that the reason for the differences was due to disturbances of some sort caused by the open lid and their nature was revealed by flow visualization. Figure 8.16 shows a sketch of the flow pattern around the collector in the direction 0°, with the lid of the collector facing into the wind. The collector lid is in the form of a sharp-edged disc, which faces nearly normally into the wind when it is open. The upper edge of the lid is then a little below the collector opening. The separation streamline from the edge of the lid passed over the collector opening, so that the area around the opening was within the separation region behind the lid. This process was assisted by the angle of the lid to the vertical when it is open, about 20°, which encouraged the vertical rise of the upper part of the separation line. The high turbulence intensities recorded over the collector in this condition were those associated with the separated flow in this region.

Apart from the varying collection efficiency with wind direction that must result from this condition, there is the equally undesirable possibility of sample contamination in the wind direction 0°. It is readily possible for water droplets accumulating on the lid to be blown upwards off the lid and into the collector, especially in strong winds.

8.4.2.3 An investigation of flow deflectors using the cylindrical ASTM gauge

The ASTM gauge investigated here was the only design using a flow deflecting device in order to control the airflow over the collector opening. This type of design has also been used on some rain gauges, for example the Swedish SMHI gauge and the Icelandic gauge. If effective, this design of collector clearly has great practical appeal. It is possible to use a simple cylindrical collector and the deflecting device is easy to make and install. However, the design tested, which used a deflector with a 45° angle, produced a low value of the acceleration but with a high value of the displacement. Because of its practical advantages and the fact that this would be an easy modification to make to the wet-only collector described above, its potential was investigated further.

Flow visualization studies of the existing design revealed that the flow pattern over the collector was not unlike that of Figure 8.16. There was a complete separation of the flow over the deflector and the upper separation line passed over the opening of the collector. The deflector did have some effect on the direction of the airflow as well, but due to its totally stalled state this was not very great. The end product, a lowered acceleration but with a greatly increased deflection was also similar to the result for the wet-only

Table 8.2 Measurements of acceleration and displacement for a cylindrical collector with a flow deflector

Shield angle (°)	*Windspeed ($m\,s^{-1}$)*	*Acceleration*	*Displacement*
Normal 45°	2.8	0.07	0.28
No shield	2.8	0.2	0.26
0	2.8	0.214	0.19
15	2.8	0.082	0.21
25	2.8	0.06	0.25
Curved	2.8	0.12	0.19
Curved + 6 mm tube	2.8	0.15	0.16
Curved + 10 mm tube	2.8	0.13	0.16

ASTM dust gauge: Cylinder — Diameter = 150 mm; Aspect ratio = 2

Shield — Diameter (top) = 450 mm; Width = 100 mm

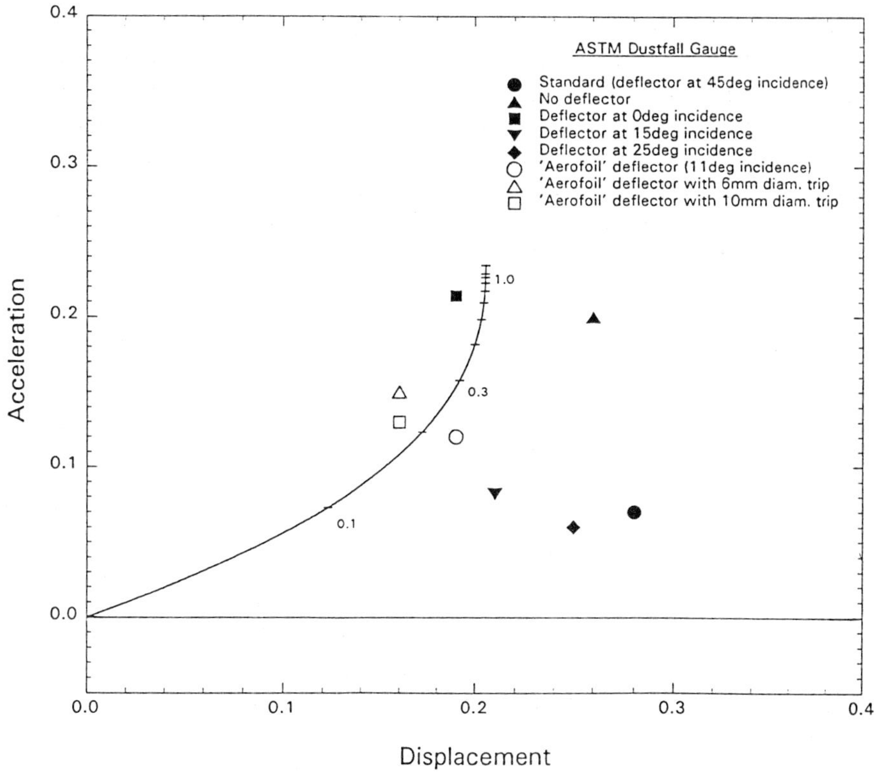

Figure 8.17 *Effect of flow deflectors on a cylindrical collector (ASTM gauge)*

collector with its lid facing into the wind. It seemed that the deflector was behaving more as a wind shield than a deflector.

A number of experiments were carried out with different deflectors at smaller incidences than 45°. The results are shown in Figure 8.17, which is a plot of acceleration against displacement for the measurements. Tabulated results can be found in Table 8.2. All the deflectors had their upper edges 5 mm above the collector opening, as for the original 45° deflector, and had the same inner and outer diameters. The plain collector, without any deflector, produced a similar acceleration but a higher displacement than the standard cylinder measurements. A deflector with zero angle, a flat disc, reduced the displacement by about 25% but did not affect the acceleration. Increasing the incidence of the deflector up to about 25° reduced the acceleration quite significantly, by up to about a factor of three, but at the expense of some increase in displacement. Flow visualization showed that these lower angle deflectors still had a poor aerodynamic performance, there was always a separated flow over their upper surfaces. This is hardly surprising, but the experiments with flat-surfaced deflectors were made on the grounds that this is an easy shape to make. Some measurements were then made with deflectors curved into a thin aerofoil form. The low Reynolds numbers at which these deflecting devices were operating (chord Reynolds numbers were of the order of 2.10^4) made it almost impossible to obtain good aerodynamic performance without boundary layer trips or other flow control devices. However, an aerofoil deflector at about 11° incidence reduced the acceleration by about 50% with an added slight reduction in displacement. A boundary layer trip wire in the form of a rounded nose was then added. This was simply a piece of split tubing (of 6 mm or 10 mm diameter) pushed on to the leading edge of the deflector. This produced a small, but not very significant, improvement in performance and flow visualization showed that the flow still separated over the upper surface of the deflector.

At this point further experiments on flow deflectors fitted to cylindrical collectors were halted. The reason was that the work so far had revealed the fundamental problem associated with this sort of collector design. There was no doubt that further development of the deflectors would have improved performance over the results of Figure 8.15, quite possibly to the point of achieving a zero acceleration over the gauge opening. The problem lay with reducing the displacement as well. The effect of a deflector is to displace downwards the airflow around it, this spreads out the streamlines above the deflector and reduces the mean windspeed in this area, which includes the flow over the collector opening. However, it has no major effect upon the basic flow pattern around the collector, which in the case of cylindrical collectors is dominated by their large aerodynamic blockage and causes the rising airflow over the collector opening. Thus the net effect of a flow deflector fitted to a cylindrical collector is to produce the basic flow pattern around a cylinder but with reduced mean windspeeds over the opening. As a result the acceleration is reduced, but the displacement is little affected. Flow visualization showed this clearly, the deflector displaced the streamlines downwards but the basic flow pattern around the cylinder remained. This problem appeared to be unavoidable.

8.4.2.4 An investigation of the effects of flow deflectors on streamlined collectors

Following the work described above, it was apparent that flow deflectors could reduce the acceleration over a collector opening but could not greatly reduce the displacement. However, with the streamlined collector shapes investigated previously it had proved possible to reduce the displacement quite markedly as a result of reducing the aerodynamic blockage. The acceleration had also been reduced, but not to such a great extent. It seemed, therefore, that a combination of deflector and streamlined collector shape might achieve the desired reduction in both parameters. The experiments described below were carried out with this intention.

The streamlined collector shapes used were the MkII and MkIII snow collectors. These are of quite similar, but not identical, shapes, the major difference being that the MkIII collector is 40% larger than the MKII in order to obtain a larger sample. Most of the experiments were carried out with the MKII collector, which is 227 mm in diameter, since its smaller size made the experiments more convenient. It was also then possible to use some of the deflectors used in the experiments on the ASTM gauge, described above. Some final experiments were also carried out on the MkIII collector. Only flat plate deflectors were used, there being insufficient time to investigate more sophisticated designs. Because the main application for the present work was to precipitation chemistry, it is desirable not only to have an aerodynamically effective collector, but also that it should be positioned so as to minimize contaminated particles or water droplets being blown off the deflector into the collector. It is thus desirable that the deflector should be as far below the collector opening as possible, ideally it should be tucked quite closely under the outer edge of the collector body. In practice it did not prove possible to do this, but these other needs were kept in mind in the design. From an aerodynamic point of view it is also desirable that, to be effective, the deflector position should allow a free flow of air between its inner edge and the body of the collector. If the gap is too small the airflow is blocked and the effectiveness of the deflector is reduced.

The results of the experiments are shown in Figure 8.18, which shows a plot of acceleration against displacement for the various deflector arrangements tried. Tabulated values of the data can be found in Table 8.3. It must be noted that this plot is of large scale, it covers only the region in the bottom left-hand corner of Figures 8.13 and 8.17. Thus the two results with the lowest combination of acceleration and displacement in Figure 8.13, which are the MkII and MkIII snow collectors, are the plotted points in the top right of Figure 8.18. It can be seen immediately in the figure that it was possible to reduce the acceleration over the collectors to zero and below. The displacement showed a rather smaller reduction and most of the deflector arrangements produced values of the displacement around 0.05. In fact the true displacement was less than this. The method of determining acceleration and displacement, using the intersect of lines drawn through the two parts of the windspeed profile over the collector, defines approximately the upper edge of the mixing layer which grows over the collector opening rather than the middle of the mixing layer. This error, which is of the order of 0.03 – 0.05 is not important for the larger values of displacement, but affects these smaller readings more seriously.

The best results were obtained with deflector incidences around 25°–35°. With the MkII collector the best results, h and k in Figure 8.18, were with deflectors of 500 mm diameter set at incidences of 25° and 33° respectively with the outer edge of the deflector set 10 mm below the top of the collector. This achieved values of the acceleration of about 0.03 and of the displacement of about 0.07. When this arrangement was scaled up pro rata to fit the MkIII gauge, a better performance resulted, n, o and p, which also showed some windspeed dependence, the performance improving as the windspeed rose. These changes were presumably due to Reynolds number effects, the performance improved as the Reynolds number increased, whether due to increased size or to increased windspeed. The best condition was obtained with a flat plate deflector of 750 mm OD at an angle of 31°, set 20 mm below the collector opening. A diagram of this arrangement is shown

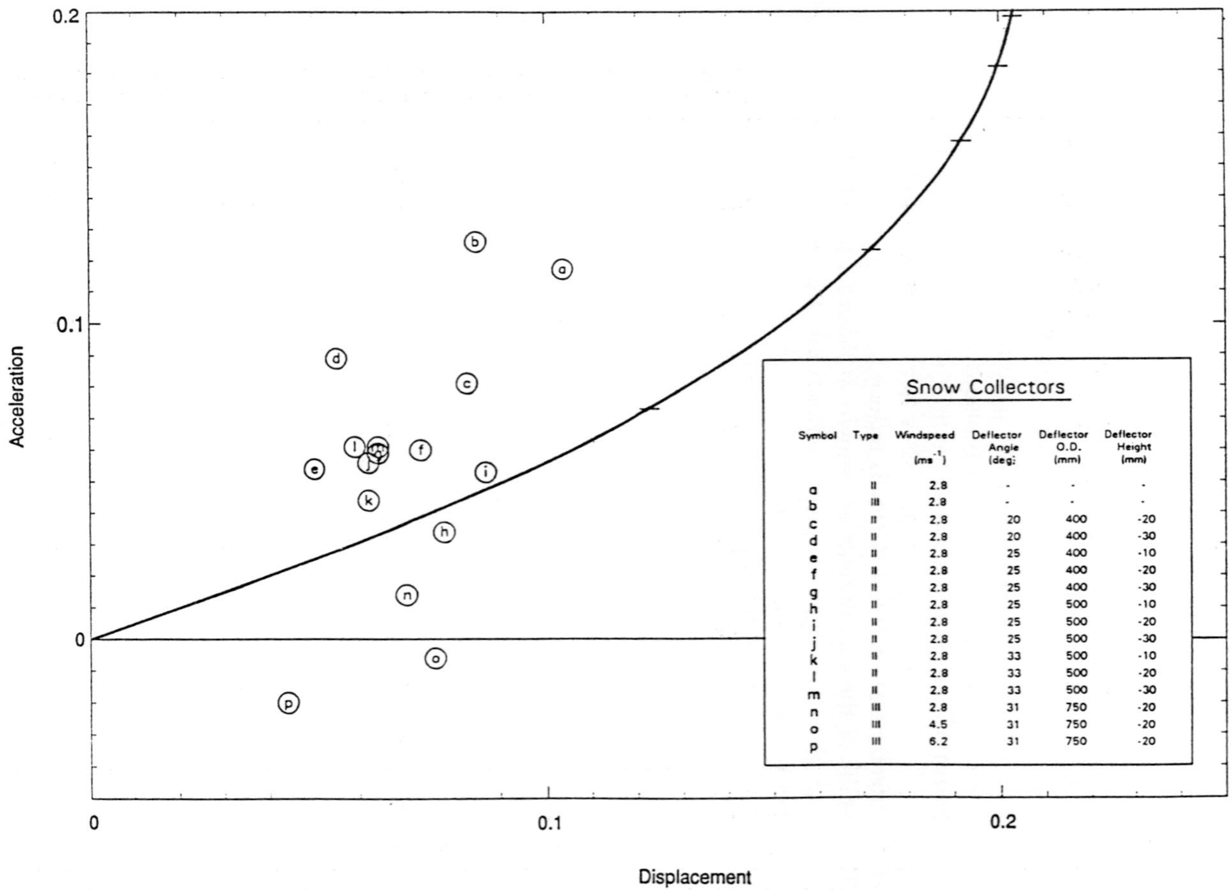

Symbol	Type	Windspeed (ms^{-1})	Deflector Angle (deg)	Deflector O.D. (mm)	Deflector Height (mm)
a	II	2.8	-	-	-
b	III	2.8	-	-	-
c	II	2.8	20	400	-20
d	II	2.8	20	400	-30
e	II	2.8	25	400	-10
f	II	2.8	25	400	-20
g	II	2.8	25	400	-30
h	II	2.8	25	500	-10
i	II	2.8	25	500	-20
j	II	2.8	25	500	-30
k	II	2.8	33	500	-10
l	II	2.8	33	500	-20
m	II	2.8	33	500	-30
n	III	2.8	31	750	-20
o	III	4.5	31	750	-20
p	III	6.2	31	750	-20

Figure 8.18 *Effect of flow deflectors on a streamlined collector (Mks ii and iii snow collectors)*

Table 8.3 Measurements of acceleration and displacement for streamlined collectors

Collector type	*Collector diameter (mm)*	*Aspect ratio*	*Windspeed ($m\,s^{-1}$)*	*Shield angle (°)*	*Shield diameter (mm)*	*Shield position (see footnote)*	*Acceleration*	*Displacement*
Snow Mk II	200	0.32	2.8	–	–	–	0.117	0.1038
Snow Mk III	310	0.29	2.8	–	–	–	0.126	0.085
Snow Mk II	200	0.32	2.8	20	400	–20	0.081	0.083
						–30	0.089	0.055
Snow Mk II	200	0.32	2.8	25	400	–10	0.054	0.05
						–20	0.06	0.073
						–30	0.059	0.064
Snow Mk II	200	0.32	2.8	25	500	–10	0.034	0.078
						–20	0.053	0.087
						–30	0.056	0.062
Snow Mk II	200	0.32	2.8	33	500	–10	0.044	0.062
						–20	0.061	0.059
						–30	0.061	0.064
Snow Mk III	310	0.29	2.8	31	750	–20	0.014	0.07
			4.5				–0.006	0.076
			6.2				–0.02	0.044

Note: For Mk II position is measured from the centre of the trip to the top edge of the shield.
For Mk III position is measured between the bowl and the bottom edge of the shield.

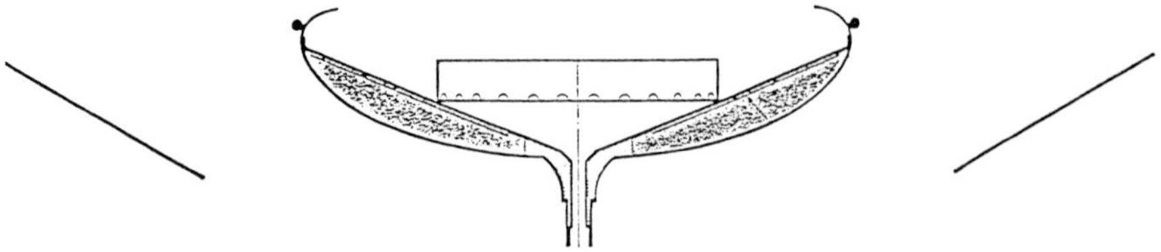

Figure 8.19 *Diagram of most effective deflector arrangement: 750 mm deflector ring around Mk 111 snow collector*

in Figure 8.19. The acceleration, with some variation, was close to zero and the displacement was around 0.04–0.07. What this value of the displacement meant in practice is shown in Figure 8.20, which shows windspeed profiles at five stations along the centreline of the collector opening for condition n, a windspeed of 2.8 $m\,s^{-1}$. It can be seen that the mixing layer grew to a depth of about 25 mm at the downwind end of the opening and that above this the windspeed profile was nearly uniform at a value close to the reference windspeed of 2.8 $m\,s^{-1}$ (this value in fact varied somewhat over the area in which the gauge was placed). Thus, as well as could be practicably managed, there was a flat, uniform and unaccelerated flow over the collector opening. This flow pattern should, in principle, produce some over-collection as, because of the reduced velocities in the mixing layer, the mean windspeed over the gauge is a little less than the undisturbed windspeed. However, when compared with the flow patterns over conventional gauge designs, this point is a minor one.

Flow visualization of condition n showed that there was a separated flow off the top edge of the deflector, but that the separation line reattached itself back on to the front of the collector before the opening. The collector opening was thus properly exposed to the main windflow, not a separated flow region. This sort of design is not entirely satisfactory, though it would be acceptable for sampling precipitation quantity. It is less acceptable for precipitation chemistry studies; there is a possibility of liquid droplets being blown off of the top edge of the deflector into the collector opening. It would be better if the deflector had an attached flow over its upper surface, so that there was no flow separation region near to the body of the collector. This should also lead to a more efficient deflector, which could then be reduced in size and set lower down. Only flat plate deflectors were investigated here due to shortage of time, but a more efficient deflector design is achievable. However, it would need careful aerodynamic design to be effective at the low Reynolds numbers at which it must operate, a deflector with a slotted-flap and boundary layer control devices would probably be needed.

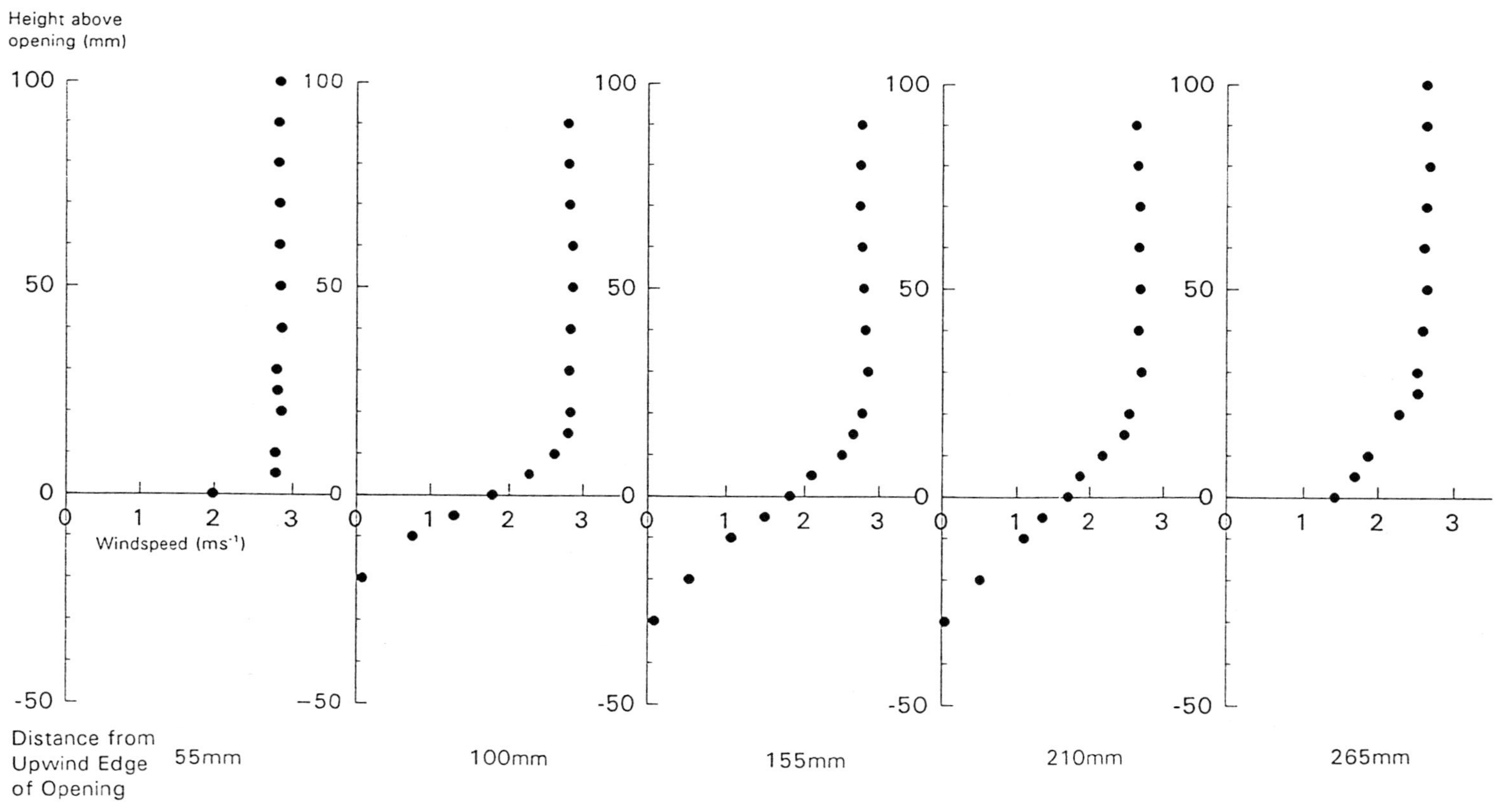

Figure 8.20 *Windspeed profiles over streamlined collector with most effective deflector (750 mm diameter, 31° incidence)*

8.4.3 Modifications to the existing wet-only collector design

The present design of wet-only collector for precipitation chemistry studies is that sketched in Figure 8.10. It is presently being modified to improve its collection characteristics. The collecting cylinder on the body of the collector is being replaced by the MkIII snow collector head, described in Hall *et al.* (1989), which has much lower values of acceleration and displacement over it opening. It also sits on a slender stem of low blockage. The lid-opening mechanism is being modified to incorporate a parallel arm movement which will keep the lid nearly horizontal when it is lowered, thus removing most of its aerodynamic blockage. It would be preferred if the collecting head could also be fitted with a deflector, as described in the section above. However, there is presently some uncertainty about these best present designs as sources of contamination. Some form of treatment of the upper surface of the deflector to reduce splashing and to discourage water droplets blowing off is being considered; this would not affect its aerodynamic performance. A more acceptable alternative would be a more aerodynamically effective deflector, which could then be smaller and set further below the collector opening in order to reduce its potential as a source of contamination.

8.4.4 Field testing of precipitation sensors

Figure 8.21 shows the cumulative frequency distribution of rainfall intensity during the period of field testing. Total rainfall during the period was 117

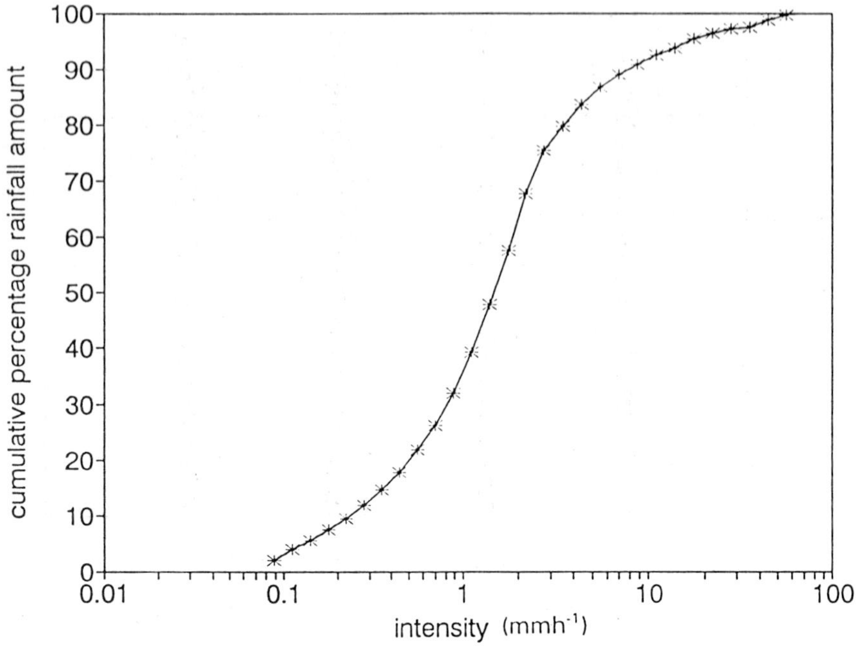

Figure 8.21 *Cumulative percentage distribution of rainfall*

mm and it is evident that over 97% of this occurred at an intensity greater than 0.1 mm hr^{-1}. In the following analyses this has been taken as the lowest intensity for consideration, the others being 0.2, 0.5, and 1.0 mm hr^{-1}; 92, 80 and 64% respectively of the rainfall falling at intensities greater than these values.

Figure 8.22 shows the percentage of time each sensor indicated rain while the ORG-700 sensor was recording intensities above these thresholds: 0.1; 0.2; 0.5 and 1.0 mm hr^{-1}. The ORG-100 shows apparently good performance, exceeding 98% agreement, when the rainfall intensity is above its detection limit of 0.1 mm hr^{-1}. The OMC and WOC sensors show similar patterns. The OMC was in agreement for 93% to 98% of the time and the WOC for 92% to 95%. At intensities between 0.07 and 0.1 mm hr^{-1} the droplet size distribution has a maximum at about 0.2 mm radius, corresponding to a wetted spot of approximate radius 0.4 mm. The diameter of wetted spots is therefore likely to be around 0.8 mm. This may explain the poorer performance of the WOC sensor which has a 1.0 mm electrode separation compared with the OMC electrode spacing of 0.6 mm. The failures to detect rain with the latter sensor may be due to the poor wettability of the silver-palladium electrodes. This could produce pearling of the small droplets, effectively reducing the diameter of the wetting spots to 0.4 mm.

On the settings employed, the ADDSOAP sensor performed less well over the whole intensity range, with a maximum agreement time of 89% at 1.0 mm hr^{-1}. This decreased as the intensity threshold was lowered to give only 48% agreement at 0.1 mm hr^{-1}. The performance under different settings will be the subject of further experiments.

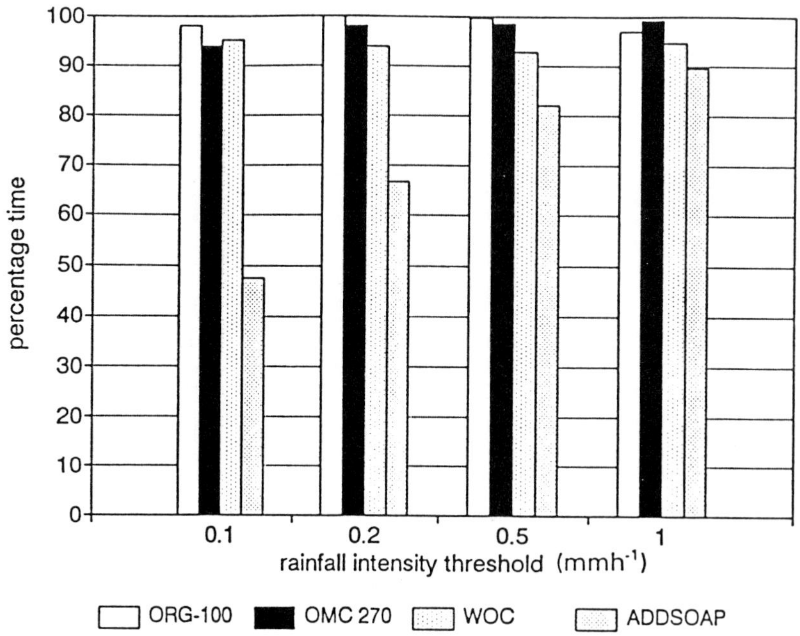

Figure 8.22 *Agreement times for 'rain' state*

As well as indicating rain when it is raining, it is also important that a sensor does not indicate rain during dry periods. Figure 8.23 shows the percentage time each sensor indicated rain while the ORG-700 recorded the 'no rain' state. This might be expected to arise mainly from delays in detecting the end of precipitation events. A sensor performing correctly should show a lower result at the 0.1 mm hr^{-1} intensity threshold and a higher result at the 1.0 mm hr^{-1} threshold.

At the important lower intensity thresholds the ORG-100 was incorrectly detecting rain for up to 30% of the time. Much of this is a result of dew forming on the instrument's lenses and increasing the output voltage. The OMC and WOC sensors again give similar results. Both performed well giving around ~2% incorrect detection at the 0.1 mm hr^{-1} threshold. This small error can be attributed to late detection of the end of events. Both show the expected increase in this percentage with intensity threshold as the ORG-700 is 'forced' to class intensities less than the selected threshold as 'no rain'.

With the humidity sensor disabled, as was the case during these experiments, the ADDSOAP sensor could conceivably suffer from frequent incorrect 'rain' state detection due to impacts from leaves, insects or other wind-blown debris. In practice, however, this appears to occur only very rarely. Employing the humidity sensor could reduce this still further.

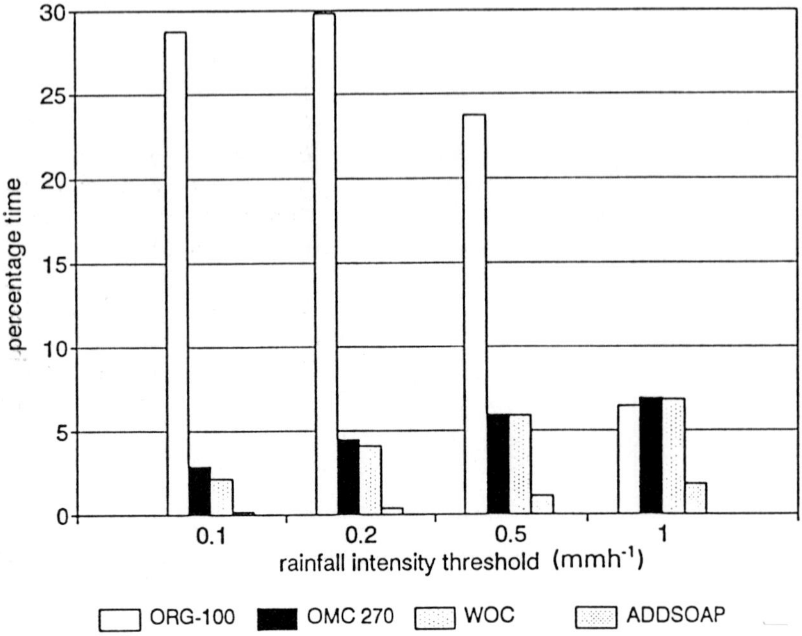

Figure 8.23 *Rain detection during 'no rain' state*

8.5 Conclusions

(1) The windflows over the openings of cylindrical precipitation collectors are directly affected by the aspect ratio (the ratio of depth to diameter) of the collector. The acceleration and displacement of the flow over the opening, the critical features which affect collection efficiency, are only reduced if the aspect ratio is below about 0.5. Above this value the flow over the opening is not greatly affected by the depth of the collector. Thus quite shallow cylindrical collectors are required if a good collection efficiency is to be obtained.

(2) The ejection of collected precipitation is also affected by the aspect ratio of cylindrical collectors. The windspeed at which ejection of precipitation of a given size occurs is at a maximum for cylinders of aspect ratios around 0.2 and for values beyond 1.5. It is at a minimum for aspect ratios around 0.8. The reasons for this behaviour are related to the internal air circulation driven inside the collector being at its strongest for aspect ratios around unity.

(3) Measurements of the acceleration and displacement of the flow over a variety of existing collector designs have shown a very wide range of values. This is particularly so for collectors with a basically cylindrical form, for which the values are not solely a function of the aspect ratio of the collectors. Funnel shaped collectors show a reduced displacement compared to the equivalent cylinder, but about the same acceleration. Streamlined collector shapes show reductions in both acceleration and displacement, but the greatest reduction is in the displacement.

(4) An investigation of the use of flow deflectors around cylindrical collectors has shown that there is a fundamental problem with designs of this sort, which is that the aerodynamic blockage of the cylindrical collector remains unaffected, as do the flow patterns associated with it. As a result, flow deflectors can reduce acceleration, but not displacement.

(5) A combination of a flow deflector and a streamlined collector shape makes possible the reduction of both acceleration and displacement over the collector opening. Using this approach it has been possible to design a collector with a flat, unaccelerated flow over the opening. Further improvement of the aerodynamic design is possible and it is intended to incorporate it into the WSL wet-only collector design.

(6) Changes in electrical resistance across intermeshed electrodes provide a reliable method for detecting precipitation state at rainfall intensities greater than 0.1 $\mathrm{mm\,hr^{-1}}$. An electrode separation of 0.6 mm or less should improve detection at lower intensities. Incorrect responses during dry periods are rare. Surface heating should be regulated to avoid dew forming and to melt and detect snow. Tin or gold plated copper electrodes may give improved wettability characteristics.

8.6 References

ASTM (1989). Method for the Collection and Measurement of Dustfall (Settleable Particulate Matter). Draft Standard D1739 (Revision).

Barrett C F, Goldsmith A L, Hall D J and Irwin J G (1985). The Variation of Precipitation Composition with Altitude: A feasibility Study. Stevenage: Warren Spring Laboratory, LR 534(AP).

Bradbury L J S and Castro I P (1971). A pulsed-wire technique for velocity measurement in highly turbulent flows. *J. Fluid Mech.*, **49**, 657-691.

BS (1969). British Standard, BS 1747, Methods for the Measurement of Air Pollution. Part 1, Deposit Gauges.

Folland C K (1988). Numerical models of the raingauge exposure problem, field experiments and an improved collector design. *Q. J. R. Meteorol. Soc.*, **114**, 1485-1516.

Goodison B E, Sevruk B, Klemm S (1989). WMO solid precipitation measurement intercomparison: objectives, methodology, analysis. Proceedings of the Symposium on Atmospheric Deposition, Baltimore, May 1989. IAHS Publication No 179, 57-64.

Hall D J and Waters R A (1986a). An improved, readily available dustfall gauge. *Atmos. Environ.*, **20**, 219-222.

Hall D J (1986b). The Precipitation Collector For Use in The Secondary National Acid Deposition Network. Stevenage: Warren Spring Laboratory, LR 561 (AP)M.

Hall D J, Cottrill S M, Goldsmith A L, Upton S L, Waters R A and Wright P (1988a). The Development and Field Use of a Snow Collector for Acid Deposition Studies. Stevenage: Warren Spring Laboratory, LR 585(PA)M.

Hall D J and Upton S L (1988b). A wind-tunnel study of the particle collection efficiency of an inverted frisbee used as a dust deposition gauge. *Atmos. Environ.*, **22**, 1383-1394.

Hall D J, Upton S L, Campbell G W, Waters R A and Irwin J G (1989). Further Development of a Snow Collector for Use in Acid Deposition Studies. Stevenage: Warren Spring Laboratory, LR 752(PA).

Ralph M O and Barrett C F (1984). A Wind-tunnel Study of the Efficiency of Three Deposit Gauges. Stevenage: Warren Spring Laboratory, LR 499(AP).

Rodda J C (1967) The Rainfall Measurement Problem. Proceedings of the International Association of Scientific Hydrology, Bern 1967. Publication No 78, 215-231.

Rodda J C and Smith S W (1986). The significance of the systematic error in rainfall measurement for assessing wet deposition. *Atmos. Environ.*, **20**, 1059-1064.

Sevruk B, Hertig J A and Spiess R (1989a). Wind Field Deformation Above Precipitation Gauge Orifices. Proceedings of the Symposium on Atmospheric Deposition, Baltimore, May 1989. IAHS Publication No 179.

Sevruk B and Klemm S (1989b). Types of standard precipitation gauge. Proceedings of the WMO/IAHS/ETH International Workshop on Precipitation Measurement, St Moritz, Switerland, pp 227-232.

Sevruk B (1990). Comments on 'numerical models of the raingauge exposure problem, field experiments and an improved collector design'. *Q. J. R. Meteorol. Soc.*, **116**, 239-242.

Stone B H (1991). An Assessment of the Collection Efficiency of United Kingdom Precipitation Collectors. Stevenage: Warren Spring Laboratory, LR 800 (AP).

— 9 —

The use of simulated rainwater solutions for analytical quality control

J L Cocksedge

9.1 Abstract

Solutions of simulated rainwaters have been prepared and subjected to interlaboratory analytical exercises. The results of two such exercises are presented and shown to validate the design and preparation of the solutions. A plan is suggested by which laboratories can use such solutions to supplement the use of standard reference materials for their quality control procedures.

9.2 Background

Ongoing monitoring of acid precipitation is needed to establish levels and trends. Surveys are carried out by many laboratories in many countries, some of the results from individual surveys feed into larger international surveys. The governments of the world then make decisions, both collectively and individually, based on the data produced. These decisions have social and financial implications for the populations of the world.

The use of analytical quality control to ensure the validity of data is crucial and, although the actual choice of analytical method will not be considered here, the use of a validated method is essential. Laboratories need to use calibration materials which are traceable and also to impose a quality control regime which includes the use of control samples and standard reference materials. The opportunity to harmonize results by taking part in collaborative trials, where the data from several laboratories is collated and compared, is essential.

Rainwater samples are analysed at Warren Spring Laboratory (WSL) from sites around the UK on behalf the Department of the Environment. Results from the primary sites in this network feed into an international network coordinated by the Norwegian Institute for Air Research (NILU) on behalf of EMEP, the co-operative programme for monitoring and evaluation of long-range transmission of air pollutants in Europe. As part of this effort NILU run intercomparison exercises. WSL is the only UK laboratory involved with the EMEP network.

As a result of the significance of this work, quality control in the laboratory is particularly important. To ensure this, the more recent batches of solutions prepared for intercomparison exercises run by the WSL (Heyes, 1986a, 1986b, 1988a, 1988b, 1989) serve the dual purpose of providing samples for an intercomparison exercise and for internal quality control. The first sample to be analysed in any analytical batch run is taken from one of these quality control solutions. From these samples, Shewart control charts are used to determine whether the instruments are operating correctly and if the data is acceptable.

9.3 WSL collaborative exercises

Early inter-laboratory exercises were arranged on behalf of the Department of the Environment to serve the primary purpose of testing the laboratories involved. The current series of inter-comparisons has been part of a project, funded by the Department of Trade and Industry, designed to bring chemical measurements on rainwater into the National Measurement System.

The objectives of the exercises are both to allow British laboratories to harmonize their analyses, and, to refine and validate methods of producing synthetic solutions for comparative and quality control purposes. Two exercises, coded A and B, have been carried out. The results of both exercises are summarized here; the results of exercise A are the subject of a report (Cocksedge, 1991).

9.3.1 Preparation of synthetic rainwaters

Synthetic rainwaters are prepared by dissolving analytical grade reagents in deionized water of good quality. The analytes of interest in collaborative trials are: pH, conductivity, chloride, nitrate, sulphate, sodium, ammonium, potassium, calcium and magnesium. Over time, the method of preparation of the WSL simulated rainwaters has been refined.

The solutions for exercise A were prepared in a similar way to those used at the end of a previous series of exercises. This method is described in more detail in a separate report (Cocksedge, 1988). A set of four solutions of synthetic rainwaters were designed each with particular target concentrations. The total volume of each solution was 10 l. For each analyte, there are two pairs of target concentrations within the set, each pair being of similar concentrations. However, the pair of solutions with comparable concentrations for one analyte is not necessarily the same as the pair for another analyte. For example: A1 and A3 were a pair for pH (4.30 and 4.22) but A1 and A4 were a pair for sodium (2.93 and 3.12 $mg\,l^{-1}$).

The solutions were made up using concentrated solutions of each component substance. Acidity was provided by sulphuric acid which was standardized by titration with sodium carbonate. Design aliquots of each of the starting solutions were combined in a series of 2 l volumetric flasks. These were then diluted to volume with water. Each solution was then poured into a 10 l container and the remaining water added by filling and draining the volumetric flask.

Before dispatch the solutions are analysed to ensure that no errors have been made in their preparation. This preliminary analysis is independent of Warren Spring Laboratory's participation in collaborative exercises. Sub-samples for dispatch are placed in new high density polyethylene bottles. Before use these bottles are washed in a laboratory washing machine which uses deionized water only without addition of detergent.

The solutions for exercise B were made by a slightly different method. Two acids were used, nitric and sulphuric. This gave two advantages: (1) the amount of sulphate was reduced thus enabling the possible concentration range to be enlarged; (2) the solutions are more similar to genuine acid rain samples.

The design aliquots were diluted to 10 l by weight rather than by volume as in previous exercises. This allowed the solutions to be prepared more quickly and overcame the problem of allowing for drainage with the volumetric flasks.

9.3.2 Storage qualities of synthetic rainwater solutions

After dispatch to the participating laboratories two sub-samples of each solution were put into long-term storage. The bottles used were from the same source as those used to dispatch the test samples. One set was stored in a refrigerator and the other set was stored in the light at ambient temperatures. The storage period was usually about 9 months. The samples were then analysed at WSL using the standard procedures.

The results for exercise A of this series were excellent, as has been the case with previous batches of synthetic rainwaters from earlier exercises. The pH values for the solutions stored at ambient temperatures were slightly low but there were no significant differences in most of the ion concentrations. The storage test results are given in Table 9.1(a).

After the analysis by WSL as a participating laboratory, some of the solutions were put into use as quality control standards in routine analytical procedures. One year after their manufacture the mean values on the control charts were still close to the theoretical, a comparison is given in Table 9.1(b).

9.3.3 Statistical evaluation of analytical results from exercise A

9.3.3.1 Calculation of standard deviations

After the receipt of their results each laboratory was supplied with the theoretical concentrations of the solutions. Subsequently, some laboratories

Table 9.1(a) Synthetic rain 9-month storage test results

Analyte	*Unit*	*Sample*	*Theoretical*	*Refrigerated*	*Ambient*
pH		A1	4.30	4.27	4.25
		A2	3.85	3.83	3.82
		A3	4.22	4.23	4.17
		A4	3.92	3.90	3.88
Conductivity	$\mu S\,cm^{-1}$	A1	76	72	73
	(at 25°C)	A2	87	83	84
		A3	52	49	50
		A4	108	101	103
Chloride	$mg\,l^{-1}$	A1	7.13	7.28	7.30
		A2	3.38	3.38	3.39
		A3	3.28	3.26	3.29
		A4	7.40	7.66	7.75
Nitrate–N	$mg\,l^{-1}$	A1	2.14	2.14	2.15
		A2	0.72	0.72	0.73
		A3	0.70	0.70	0.72
		A4	2.19	2.24	2.18
Sulphate–S	$mg\,l^{-1}$	A1	1.57	1.60	1.62
		A2	3.01	3.04	3.03
		A3	1.73	1.78	1.79
		A4	2.69	2.77	2.80
Ammonium–N	$mg\,l^{-1}$	A1	1.96	1.90	1.99
		A2	0.86	0.86	0.88
		A3	0.79	0.74	0.84
		A4	2.12	2.05	2.10
Sodium	$mg\,l^{-1}$	A1	2.93	2.93	3.08
		A2	1.99	2.04	2.18
		A3	1.91	1.93	2.13
		A4	3.12	3.19	3.22
Potassium	$mg\,l^{-1}$	A1	0.26	0.28	0.26
		A2	0.73	0.74	0.79
		A3	0.66	0.66	0.71
		A4	0.31	0.30	0.32
Calcium	$mg\,l^{-1}$	A1	1.47	1.47	1.47
		A2	0.24	0.24	0.25
		A3	0.29	0.29	0.30
		A4	1.37	1.36	1.38
Magnesium	$mg\,l^{-1}$	A1	0.66	0.67	0.68
		A2	0.18	0.19	0.19
		A3	0.24	0.25	0.25
		A4	0.60	0.62	0.61

Table 9.1(b) Comparison of mean values on current control charts with theoretical concentrations ($mg\,l^{-1}$)

Analyte	*Mean – current chart*	*Theoretical concentration*
Chloride	3.25	3.28
Nitrate-N	0.71	0.70
Sulphate-S	1.74	1.73
Sodium	3.14	3.12
Ammonium-N	2.11	2.12
Potassium	0.31	0.31
Calcium	1.46	1.47
Magnesium	0.663	0.66

reported corrected results which, formerly, had some error in the calculation or in the result transcription. These amended results were used for the statistical evaluation. Other laboratories reanalysed samples after receiving the theoretical figures, these results were noted but the original data was used in the evaluation.

The standard deviation for each analyte was calculated after outliers had been excluded. The Dixon Q-test was used to identify outliers:

$$Q = \frac{\text{nearest neighbour-suspect value}}{\text{range}}$$

Tables of Q-values give guidance for rejection at a particular level of confidence. For example, the value of Q for 10 measurements at the 90% confidence level is 0.41. Values of Q higher than this mean that a result can be rejected with this level of confidence.

Some laboratories reported a 'less-than' value for calcium and magnesium. These solutions contained concentrations below, or apparently below, the detection limit of their method. These were not included in the calculations.

9.3.3.2 Youden plots

The results for collaborative analytical exercises can be displayed graphically by a method described by Youden (1959). For each determinand the results of a pair of solutions are plotted with one solution on each axis. For example, the pH for sample A1 by laboratory x is plotted against the pH reported for sample A3 by laboratory x. The theoretical and median concentrations (or measurements) are also indicated on each figure and have the effect of dividing the figure into quadrants. The intersections give the points at which both results represent the 'true' result or the median pair of results.

Systematic bias can be observed in the scatter in the 45° direction across the bottom left and top right quadrants. This is because a laboratory which obtains a low result for one solution because of a systematic error will obtain a low result for the other (provided the systematic error is greater than any

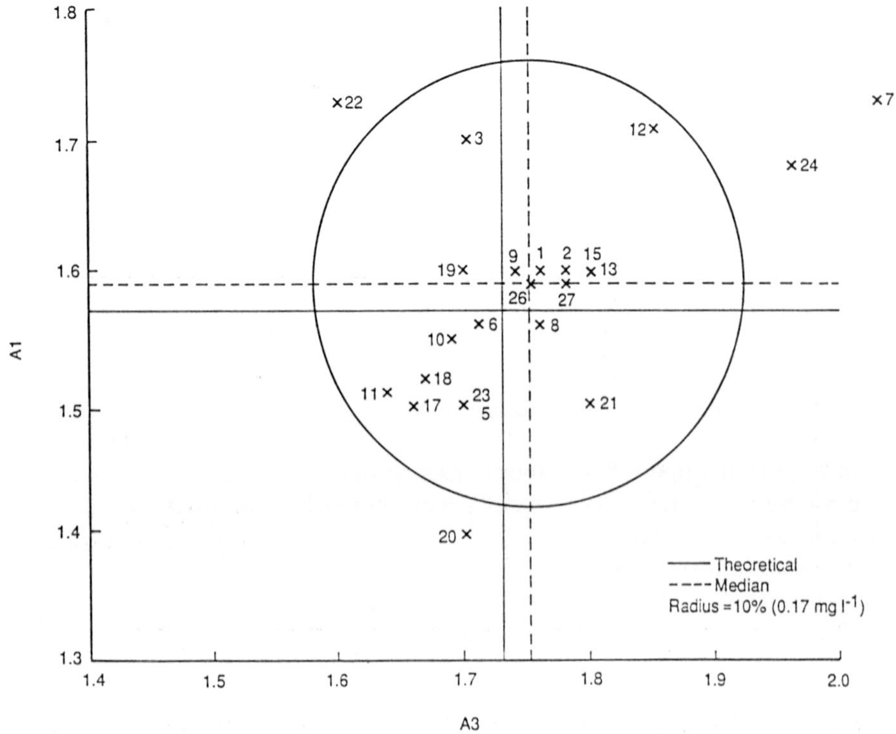

Figure 9.1 *Synthetic rain Youlden plot for sulphate, solution A1 against solution A3 (mg l^{-1} SO_4 as S)*

random error present). Random error does not favour any quadrant and therefore is demonstrated in the distance of the points from the 45° line.

A circle around the intersection of two median lines is constructed to indicate a target area for results. For the purpose of this exercise it was decided that the circle would enclose all those laboratories which had achieved results within 10% of the median for both the conductivity and the ion concentrations. The circles on the pH plots were drawn at 0.1 pH units.

A selection of the plots from the first exercise can be seen in Figures 9.1 to 9.3. Figure 9.1 (for sulphate A1 versus A3) shows the majority of the laboratories within the target and a fairly random scatter of results. Figure 9.2 (nitrate A2/A3) shows far less random scatter and a more pronounced systematic error, some laboratories are some distance along the diagonal from the theoretical and median point. Finally, Figure 9.3 (potassium A1/A4) shows significant amounts of both types of error. Relative standard deviations for potassium determinations on this pair of solutions were of the order of 40%, after the exclusion of outliers.

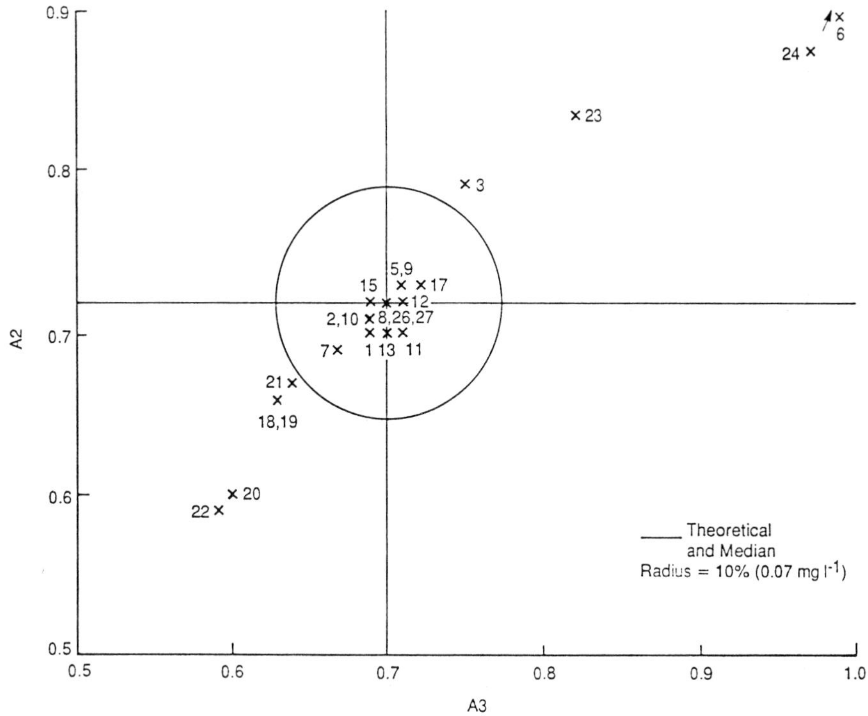

Figure 9.2 *Synthetic rain Youlden plot for nitrate, solution A2 against A3 (mg l^{-1} NO_3 as N)*

9.3.4 Comparison of measured and theoretical concentrations

Tables 9.2(a) and 9.2(b) give summaries of the results of exercises A and B respectively. The calculated theoretical concentrations for each determinand are listed together with the mean and median concentrations. The standard deviation for each set of results is also given.

In the case of exercise A, the median pH results were within 0.01 pH units of the theoretical for all solutions. The mean and median concentrations of ions were all within 10% of the theoretical concentrations and most were within 5%. One striking detail of the Youden plots was that, even when there was a wide scatter of results, the theoretical and median intersecting lines were remarkably close.

In general, the results for exercise B are similarly impressive. Median pH values are slightly further from the theoretical, the maximum difference being 0.05 units. This may be a factor due to the smaller number of participating laboratories. Twelve laboratories determined pH on this occasion whereas 21 did so in exercise A.

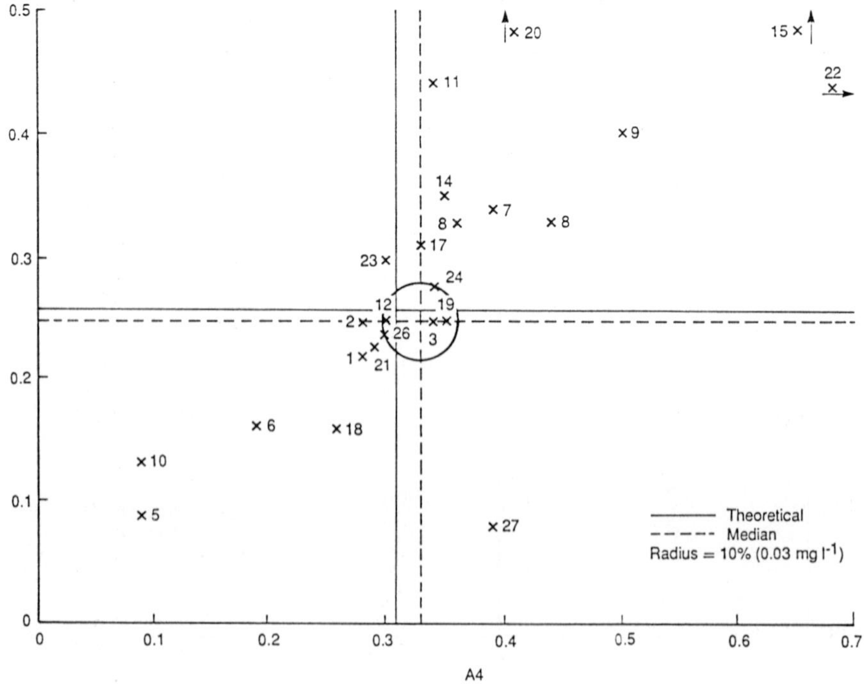

Figure 9.3 *Synthetic rain Youlden plot for sulphate solution A1 against solution A3 (mg l^{-1})*

9.4 Suggested method for the utilization of simulated rainwater samples for quality control

Naturally, not every laboratory which requires a quality control sample can prepare one and get it analysed by a number of other laboratories. On the other hand standard reference materials are expensive to use on a day-to-day basis. In addition, the certificate is usually valid only on the day of opening. Laboratories could devise a scheme whereby a synthetic solution of this type could be incorporated as a valid Control sample. An example of such a scheme is as follows:

1. A synthetic solution must be carefully prepared, independently of all calibration standards, according to a method validated by its use in laboratory intercomparison exercises.
2. The analytical methods of choice should be optimized using existing standards and the synthetic solutions so that reproducible results are obtained with acceptable precision. These precision figures may be able to be used to prepare control charts if standard reference material results are good (see Stage 3).

Table 9.2(a) Comparison of observed and theoretical concentrations in synthetic rain solutions (Exercise A 1991)

Analyte / unit	*Solution*	*Concentrations* *Theoretical*	*Mean*	*Median*	*Standard Deviation*
pH	A1	4.30	4.28	4.30	0.273
	A2	3.85	3.88	3.86	0.396
	A3	4.22	4.20	4.22	0.347
	A4	3.92	3.92	3.92	0.287
Conductivity	A1	76	72.1	72	4.22
$\mu S\,cm^{-1}$	A2	87	80.6	82.5	6.21
at 25°C	A3	52	49.4	50	2.86
	A4	108	102.5	102	4.02
Chloride	A1	7.13	7.07	7.07	0.392
$mg\,l^{-1}$	A2	3.38	3.30	3.29	0.396
	A3	3.28	3.12	3.22	0.320
	A4	7.40	7.45	7.44	0.809
Nitrate-N	A1	2.14	2.23	2.10	0.369
$mg\,l^{-1}$	A2	0.72	0.73	0.72	0.092
	A3	0.70	0.71	0.70	0.101
	A4	2.19	2.30	2.19	0.365
Sulphate–S	A1	1.57	1.59	1.59	0.082
$mg\,l^{-1}$	A2	3.01	3.03	3.01	0.277
	A3	1.73	1.75	1.75	0.097
	A4	2.69	2.71	2.71	0.229
Ammonium–N	A1	1.96	1.96	1.97	0.135
$mg\,l^{-1}$	A2	0.86	0.86	0.86	0.074
	A3	0.79	0.80	0.80	0.083
	A4	2.12	2.12	2.12	0.179
Sodium	A1	2.93	2.90	2.90	0.389
$mg\,l^{-1}$	A2	1.99	2.00	2.00	0.276
	A3	1.91	1.95	1.90	0.347
	A4	3.12	3.14	3.10	0.599
Potassium	A1	0.26	0.28	0.25	0.110
$mg\,l^{-1}$	A2	0.73	0.72	0.74	0.097
	A3	0.66	0.65	0.64	0.134
	A4	0.31	0.32	0.33	0.120
Calcium	A1	1.47	1.39	1.38	0.194
$mg\,l^{-1}$	A2	0.24	0.24	0.24	0.069
	A3	0.29	0.27	0.28	0.064
	A4	1.37	1.34	1.36	0.195
Magnesium	A1	0.66	0.64	0.64	0.043
$mg\,l^{-1}$	A2	0.18	0.17	0.17	0.033
	A3	0.24	0.23	0.23	0.035
	A4	0.60	0.59	0.60	0.037

Table 9.2(b) Comparison of observed and theoretical concentrations in synthetic rain solutions (Exercise B 1991/2)

Analyte / unit	*Solution*	*Concentrations* *Theoretical*	Mean	*Median*	*Standard deviation*
pH	B1	4.48	4.48	4.46	0.122
	B2	4.41	4.37	4.38	0.070
	B3	3.96	3.96	3.96	0.057
	B4	4.07	4.04	4.02	0.060
Conductivity	B1	48	50	46	7.7
$\mu S\,cm^{-1}$ at	B2	33	35	32	8.6
25°C	B3	77	78	76	7.0
	B4	86	87	85	10.3
Chloride	B1	6.56	6.52	6.51	0.248
$mg\,l^{-1}$	B2	2.23	2.22	2.20	0.279
	B3	2.40	2.37	2.40	0.280
	B4	6.88	6.71	6.68	0.281
Nitrate–N	B1	0.69	0.67	0.68	0.050
$mg\,l^{-1}$	B2	0.64	0.63	0.64	0.046
	B3	1.95	1.96	1.95	0.114
	B4	1.92	1.93	1.91	0.110
Sulphate–S	B1	0.73	0.77	0.74	0.168
$mg\,l^{-1}$	B2	0.79	0.84	0.81	0.189
	B3	2.18	2.30	2.18	0.415
	B4	2.02	2.05	2.03	0.092
Ammonium–N	B1	1.18	1.12	1.13	0.236
$mg\,l^{-1}$	B2	0.39	0.35	0.37	0.085
	B3	0.51	0.47	0.47	0.120
	B4	0.98	0.92	0.92	0.099
Sodium	B1	0.99	0.99	0.98	0.066
$mg\,l^{-1}$	B2	0.79	0.78	0.76	0.057
	B3	2.97	2.95	2.93	0.180
	B4	3.17	3.17	3.16	0.217
Potassium	B1	0.58	0.66	0.61	0.195
$mg\,l^{-1}$	B2	0.21	0.24	0.22	0.100
	B3	0.26	0.29	0.27	0.103
	B4	0.52	0.55	0.54	0.091
Calcium	B1	0.37	0.35	0.36	0.049
$mg\,l^{-1}$	B2	0.44	0.41	0.42	0.067
	B3	0.73	0.73	0.72	0.089
	B4	0.81	0.79	0.79	0.067
Magnesium	B1	1.05	1.04	1.05	0.061
$mg\,l^{-1}$	B2	0.36	0.36	0.35	0.039
	B3	0.30	0.31	0.30	0.044
	B4	1.35	1.35	1.33	0.057

3. Method accuracy must now be established using standard reference materials.
4. Using statistical tests the 'true' and measured results for the reference solution should be compared. If a significant difference arises the cause, for example suspect calibration standards, should be found and rectified.
5. The synthetic samples are now re-analysed and, provided theoretical and measured concentrations are acceptable, mean values and standard deviations used to construct control charts.
6. Storage tests on the synthetic solutions should be carried out, together with periodic checks using reference materials.

9.5 Conclusions

The method of production of simulated rainwater samples has been validated by the excellent agreement between the mean, median and theoretical measurements for all determinations in the inter-laboratory comparisons. These were achieved despite, in some cases, large standard deviations in the individual figures which make up the group.

Stability tests on batch A indicate that the solutions are very stable, even when no particular measures have been taken to preserve them. This confirms storage information which has been found with all batches of low pH solutions, pH 3.5 to 4.5, used in previous exercises (Heyes, 1986a, 1986b, 1988a, 1988b, 1989). If solutions of higher pH, >5, were to be required precautions would have to be taken to exclude the carbon dioxide from the atmosphere. For example, solutions currently under investigation by the Community Bureau of Reference are prepared under nitrogen and stored in sealed quartz vials.

Backed up by initial and periodic confirmation using reference materials, laboratories may use such solutions to monitor their analytical quality. This measure would assist them to be able to achieve and support NAMAS accreditation.

9.6 References

Cocksedge J L (1988). The Design and Production of Synthetic Rainwater Samples. Stevenage: Warren Spring Laboratory, 1988, LR 684 (CS).

Cocksedge J L (1991). Chemical Analysis of Precipitation Samples Intercomparison Exercise A 1990. Stevenage: Warren Spring Laboratory, 1991, LR 855 (CA).

Heyes, C J (1986a). Interlaboratory Comparison of Precipitation Analysis – April 1985. Stevenage: Warren Spring Laboratory, LR 580 (AP)M.

Heyes, C J (1986b). Interlaboratory Comparison of Precipitation Analysis – November 1985. Stevenage: Warren Spring Laboratory, LR 581 (AP)M.

Heyes, C J (1988a). Interlaboratory Comparison of Precipitation Analysis - November

1986. Stevenage: Warren Spring Laboratory, LR 674 (AP)M.
Heyes, C J (1988b). Interlaboratory Comparison of Precipitation Analysis within the National Materials Exposure Programme–January 1988. Stevenage: Warren Spring Laboratory, LR 695 (AP).
Heyes, C J (1989). Interlaboratory Comparison of Precipitation Analysis - December 1987. Stevenage: Warren Spring Laboratory, LR 701 (AP).
Youden W J (1959). Graphical diagnosis of interlaboratory test results. *Industrial Quality Control,* **15**, 24-28.

Questions asked during discussion session I

To A van der Meulen

D Mark and S Upton:

Thank you for letting us see this paper which we found very interesting. As Dr van der Meulen was unfortunately unable to present the paper we would like to make some written comments.

(1) The ISO thoracic aerosol convention and the PM10 convention do not describe the same curve. The ISO curve is a cumulative lognormal curve with $d_{50} = 10$ μm and $\sigma_g = 1.5$, reaching 0 at about 30 μm, whilst the PM10 curve with the same d_{50} goes to 0 at 16 μm (see page 24734 of the US Federal Register, 1987). This difference may be important because, whilst the currently-approved PM10 samplers conform very well to the PM10 convention, new samplers operating to different principles may not, but they may instead conform well to the ISO convention.

(2) In the paper the author suggests that, because of some problems that he perceives with the wind-tunnel testing of samplers, the EC Method for demonstrating compliance with the convention should be based on elaborate ambient field testing rather than the more systematic and comprehensive approach adopted by the US comprising both wind-tunnel and field testing. However, we have difficulty in finding the evidence stated in the first two reasons for neglecting the wind-tunnel tests and would like to suggest that only in wind-tunnels can the sampling performance be systematically determined. Field determination of sampling performance is complicated by the lack of reliable information about the true particle size distribution and concentration of the aerosol to be sampled. Furthermore, in the wind-tunnel tests carried out at Warren Spring Laboratory, attempts are made to condition the airflow to simulate the atmospheric boundary layer. Whilst it is not possible to generate large turbulent eddies, most samplers would see these as variable windspeed or flow direction, which can be tested separately in the wind-tunnel. In addition, the prototype IOM ambient inhalable aerosol sampler would probably have become the European

Reference Sampler for inhalable aerosol if just the field measurements were taken into account. Wind-tunnel experiments carried out at Warren Spring Laboratory showed that the sampler suffered serious internal wall loss problems, especially when sampling large sticky particles. The field measurements had shown good agreement with the WRAC reference instrument.

(3) We are concerned about the philosophy of using the Aerodynamic Particle Sizer (APS) to enable the size selective characteristics of the candidate sampling instrument to be determined in the field. There are two problems: (a) it will be very difficult to ensure that a representative sample for all particle sizes is obtained by placing the entry of the APS in the entry of either the Andersen or the Wedding inlet; (b) the size selective performance is not obtained by comparing the mass size distributions penetrating to the filter with and without the PM10 entry. This is because the entry efficiency of the bare pipe when the PM10 device is removed will not be unity and will vary with particle size and windspeed.

(4) In conclusion, we are pleased to see a draft protocol for the field evaluation of samplers for PM10 aerosol, but would urge that the protocol includes both wind-tunnel and field testing so that an unambiguous assessment is obtained. This does not mean extensive wind-tunnel tests as most PM10 samplers currently available have already been approved by the US EPA protocol.

A Van der Meulen

Thoracic aerosol conventions

In the table below various thoracic sampling efficiency curves (as percentages of the total airborne particles) are compared. These are: the 'old' 1983 ISO thoracic convention; the 'new' 1991 ISO thoracic convention; results from the PM10 formulae with a 50% cut-point, d_{50}, = 10 μm aerodynamic diameter, and geometric standard deviation $\sigma = 1.5$; the EPA PM10 tabulated curve; and the recent (1991) EPA approved wind-tunnel result for a PM10 low volume Andersen type impactor device.

As can be inferred from the table, both the PM10 sampling formulae and the wind-tunnel tested PM10 device reach zero at about 30 μm; however, the EPA tabulated sampling efficiency approaches zero for particle sizes of about 16 μm. Hence, currently approved PM10 samplers do conform very well to both the old 1983 ISO thoracic convention and the new 1991 one.

Note that the total airborne particles are made up for the larger part (well over 70%) of particles < 20 μm. The sampling efficiency for particles over 20 μm is well below 10%. Hence, it is to be expected that the differences in large particle sampling characteristics will not become important.

Diameter (μm)	*ISO 1983 (%)*	*ISO 1991 (%)*	*EPA PM10 Formula (%)*	*EPA PM10 Tabulated (%)*	*PM10 wind-tunnel (%)*
1	95.6	97.1	100	100	
2	91.8	94.3	100	94.2	
3	88.4	91.7	100	92.2	
4	84.7	89.0	99	89.3	
5	79.5	85.4	96	85.7	82–98
6	72.2	80.5	90	81.2	
7	63.7	74.2	82	75.9	64–78
8	54.4	66.6	72	69.7	54–61
9	44	58.3	61	62.8	41–45
10	36.7	50.0	50	55.1	
11	30	42.1	41	46.5	30–32
12	22.9	34.9	32	37.1	
13	19	28.6	26	26.9	
14	13.7	23.2	20	15.9	
15	11	18.7	16	4.1	
16	8.0	15.0	12	0	
18	4.6	9.5	7.5		
20	2.7	5.9	4.5		
22	1.5	3.4	2.6		
25	0.7	1.8	1.2		0.9–4.6
30	0	0.6	0.3		
35		0.2	0.1		
40		0.1	0		
50		0			

Field vs wind-tunnel testing to show compliance

It has been widely acknowledged that there are systematic differences up to 30% in ambient field experiments between high and low volume PM10 instruments both approved by the US-EPA wind-tunnel procedure. Obviously, such differences are well beyond the required sampler performance specifications. The pertinent US-EPA requirements are that the comparability of collocated samplers should be better than 15% and that the expected mass concentrations calculated from the sampling effectiveness curves must be within $\pm$ 10% of that calculated for the ideal sampler. Hence, it is felt that (wind-tunnel) test procedures allowing for such sizeable deviations should be considered with due care.

Obviously, wind-tunnel experiments serve as a most useful and systematic design guide to mimic the target sampling efficiency. Therefore, it is worthwhile to investigate wind-tunnel procedures to elucidate the various aspects, allowing for the aforementioned differences.

Among others, it is proposed by the EC that Member States have to use the PM10 reference method under routine conditions for the survey of ambient air quality, or methods which are equivalent to the reference method. This calls for appropriate EC-PM10 reference equivalence procedure in real-world measuring conditions (covering a wide range of characteristic ambient situations in EC). Furthermore, it is thought that an EC-PM10 reference equivalence procedures should not be burdened by a discussion on the applicability of wind-tunnel procedures; apparently the applicability of wind-tunnel findings to ambient situations remains a controversial issue. As to the IOM sampler, this clearly demonstrates the need to perform the reference equivalence experiments over a wide range of characteristic ambient situations indeed. The IOM sampler has only been subjected to a limited field test covering a limited range of ambient situations.

APS size selective testing

It is not proposed that the APS procedure should serve as an integral part of any size selective testing procedure. It is proposed only to investigate the feasibility of such a procedure to test the cut-off characteristics of a candidate PM10 sampler. Obviously, the remarks made by David Mark and Stuart Upton will be taken into due consideration. As they point out, indeed the PM10 characteristics are not obtained by comparing the complete PM10 assembly with a bare inlet pipe only. It is thought worthwhile to test the various parameters influencing the PM10 size selective characteristics by comparing with the PM10 entry assembled, for example, with and without the insect screen in the very inlet itself, with and without the PM10 impactor device, etc.

To J Wilken

R Storeton-West: Did you have an ozone filter on the Luminol instrument?

J Wilken: No, as some reports say that these also remove NO_2 to some extent.

G Marshall: The majority of the measurements were at low concentrations. Is the Teco analyser sufficiently sensitive to measure these concentrations?

J Wilken: Yes the Teco 42S has a detection limit of 50 ppt, however, it would probably have been better to switch the instrument to a lower operating range of say 0-50 ppb.

To S Eaton

J Scawin: How do you ensure that the zero reference gases you use are correct?

S Eaton: Zero gases are BOC metrology grade.
W Rudolf: We put gas through a carbon filter then a molecular sieve to dry it. The instruments can then determine 0.3 ppb from zero.

R Coombes: Have you looked at the efficiency of the NO_2 to NO converter?
S Eaton: Yes, we measured it using gas phase titration.

W Rudolf: The error you stated in static dilution was 4%. We usually get a 1% error. Why is yours 4%?
S Eaton: The greatest error is in the syringe used for the concentrated gas.

To B Sevruk

K Nicholson: Evaporative losses do not affect deposition flux but wind induced losses do. Have you done work on this?
B Sevruk: Small particles taken away by wind often contain the most pollution. It would be useful to chemically analyse material from several collectors at the same site.

To J Irwin and D Hall

K Playford: Have deflectors other than solid ones been used, for example the fence described by Boris Sevruk?
D Hall: It is better to design a good collector than to use a windbreak fence as, if precipitation is to be analysed, contaminants may be picked up from the fence.

A Webb: Does a flat design encourage splashing? Have you looked at the chemical composition of small rain-drops?
D Hall: While flat designs have improved aerodynamics, they can suffer from problems with splashing. From that point of view a deeper shape would be better but, on balance, we have found the shallow shapes to be preferable.
J Irwin: Although there are not a great many data, concentrations in small droplets are generally greater than in larger drops. For example, concentrations in cloud-water are often several orders of magnitude larger than in rain drops where growth and dilution have

taken place. While this is important in event studies, over longer periods the effect is less significant as the smaller drops contribute much less to total rainfall as volume is related to radius cubed.

K Wrekolsa: Rain collectors are contaminated by resuspended material from the ground, such as radionuclides. It would be useful to have a collector that samples above the suspended particulate layer, how high would this have to be?

D Hall: There is no effective height, as windspeed increases the number of particles decreases. We must make sure that we know what we want to measure; to estimate the effect of rain on crops we would need the collector near the ground.

Questioner: How would you sample above the North Sea?

B Sevruk: The problems of sea spray make measurements using conventional collectors difficult. Some tests have been carried out using Doppler radar detectors which should become available at reasonable prices in the not too distant future.

J Irwin Doppler radar is certainly a good approach. Alternatively it may be appropriate to measure precipitation at sites around the North Sea and then use models to predict rainfall or deposition over water.

To J Cocksedge

K Goulding: Can you associate outliers with a particular method?

J Cocksedge: Survey A did not ask for detailed information on the analysis methods used, however, in Survey B participants where asked for information such as which columns used for IC standards used.

Part 2

Source assessment

— 10 —

Introduction

C. Schofield

Assessments of air pollution sources are carried out for a number of reasons.

1. *Compliance testing*: Regulations concerning air pollution emissions from specific processes are often expressed in terms of emission concentrations or rates of specified substances or classes of substances. Compliance with the regulations is therefore tested by measuring the appropriate parameter under known circumstances.
2. *Inventory compilation*: EC Member States are required to compile inventories of emissions of the principal pollutants from all possible sources. Measurements of all sources is impracticable and emission factors are used which relate emissions to a level of activity for a class of sources. Such emission factors require rigorous measurement.
3. *Diagnostic studies*: Measurements of emissions are carried out to understand the relationships between emissions and process design or operation.
4. *Performance testing*: The performance of abatement or suppression systems can only be assessed rigorously by appropriate measurement.

In all of these cases, assessments need to be carried out in such a way as to provide traceable measurements of assured quality, in order to have confidence in the judgements to be made based on those results. The papers in this session address a number of important pollution issues and the methods of source assessment.

Where the emission is confined in a stack or duct, measurements are usually made by extracting a representative sample and analysing that sample in the laboratory or continuously close to the stack. Sample integrity is an important consideration, especially in the case of dioxins at nanogram concentration, where, during sampling, the sample is exposed to changing conditions which could affect its chemical constitution. Peter Coleman's paper addresses this problem whilst Alwyn Fernandes's paper deals with the extremely demanding analytical task of analysing for dioxins and furans at sub-nanogram levels. In the case of continuous measurements of acid gas emissions, confidence in the results depends, in part, in the instrument calibration which is discussed by David Walker.

Fugitive or unconfined sources have to be assessed by measurements in the ambient atmosphere remote from the source. A number of papers deal with the issue of particle or dust source assessments in this way. David Hall describes recent improvements in the design of dust deposit gauges and particle flux gauges, taking account of their aerodynamic characteristics. Peter Clayton describes how more convential measuring systems can be used in a directional mode, taking account of wind direction, in order to determine the contribution from specific sources. Mark Kibblewhite's paper relates to procedures for analysing lead and cadmium in samples from stack or ambient samplers.

— 11 —

Improvements in dust gauge design

D J Hall, S L Upton and G W Marsland

11.1 Abstract

Large particles in the atmosphere, commonly called dust or grit, are a major cause of nuisance and can also be a cause of health and other problems, either from direct ingestion or through secondary pathways after they have deposited. Two sorts of measurements of dust, mainly for nuisance purposes, are commonly made using passive samplers. These are of deposition and of the horizontal flux of particles past a sampling station. Both deposition and flux gauges of the sort commonly in use show seriously deficient catch characteristics which are strongly wind dependent, typically collecting only a small fraction of what they should. These deficiencies are mainly due to the aerodynamic characteristics of the gauges. Wind induced flows around the gauge either prevent the entry of particles into the gauge opening, or remove them due to a wind generated circulation inside the gauge. Deposition gauges have been the subject of some previous investigation of this sort, but flux gauges have received very little attention. The paper describes investigations of the behaviour of both deposition and flux gauges with a view to improving their performance.

11.2 Introduction

Atmospheric dust (that is, large particles beyond about 20 μm aerodynamic diameter) has traditionally been considered mainly as a cause of nuisance (it is one of the two major causes of complaint about air pollution, alongside odours) or of health effects due to secondary pathways following its deposition to the ground. More recently, the realization of the effectiveness of the human respiratory tract in inhaling particles up to and beyond 100 μm (Vincent, 1989) has led to a greater interest in the primary health effects of large particles. The present paper is concerned with the traditional problem of dust deposition and its measurement, which remains of great importance for the reasons stated.

The two measurements of interest in monitoring deposited dust are of deposition to the ground and of the horizontal flux of particles past a measuring station. Both types of measurement are normally made with passive gauges. The deposit gauge is usually a container of some sort with a horizontal opening and the flux gauge is a container with a vertical opening. Figures 11.1 and 11.2 show the respective current British standard realizations of these gauges. The deposit gauge, Figure 11.1, is a bowl on a stand with a 'bird guard' and collecting bottle. It is typical of the variety of deposit gauges used in various countries. Many bear a resemblance to rain gauges, from which most were derived. The main distinction between existing dust and rain gauges is in the elevation of dust gauges to sufficient heights, typically between 1 and 2 m, to avoid collection of saltated material from the surface. The flux gauge, Figure 11.2, is described as the 'British Standard Directional Gauge' as it uses four flux gauges set 90° apart in order to determine the source direction of the dust. The basic collection unit is a vertical cylinder with a vertical slot cut in it.

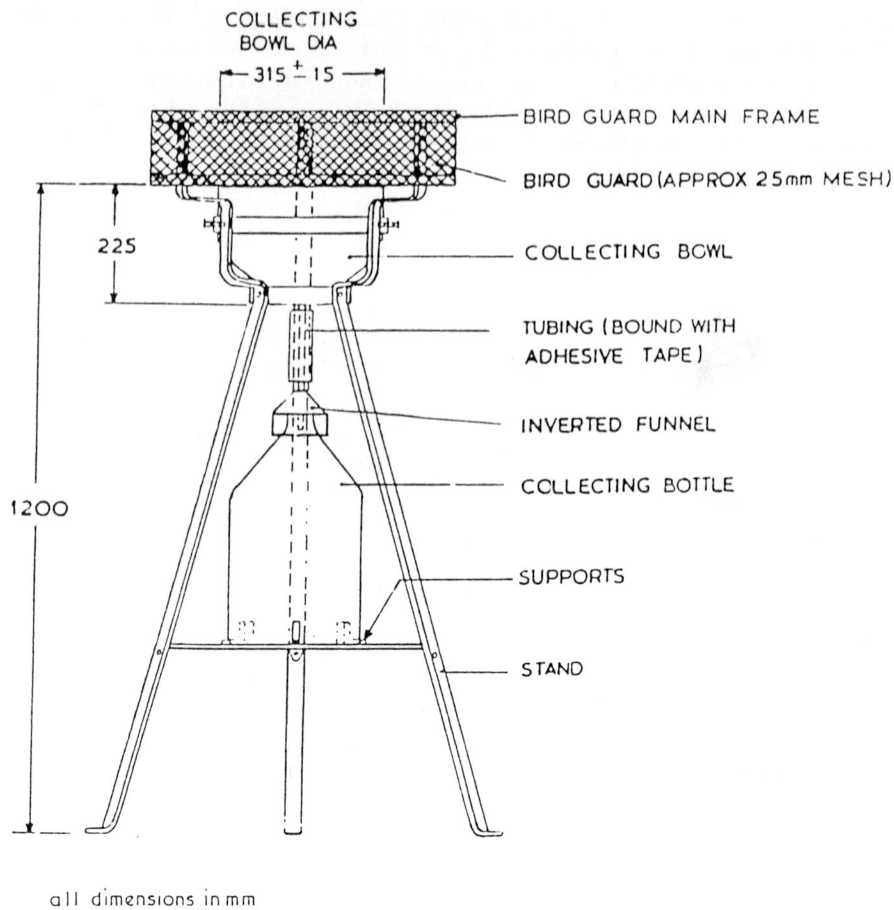

Figure 11.1 *British Standard (BS 1747 part 1) horizontal dust deposit gauge*

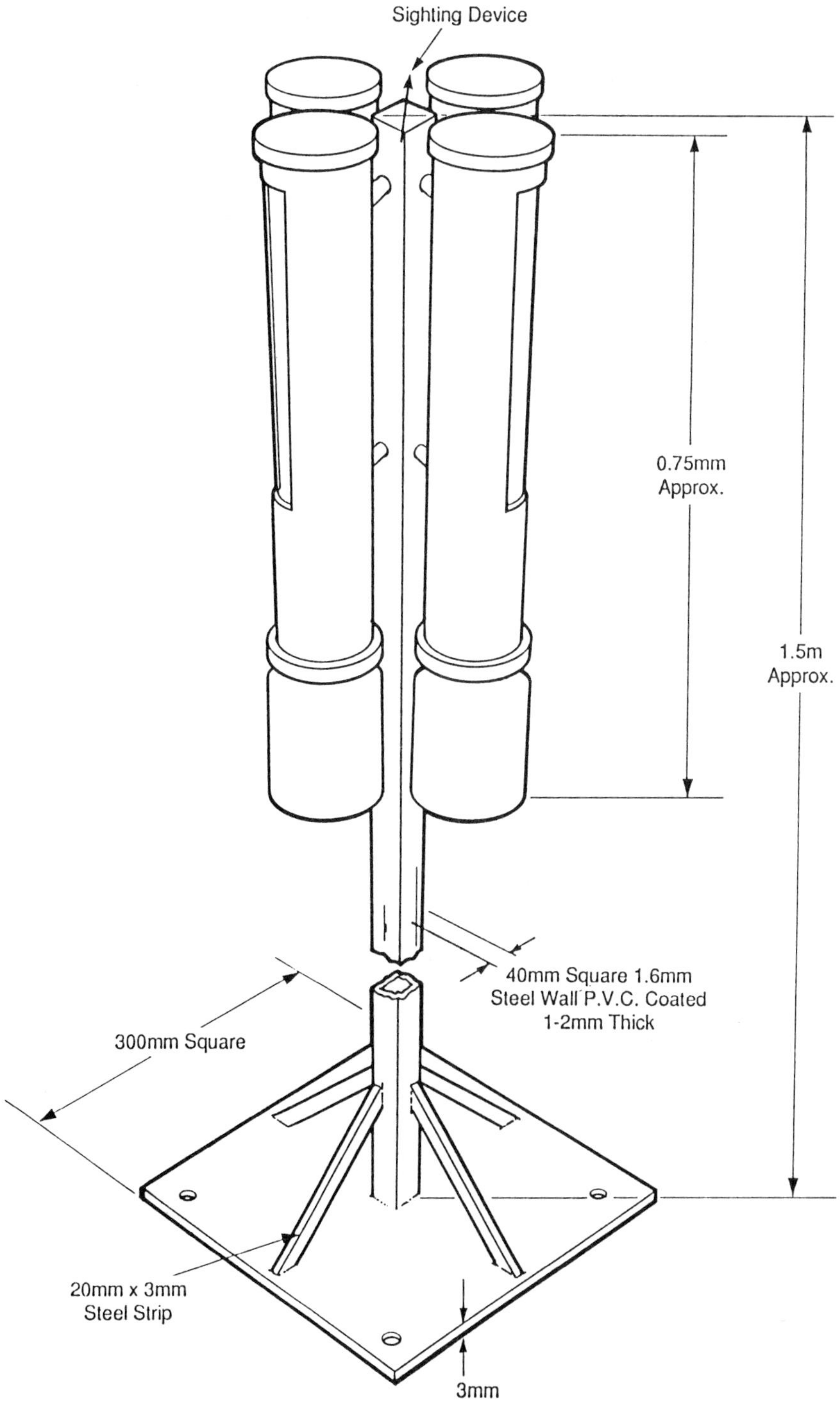

Figure 11.2 *British Standard directional dust gauge (BS 1747 part 5)*

For large particles, the mass deposition to the ground, D, is given by,

$$D = A \int \chi V_s \mathrm{d}t \tag{1}$$

where A is the surface area over which deposition occurs, χ is the ambient mass concentration of the particles and V_s is the falling speed of the particles.

Similarly, the mass flux of particles past a measuring station, F, is given by,

$$F = A \int \chi U \mathrm{d}t \tag{2}$$

where U is the windspeed.

For a gauge, the surface area, A, is the area of the gauge opening, whether horizontal as with the deposition gauge, or vertical as with the flux gauge. Note that the particle masses collected by the two different types of gauge are not directly related, deposition is dependent on the falling speed of the particles while flux is dependent on the windspeed. There is no simple relationship between these different collected masses. The possibility of inferring one type of measurement from the other is too complex to be made use of in practice.

In practical dust sampling, deposit gauges are used at points of complaint or areas where information on deposition is required. Flux gauges may be used at similar locations in order to assess the source direction of depositing material, but they do not directly indicate the local deposition. Flux gauges can also be positioned close to a known source in order to assess its scale by monitoring the flux of material from it. Similarly, a line of flux gauges spaced away from a known source can indicate the depletion rate of material from that source with increasing distance. It is also possible, using a flux gauge, to determine the average particle concentration if the run of wind past the gauge is known. This can be measured with a simple cup counting anemometer.

So far, the discussion has not considered the performance of deposition and flux gauges, it has been assumed that they collect according to equations (1) and (2), above. In practice this is far from the case. Measurements of the collection efficiencies of the British Standard Deposit gauge, together with some other deposit gauge designs, by Ralph and Barrett (1984), and for the British Standard Directional gauges, by Ralph and Hall (1989), have been made using reduced scale models under carefully controlled conditions in a wind-tunnel. These have shown the collection efficiencies of dust gauges to be very variable, depending upon windspeed and particle size, and at times to be very low, values below 10% being readily achieved. Figures 11.3 and 11.4 show collection efficiency measurements for the two British Standard gauges taken from this work.

The deposit gauge, Figure 11.3, shows collection efficiency reducing rapidly with increasing windspeed and decreasing particle size. At windspeeds beyond the lightest and for the particle sizes which travel any

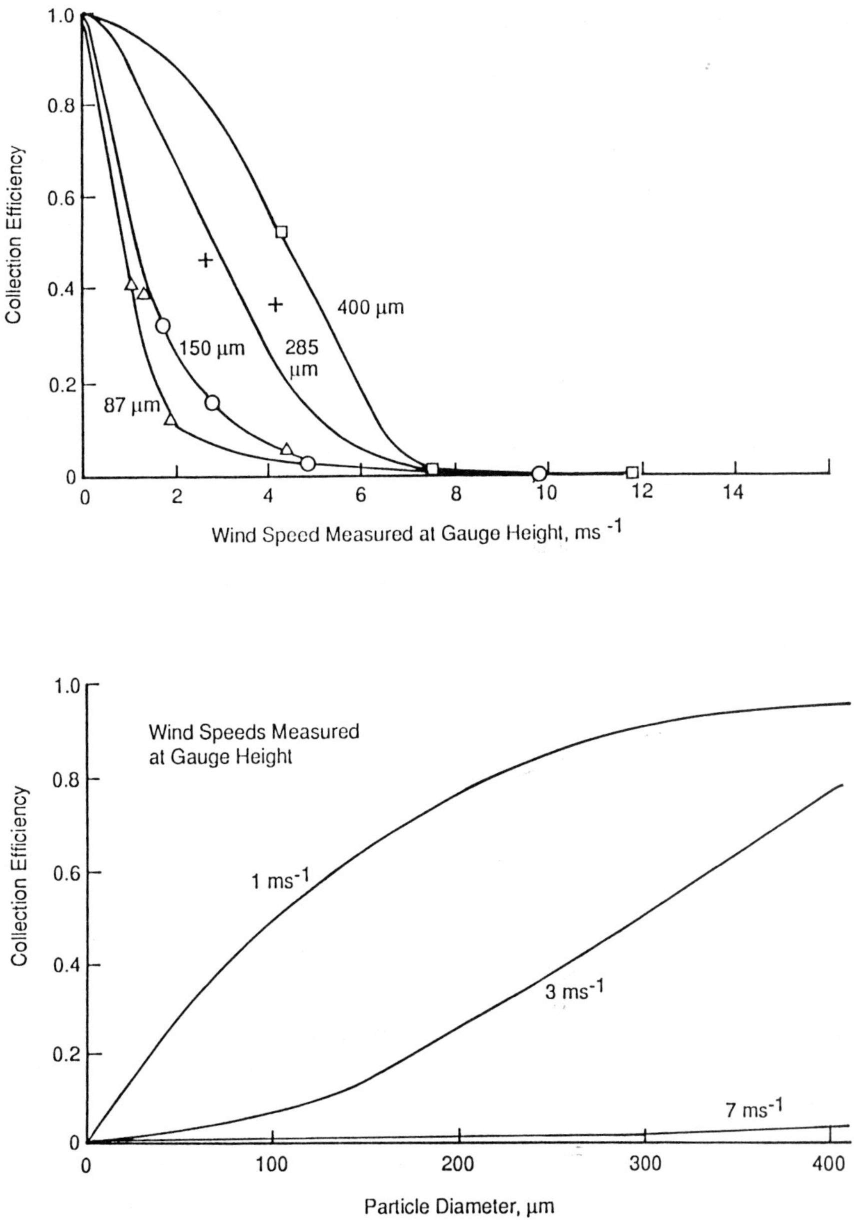

Figure 11.3 *Collection efficiency of the BS horizontal dust deposit gauge for varying windspeeds and particle sizes*

distance in the wind (mainly those below 200 μm) the measurements show a quite poor collection efficiency, below 20%. Ralph and Barrett showed that the 'bird guard', which is currently a plastic mesh with 60% open area, contributed to this poor performance and that removing this, together with adding a baffle inside the gauge, produced a distinct improvement in

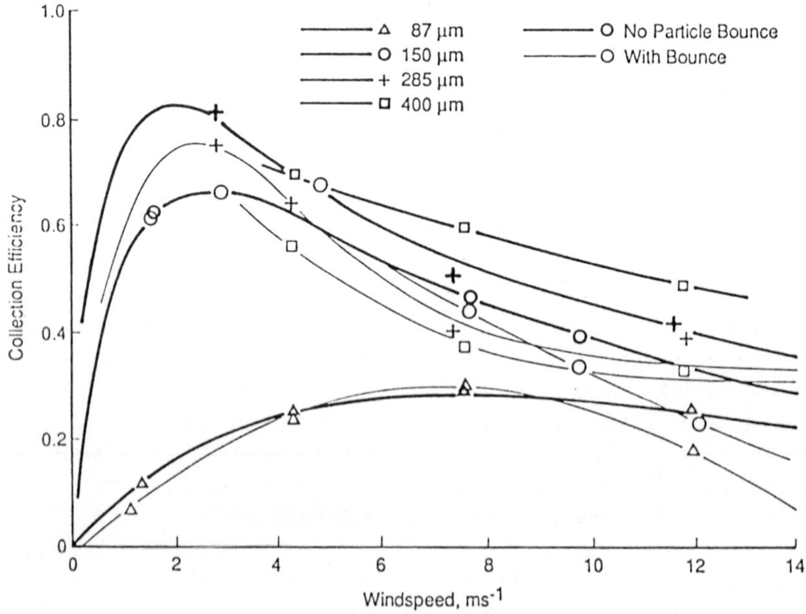

Figure 11.4 *Variation in collection efficiency with windspeed and particle size for BS directional gauge*

collection efficiency. However, even with these improvements the collection efficiency remained poor, largely below 20% for the smaller particle sizes and higher windspeeds.

The directional (flux) gauge, Figure 11.4, shows a maximum in collection efficiency at a windspeed which varies with the particle size. However, this maximum mainly occurs at low windspeeds of a few metres per second, so that at the higher windspeeds collection efficiency still falls with increasing windspeed. Overall, collection efficiency is not good, being mostly below 50% for the smaller particle sizes, and below 30% for sizes below 100 μm. The directional gauge does, however, have quite good directional properties, so it can be a reliable indicator of the source direction of wind-blown particles.

It is interesting to compare this sort of dust gauge performance with the conditions associated with the majority of practical dust nuisance problems. Here, it is commonly the case that the amount of dust released to the atmosphere increases with increasing windspeed, with the handling of powdery materials in the open, for example. A similar situation occurs with dust carried away from stockpiles. The nature of the saltation process which lifts particles from the surface is such that below a critical windspeed there is no loss, while above it there is a very rapid increase in the flux of airborne dust with small further increases in windspeed. Thus with present gauge designs, the collection efficiency falls away as, usually, the concentration of airborne dust increases, thus exaggerating the errors in estimating deposition or flux. On the whole, it would be more satisfactory if dust gauges were designed to have the best collection efficiency at high rather than low windspeeds. The strong dependency of gauge performance on windspeed and particle size is also unhelpful, as it results in unrepresentative sampling. It would be acceptable if a gauge had a reduced performance (of, say, 60%-70% collection efficiency), provided that it was independent of windspeed and particle size. The error could then be easily accounted for.

The reasons for this poor performance of dust gauges are due to their aerodynamic characteristics. In the case of the deposit gauge there are two specific problems. Firstly, aerodynamic blockage caused by the substantial bulk of conventional gauge designs causes a pronounced rise of the separation streamline over the gauge opening, together with an associated acceleration in the mean windspeed just outside the separation region. As a result of this, particles descending under gravity in the wind which would be expected to enter the gauge are displaced away from the opening, thus reducing the catch. The effect becomes more pronounced as the windspeed increases and the falling speed and the inertia of the particles reduces (essentially as they reduce in size). Secondly, the wind drives a circulation inside the gauge which can remove particles already collected.

In the case of the flux gauge there are also two specific problems. Firstly, there is no flow into the closed container of the gauge, so that the flow stagnates at the entrance slot. Particles can therefore only enter the gauge by their inertia carrying them into it. Since typical gauges are relatively large compared with the scales of inertial particle motions this effect significantly limits the catch; it is also both windspeed and particle size dependent. Secondly, there is, as with the deposit gauge, a wind driven circulation inside the gauge which tends to remove collected particles. The first effect is to some extent predictable (Bush *et al.*, (1976)) and largely accounts for the fall in collection efficiency on the low windspeed side of the maximum in Figure 11.4. The second effect is not readily predictable and probably accounts for most of the fall in collection efficiency on the high windspeed side of the maximum in Figure 11.4.

Similar problems to those occurring in deposit gauges occur in rain and snow gauges, though due to the generally higher falling speeds of rain and snow compared with dust, the errors so induced are not so severe. The problem has been well recognized in the meteorological world; references to wind-induced errors in rain collection exist as far back as 1865. The subject is discussed in more detail by Hall *et al.* in this volume. Correcting for wind-induced errors in snow and rain sampling has received some attention, mainly related to snow collection where the errors are larger. Most of the work has been concerned with the design of wind shields to reduce the effective windspeed around the gauge, with some limited attention being given to the design of more effective gauge shapes (Folland, (1988), Hall *et al.*, (1988a, 1989)). Curiously, virtually no attention had been given to this problem in relation to dust sampling until the wind-tunnel work carried out at Warren Spring Laboratory by Ralph and Barrett (1984). This is despite the much more severe problems associated with dust collection due to the generally much lower falling speeds of dusts compared to rain or snow, and also despite the British Standard deposit gauge (or its earlier equivalent) having been in use since the turn of the century. The position with the flux gauge is similar. There has been very little consideration of its performance either, except for that noted above, though it is of a more recent design (Lucas and Moore, 1964) than the deposit gauge.

Following the earlier work carried out at Warren Spring Laboratory, some attention was given to the development of improved dust deposit gauges. This was in parallel with the work on improving precipitation gauges for acid precipitation studies noted above (Hall *et al.*, 1988a, 1989). There are, broadly, two ways of improving the airflow over a gauge opening. These are, firstly, to reduce the aerodynamic blockage of the gauge by reducing its depth and, secondly, by modifying the flow around the gauge opening either by shaping the gauge appropriately or by fitting flow deflectors of some sort around it. The work at Warren Spring was aimed at both reducing the depth of the gauge and at shaping it in order to modify the flow over the opening. At the same time, attention was given to reducing the blowout of material which had already entered the gauge. The work on dust deposit gauges was mainly on a gauge shape (an inverted frisbee) with minimal aerodynamic blockage (Hall and Waters, 1986, Hall and Upton, 1988b). That on precipitation gauges (Hall *et al.*, 1988a, 1989) was on a mixture of reduced blockage and gauge shaping.

The gauge shape initially investigated for dust collection was an inverted frisbee, chosen simply because it was a readily available shape with low aerodynamic blockage which might have promise as an effective dust gauge. On further investigation, this proved to be so. Wind-tunnel measurements showed the collection efficiency of a gauge made sticky (to improve its retention) not falling below 30%. These results, taken from Hall and Upton (1988b) are shown in Figure 11.5. Though not perfect, this represented a distinct improvement over more conventional designs. Furthermore, this gauge of limited depth (the depth/diameter is about 1/7) showed, contrary to expectations, a greater resistance to particle loss due to blowout than deeper conventional gauges. The reason for this is that the strongest internal circulation occurs in containers of approximately square cross-section, that is of equal depth and diameter. Other shapes show weaker internal circulations.

The experiments described here are an extension of the previous work on deposition gauges, together with an investigation of and some development work on flux gauges. The work on deposition gauges concentrates mainly on the inverted frisbee shape, since this showed such good potential. The work on flux gauges investigates both simple closed container designs similar to the existing gauge, and passive through-flow devices. Since the programme is still very active, none of this work is complete, so that the present report is more of a progress report than a final conclusion. This applies particularly to the field experiments, where data is collected only slowly and interpretation is thus delayed.

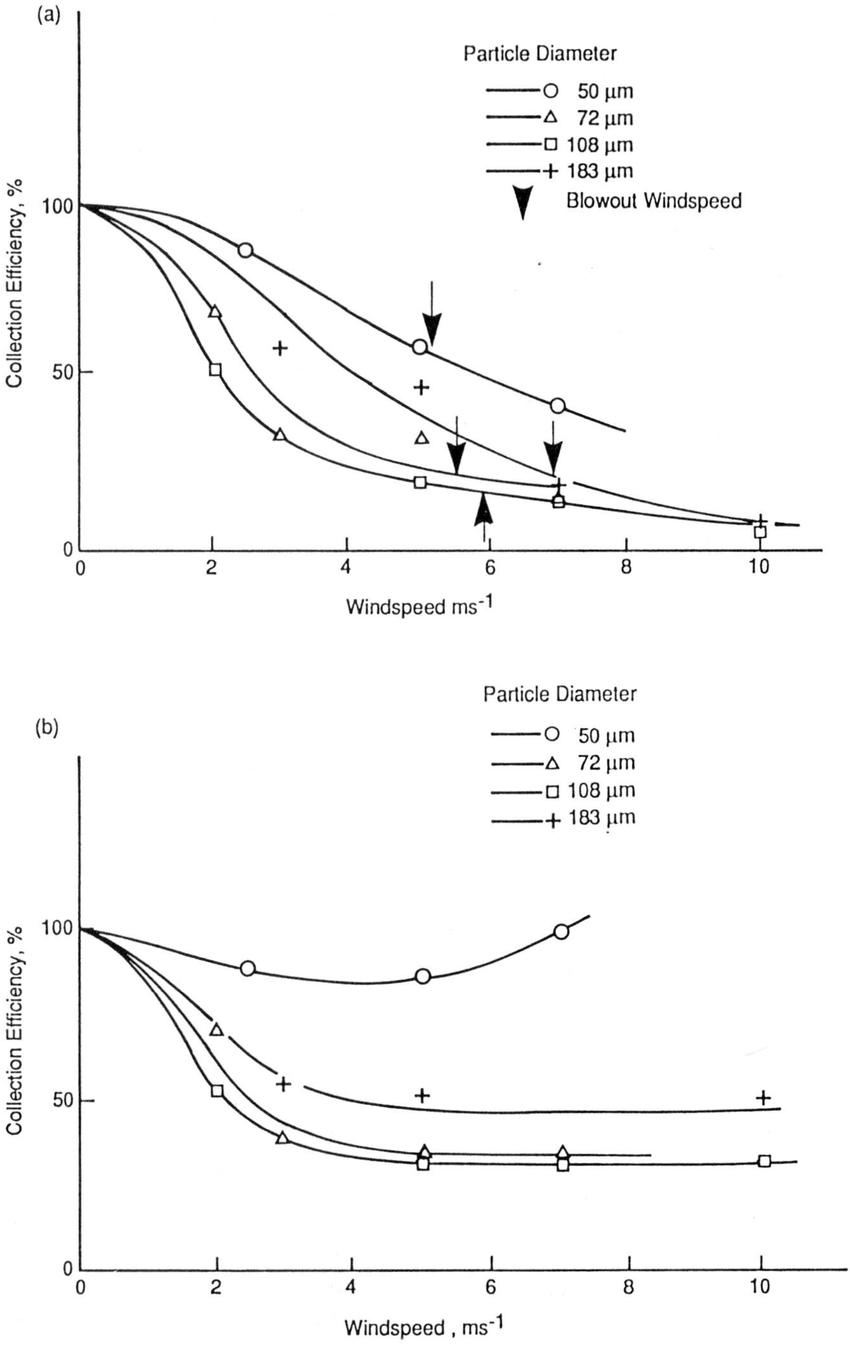

Figure 11.5 *Particle collection efficiency of inverted frisbee dust gauges: (a) dry frisbee; (b) 'sticky' frisbee*

11.2 Experimental methods and field testing

11.2.1 Wind-tunnel experiments

The detailed investigations of gauge performance described here have been carried out in wind-tunnels under carefully controlled conditions. Details of the experimental methods have been described elsewhere, in Ralph and Barrett (1984) and in Hall and Upton (1988b), so a brief description only is given here. Measurements of dust collection efficiency are made in a wind-tunnel by setting up a flow of known windspeed containing uniformly distributed particles of known size and concentration. From equations (1) and (2), it is known what mass of particles an ideal gauge should collect. The gauge under test is placed in the flow and its actual collected mass found by weighing. The ratio of these two masses is the collection efficiency.

Most of the experiments described here were carried out in the Pollution Abatement Division's Number 2 wind-tunnel. This has a long working section of 0.6 m height and 1.2 m width. It is set up with a turbulent flow intended to reproduce some degree of atmospheric turbulence. The monodisperse particles were sieved glass spheres. The smaller gauges, such as the inverted frisbee, were tested at full scale. The larger deposit gauges, such as the British Standard Deposit gauge and the proposed ISO gauge, were tested at one-third scale. The British Standard Directional Gauge was initially tested at one-third scale (Ralph and Hall, 1989). The subsequent investigations on alternative flux gauge shapes described here were carried out at half scale. In reduced scale gauge experiments, both the windspeed and particle size were also appropriately scaled, as described by Ralph and Barrett (1984). All windspeeds and particle sizes referred to in this paper are of full-scale values or full-scale equivalent values. Quoted particle sizes are of aerodynamic diameters.

Some experiments were also carried out in the Division's No 1 wind-tunnel. This is a much larger wind-tunnel, intended mainly for dispersion modelling. However, it is also used for testing the larger samplers, especially ambient aerosol samplers. The measurements of small particle collection efficiency of the frisbee gauge were carried out in this way. This wind-tunnel is also used for velocity measurements over gauges.

11.2.2 Field measurements

Most of the field measurements have been carried out by York University, assisted by Selby District Council, using sites in the North Yorkshire area. This work is described separately in detail by Vallack and Chadwick (1990, 1991a, 1991b). Though a number of sites have been used, most of the recent work has concentrated on two sites. One, near a power station, suffers from fairly high dust deposition rates. The other, in largely open country has the low deposition rates consistent with rural areas. British Coal have also assisted in testing samplers on a reciprocal information exchange basis. A few other unpublished data sets from various sites have also been made available to the authors.

11.3 Dust deposition gauges

11.3.1 Summary of programme

Work on deposition gauges has concentrated mainly upon the frisbee type of shallow gauge. This showed good promise as a deposition gauge and some further investigation of its properties were felt to be worthwhile. Wind-tunnel and laboratory investigations covered three aspects of the gauge's performance. These were, particle retention, the collection efficiency of small particles and the effect of eddy shedding devices as performance improvers. Particle retention problems became a matter of interest following the first field experiments. The other two items followed on from the previous wind-tunnel work on the performance of this gauge (Hall and Upton, 1988b). Field experiments were mainly concerned with comparisons between gauges, including the frisbee type, and with investigating the practical effectiveness of particle retention devices.

11.3.2 Modified deposit gauge design

The design of gauge used in the experiments and field trials (and which has been in more general use recently) was of the frisbee type, but no longer used a plastic frisbee. Plastic frisbees were originally used for quite short-term specific investigations, typically over a few days, for which they seemed to be generally satisfactory. When used in longer-term trials two problems arose. Firstly, there were difficulties with dimensional stability and degradation of the plastic. Since the gauge was very cheap, the latter problem was not of dominant importance, but the gauges also tended to warp and split. Secondly, partly due to the plastic frisbee having a flat bottom and partly as a result of the warping, pools of water tended to accumulate in the gauges in wet weather. There was some evidence that deposition was reduced in these conditions and it was thought that a possible cause might be losses due to splashing from further rainfall into the pooled water already in the gauge, which removed material previously deposited.

The gauge shape used is shown in Figure 11.6. The outer rim was of the same cross-section as a frisbee but the bottom had a shallow curvature (falling about 10 mm from edge to centre) which was sufficient to allow drainage of rainfall without greatly modifying the basic shape. Wind-tunnel velocity measurements over the opening showed that the displacement and acceleration of the windflow over the gauge opening were little affected by this additional depth. The gauge used in the present studies was spun in aluminium, anodized and fitted with the mounting stem shown in the

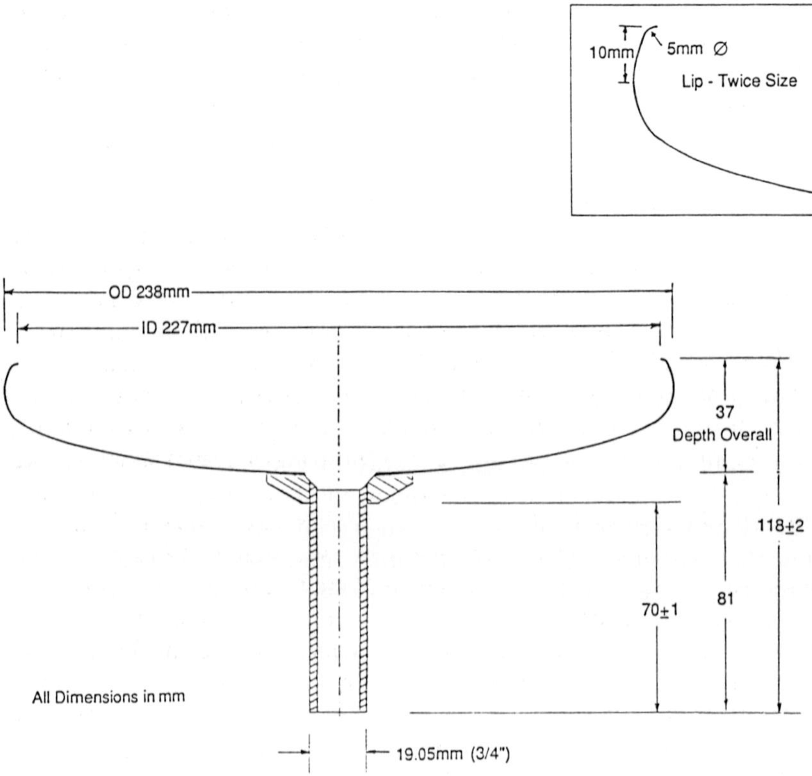

Figure 11.6 *Section through metal frisbee*

drawing. This was designed to fit inside a standard 3/4" plastic pipe which both acted as a drain tube and could be used to mount the gauge in any convenient way. In practice it is also possible to use a plastic frisbee in this way. An earlier design used a plastic frisbee pulled down at its centre on to a shallow aluminium dish of about two-thirds of the diameter of the frisbee (this was the only fixing). This both provided the curvature of the bottom of the gauge and greatly stiffened its form, so that warping was largely eliminated. It did not, however, prevent degradation and splitting of the plastic, but in view of the lower cost of the plastic frisbee it remained a viable alternative to the spun metal model. For the present work, where a high level of stability in the shape was needed for basic experiments, the spun metal model was used.

In addition to this, two internal modifications to the frisbee have been investigated, both with the intention of controlling splashing and improving particle retention. The first was an anti-splash screen fitted inside the frisbee. This was simply a nylon mesh fixed to a stainless steel ring of 6 mm diameter, just small enough to drop inside the frisbee, thus the mesh was about 6 mm above the surface at the edge of the gauge and about 16 mm above at the centre. The mesh was intended to break up raindrops without generating excessive splashing, preventing large drops reaching the gauge surface. Also, any splashing from smaller drops reaching the gauge surface would have to pass back through the mesh, further reducing the probability of material being carried out of the gauge. Experiments were carried out on a number of mesh sizes using a watering can to provide a heavy rain on to the mesh, poured from a height of about 1.5 m. The degree of splashing was judged subjectively. With no mesh and a flat bottomed frisbee, pools of water formed in the gauge and a great deal of splashing occurred, with large numbers of water drops passing out of the gauge. The modified gauge with a curved bottom surface improved matters, since the pools of water were removed, but a significant amount of splashing was still evident. The meshes produced a distinct further improvement, but there was some variation with the type of mesh. Fairly open meshes were investigated, all with open areas exceeding 30%, the choice being dictated mainly by what was readily available. All the meshes tried produced some reduction in splashing, but the most effective was a mesh of 2.4 mm pitch and 50% open area (the aperture was 1.7 mm and the thread size 0.7 mm). A finer mesh of about half the pitch and the same open area was not as effective. With the most effective mesh the splashing was greatly reduced, though not completely eliminated. It was expected that the mesh would also reduce the recirculation velocities within the gauge, which should at least have improved the resistance to blowout. However, this did not seem to occur in any significant amount. Some brief measurements of the collection and blowout performance of the frisbee shape with the curved bottom (without a sticky coating) and fitted with the most effective anti-splash mesh showed no difference in the collection or retention properties within the limits of accuracy of the experiments.

The second modification investigated was the use of a foam lining to the frisbee. This was intended to act both as an anti-splash device and to eliminate the need for a sticky coating in the gauge to assist particle retention. Though the sticky coating (liquid paraffin was recommended but other oils would be equally effective) appeared to be quite successful in retaining particles, it is a little inconvenient in practical use. It must be washed off using an organic solvent, which requires the use of a fume cupboard in normal laboratory practice. It was thought that a layer of foam on the bottom of the gauge might serve both as an effective particle retention device, instead of the sticky layer, and as a splash eliminator. The foams investigated were of the open cell type, which have had a growing use over recent years as filters and as particle traps for aerosol samplers. The coarser foams, of 10 and 20 pores per inch (PPI) have been investigated in a sheet of 10 mm thickness covering the whole inner area of the gauge. Water percolates through them quite readily, so that they do not inhibit drainage of the gauge. Since this matter has only been taken up quite recently, it has not been possible to check whether the foam affects the collection efficiency of the gauge. This is possible not only because of any improved retention that might occur, but also because the internal depth of the gauge is effectively reduced by some proportion of the depth of the foam. This latter effect is somewhat uncertain, but is probably not large as the newer gauge shape with the curved bottom surface has an internal depth which is greater by about the thickness of the foam used. Wind-tunnel measurements made of the blowout of a range of particle sizes, using glass spheres sprinkled on to the surface of the foam, showed that the 10 PPI foam was more resistant to blowout than 20 PPI, it was presumed because the test dust was able to penetrate further into the coarser foam and was thus better sheltered from the airflow. The windspeeds at which blowout occurred with the 10 PPI foam were about 50% higher than those for an uncoated gauge, reported in Hall and Waters (1986). Splashing of rainfall on to the foams was checked using the same test as described above for the mesh inserts. Both the 10 PPI and 20 PPI foams proved very effective, largely eliminating splashing and showing a superior performance to the meshes.

Field experiments are currently in progress with the new design of frisbee gauges with the curved bottom surface, fitted with anti-splash screens or with foam inserts. There was some initial concern that it might prove difficult to recover collected particulate from the foam, but present experience suggest that this is in fact fairly readily manageable by simply washing it in water.

11.3.3 Wind-tunnel investigations

11.3.3.1 Background

The last published paper on the performance of the inverted frisbee deposit gauge (Hall and Upton, 1988b) provided measurements of its collection efficiency as a function of windspeed and particle size. These showed a number of interesting features, two of which were felt to be worth further investigation. Firstly, the collection efficiency of small particles showed an unusual variation. Rather than just reducing with decreasing particle size as occurred with other gauges, the collection efficiency passed through a minimum for particle sizes around 100 μm. Below this size collection was enhanced by the turbulent diffusion of particles through the shear layer over the gauge opening. This is illustrated in Figure 11.7, which shows Figure 7 of the earlier paper, containing the measurements of collection efficiency as a function of particle size. The collection efficiency was still rising at the lowest aerodynamic particle size that could be used in the experiments, 50 μm (this corresponded to a physical size of 20 μm, for the glass spheres used in the experiments, these were the smallest that could be sieved). The behaviour of smaller particles has now been investigated using a different experimental technique. This work is described here. Secondly, the paper also showed that it was possible to enhance the collection in the region of the minimum in collection efficiency by increasing the turbulence over the gauge opening using an eddy shedding bar placed upstream. This effect is illustrated in Figure 11.8, which shows the relevant result from Figure 14 of the earlier paper. This technique is clearly of great interest as a method of enhancing

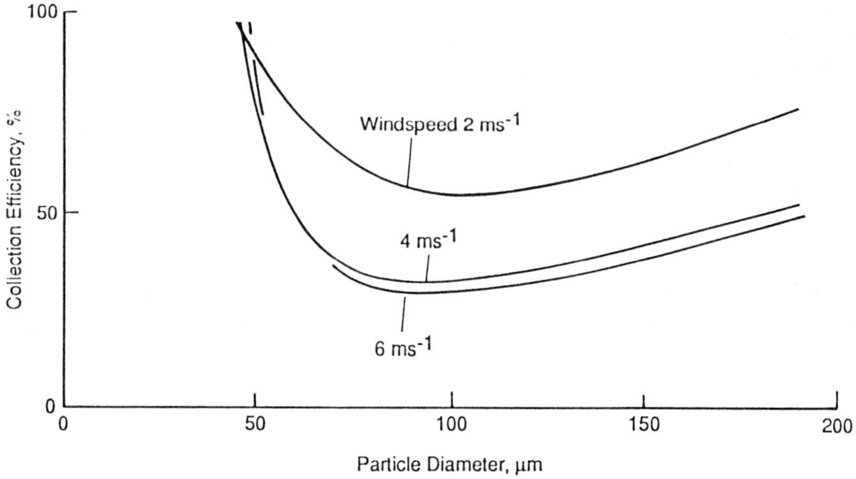

Figure 11.7 *Collection efficiency of a sticky frisbee as a function of particle size*

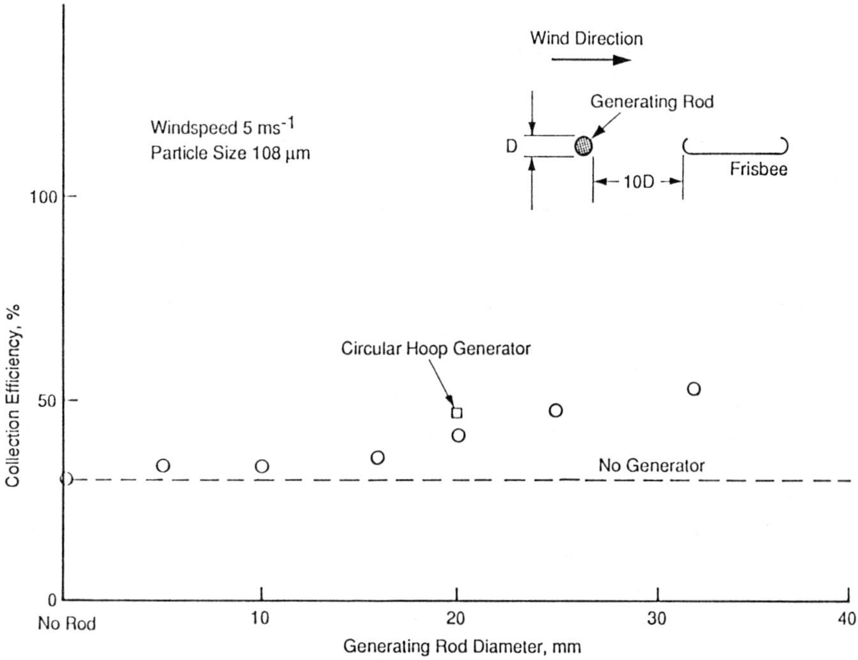

Figure 11.8 *Effect of turbulence from eddy shedding rods on frisbee particle collection efficiency*

performance, but it was not possible to proceed further with it at the time. This has also now been done and the results are reported here.

11.3.3.2 Collection efficiency of small particles (<20 μm) by an inverted frisbee

The measurements were carried out in the Laboratory's Number 1 wind-tunnel using liquid and solid aerosols marked with fluorescent tracers. The techniques used were the same as those for testing aerosol samplers, which has been a major activity of the Laboratory. The measurements were carried out at the same time as another aerosol sampler testing programme, that described in Hall *et al.*, (1990a), using some of the larger particle sizes in these experiments, between 7 μm and 20 μm. The windspeeds and particle sizes tested, and the results of the measurements, are laid out in Table 11.1. The measurements used both solid (uranine) and liquid droplet particles, but in only one case, for 19 μm particles at 9 $m\,s^{-1}$ windspeed, was a measurement made with both types of particle. The choice of particle sizes and windspeeds is a little arbitrary for the present purposes, since they were those chosen for the other experiment, so that they do not precisely match the test conditions used by Hall and Upton (1988b), but they cover the required range of test conditions fairly well. The performance of the gauge is expressed in two ways in the table. Firstly as a collection efficiency in the form normally used for deposition gauges, where it is assumed that deposition to the ground is governed by the particle falling speed, as indicated in equation (1) earlier. This is usually the case for large particles, beyond about 20 μm. Secondly, it is expressed as an effective particle

Table 11.1 Results of collection efficiency measurements of small particles into frisbee gauge

Test conditions			*Gauge performance*	
Test windspeed $m\,s^{-1}$	*Particle size*[*] *μm*	*Particle falling speed* $cm\,s^{-1}$	*Collection efficiency (based on eqn (1))* %	*Effective particle deposition velocity* $cm\,s^{-1}$
1	8.9[s]	0.24	130	0.3
1	18.5[s]	1.02	150	1.6
2	6.6[s]	0.13	190	0.25
2	8.9[s]	0.24	150	0.36
4	17.8[s]	0.94	460	4.3
6	19[d]	1.08	500	5.3
9	8.5[s]	0.22	2800	6.2
9	13.8[d]	0.57	1400	8.1
9	18.1[s]	0.98	560	5.5
9	18.5[d]	1.02	430	4.4

[*] [d] = droplet aerosol, [s] = solid uranine aerosol

deposition velocity into the gauge. For smaller particles deposition velocities are usually in excess of the falling speed due to the action of a number of additional deposition processes and deposition to the ground is normally expressed in terms of an effective deposition velocity. The results are also shown in Figures 11.9 and 11.10 which show, respectively, the collection efficiency and effective particle deposition velocity as a function of particle size; note the log scales. The earlier measurements of Hall and Upton (1988b) for larger particles are also shown on the figures for comparison. The results shown are for a frisbee with a sticky surface, which is most appropriate for comparison with the aerosol measurements as in both cases contact with the surface should result in complete retention. In addition, Figure 11.10 shows a line for a deposition velocity equal to the particle falling speed, which would correspond to a collection efficiency of 100%.

The most significant feature of the small-particle measurements is the high rates of deposition into the gauge. The lowest recorded collection efficiency is 125% and the highest is 2800%. It can be seen in Figure 11.9 that the

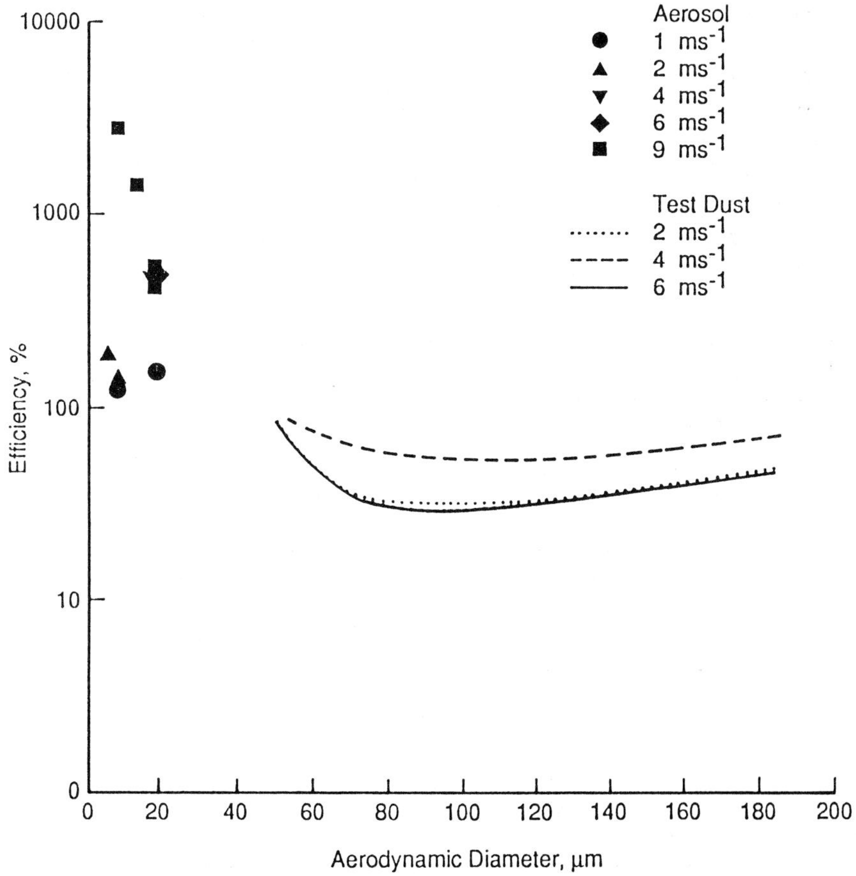

Figure 11.9 *Collection efficiency of frisbee gauge for large and small particles*

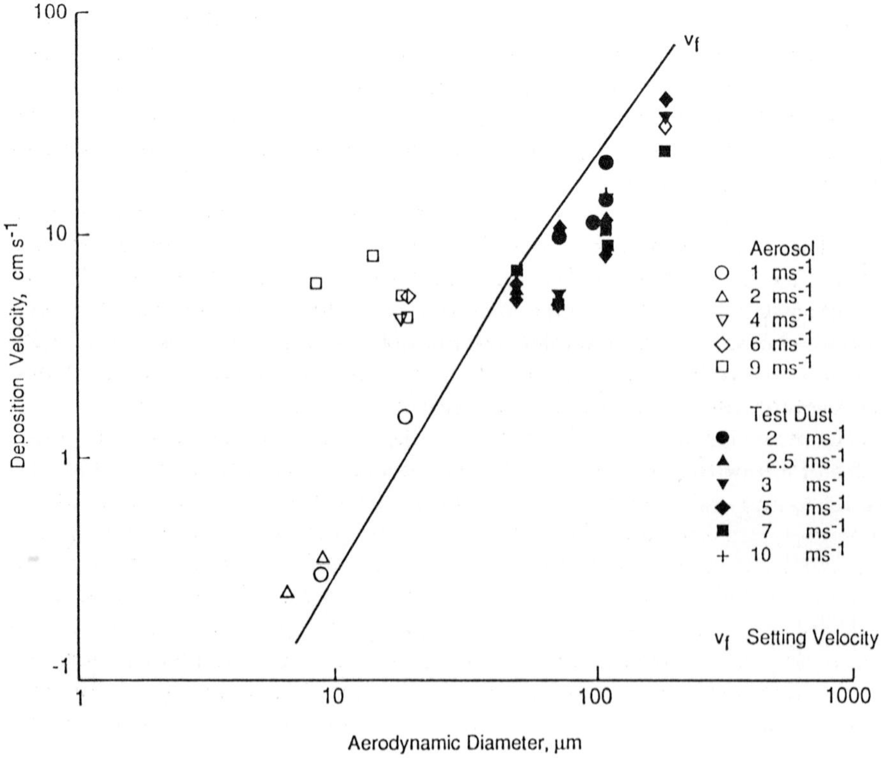

Figure 11.10 *Effective deposition velocity of particles collected by a sticky metal frisbee dust gauge*

collection efficiency continued to rise with decreasing particle size, generally showing a continuous progression from the earlier measurements with large particles. The aerosol measurements also showed a windspeed dependency, with the collection efficiency increasing with increasing windspeed. This effect was particle size dependent. Thus the particle sizes around 8-9 μm showed an increase in collection efficiency from 130% to 2800% between windspeeds of 1 m s^{-1} and 9 m s^{-1}, while the larger particle sizes around 18-19 μm showed a more limited increase from 150% to about 500%. This result is not surprising; the deposition of particles by turbulent transfer will increase as the turbulent velocities increase (with increases in windspeed) and as the inertia of the particles decreases (with reductions in size).

With the collection expressed as a deposition velocity, in Figure 11.10, the large- and small-particle results divide, firstly, either side of the line on the figure for a deposition velocity equal to the particle falling speed. The large-particle measurements of Hall and Upton (1988b) lie on the right-hand side of the line, indicating the (known) inefficiency of the gauge at collecting large particles. The small-particle measurements described here lie on the left side of the line, indicating deposition velocities into the gauge in excess of the particle falling speed. This is, of course, consistent with the high collection efficiencies, beyond 100%, shown in Figure 11.10. Secondly, the small-particle deposition velocities show the windspeed dependency previously mentioned. Below 5 $m\,s^{-1}$ windspeed the deposition velocities are relatively close to the particle falling speed and therefore to the line in the figure. At higher windspeeds the particle deposition velocity shows a limited variation between about 4 and 8 $cm\,s^{-1}$. The different forms of behaviour seem to divide either side of a windspeed of about 3 $m\,s^{-1}$ and the measurements otherwise show no other clear signs of systematic effects due to windspeed.

The question of what characteristics a dust deposit gauge should ideally have for collecting small particles has not previously been considered. When used for sampling deposited dust the collected mass tends to be dominated by the larger sizes, so that the small contribution from the smaller sizes is not so important and on this basis the precise collection characteristics in this size range would not seem to be very critical. However, there has been a growing trend to analysing the contents of deposit gauges, especially for low-level contaminants like heavy metals and dioxins. In these cases it is not always certain which size fraction contains the bulk of the contaminant, so if it is the smaller fraction then the collection characteristics of this fraction become more important. Ideally the gauge should mimic the collection processes of the ground over which it is situated. This requires that the deposition rate should pass from a value equal to the falling speed, for large particles, to the enhanced deposition velocity associated with smaller particle sizes. The size division is not a precise one, but starts to occur at particle sizes around 20 μm. Similarly, there is a wide range of effective deposition velocities for small particles, depending mainly upon the particle size and the nature of the surface. Sehmel's (1984) recent review shows the range of deposition velocities that can occur, but for particle sizes around 10 μm values are typically of the order of $cm\,s^{-1}$. A commonly quoted range is 1-3 $cm\,s^{-1}$. This is more or less what the frisbee gauge seems to do. The values measured here are of this order, ranging from 0.3 to 8.1 $cm\,s^{-1}$. Within the order of accuracy to which deposition estimates are usually made, the frisbee gauge therefore seems to mimic small particle deposition acceptably.

There is no similar data for the traditional shapes of deposition gauge, so their collection behaviour for small particles is not known. However, Hall and Upton (1988b) pointed out that deposition into these gauges by turbulent diffusion must be significantly less than for the shallower, frisbee type of gauge simply because of the greater height of the separated shear layer over the opening of the gauge and the greater distance over which turbulent transfer must therefore occur. In support of this, Ralph and Barrett's (1984) measurements of conventional gauges show no evidence of enhanced collection of smaller particles.

11.3.3.3 The effects on collection performance of enhanced turbulence over the gauge opening

In the earlier measurements of Hall and Upton (1988b), the enhanced collection performance of the smaller particle sizes by an inverted frisbee was investigated by generating additional turbulence over the gauge opening using the eddy shedding of a cylindrical rod placed upwind. The results of that experiment (Figure 14 of the original paper) are reproduced here in Figure 11.9. Rods of a range of diameters were used, placed ten individual diameters upwind of the gauge opening so that by this distance any mean velocity perturbation had disappeared and the scale and turbulence intensity of the eddies generated by the rod were in proportion to its diameter. It can be see that as the rod diameter increased up to 30 mm, the largest size used, the collection efficiency of the gauge also increased, from 30% to 52%.

Besides demonstrating the reason for the enhanced collection of small particles, this device also clearly had some practical potential for improving the collection efficiency of the gauge. However, the measurements for the single particle size and windspeed, 108 μm and 5 m s^{-1}, described above, were insufficient to develop a suitable design. Firstly, it was not clear whether the earlier results were close to the optimum improvement obtainable. Secondly, it is not adequate merely to increase the collection efficiency of the gauge, but to increase it appropriately in those areas where it is deficient, mainly the particle sizes around 100 μm, where the collection efficiency was at a minimum. This needed to be done without excessively increasing the collection at both larger and smaller particle sizes. Larger particle sizes, having relatively high inertia and steep falling angles, were thought unlikely to be much affected by enhanced turbulence over the gauge opening. However, there seemed likely to be a problem in avoiding excessive enhancement of the collection of smaller particles which, from the preceding discussion, can be seen (in Figure 11.10) to be already quite high.

One possible approach was to use the rod additionally as an impaction trap for the smaller particles. The smaller particles have a very shallow trajectory (for 50 μm particles it is at most a few degrees) and in the original experiment the rod was deliberately placed with its upper edge no higher than the top of the gauge in order to keep it clear of the paths of the smaller particles. By raising the rod a little, it was hoped that it would have some effect as a size selective particle separator, preferentially blocking the low angle trajectories of smaller particles whilst still enhancing the catch of the larger sizes around 100 μm, which having higher trajectories would be more likely to clear the rod.

The experimental programme was carried out with these two aims; to see what improvements in catch due to an eddy shedding device were possible and to see whether it could be used to enhance deposition into the gauge preferentially as desired. The rod was replaced by a flat bar of 25 mm height. On later reflection, there seemed no good reason for using a rod originally, it is no better an eddy shedder and suffers more readily from Reynolds number effects. A bar produces a slightly wider wake than a rod, so a given height of bar is equivalent to a larger diameter rod. It was also thought that a bar would act as a better particle trap. It seemed from the first experiments that the effectiveness of the bar increased with size and the height of 25 mm chosen was the largest that was felt to be consistent with anything resembling a streamlined flow over the gauge.

Figure 11.11 shows measurements of the effect of the position of the bar on the collection efficiency of 72 μm particles at a single windspeed of 5 m s^{-1}. The position of the bar was varied between 3 and 15 bar heights upwind of the upwind edge of the frisbee, and its height between level with the opening of the frisbee and 25 mm above this point. It can be seen from Figure 11.11 that the highest collection efficiency occurred with the bar between 5 and 10 bar heights from the frisbee and with its upper edge close to the opening. At the closer distances, the bar was more effective when set a little below the opening and at the further distances when set a little above. Since it was desired to set the bar above the level of the opening, the greater distance

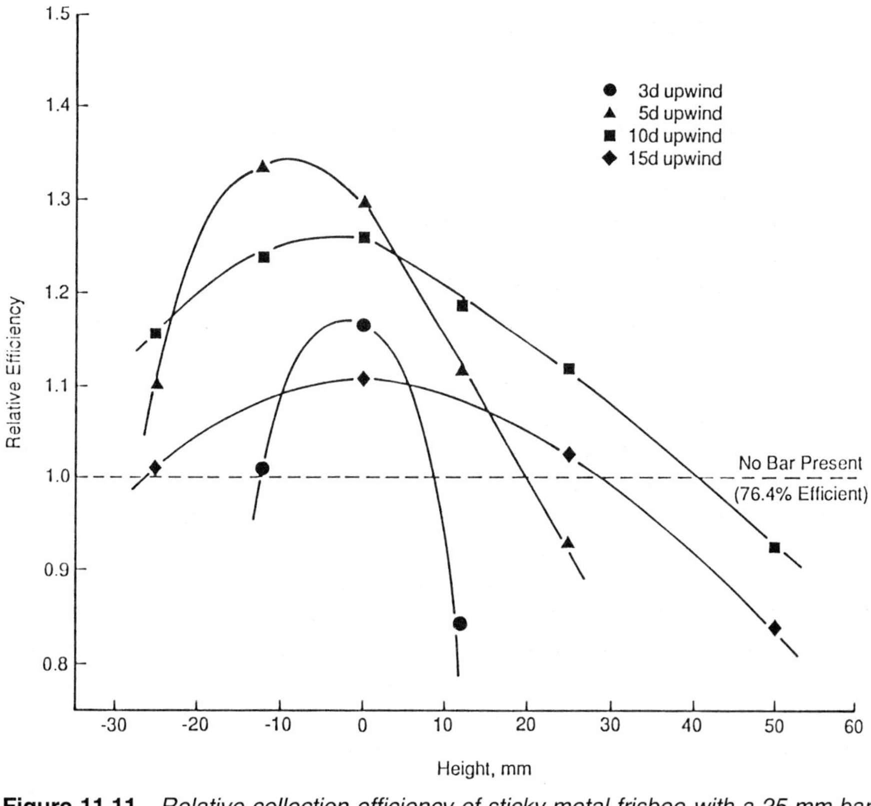

Figure 11.11 *Relative collection efficiency of sticky metal frisbee with a 25 mm bar at several positions upwind: windspeed 5 m s^{-1}, particle size 72 μm*

seemed preferable. This was the bar position originally selected, which thus seems to have been close to the optimum.

Figure 11.12 shows the effect of the vertical position of the bar on the collection efficiency. Three particle sizes were used, 50 μm, 72 μm and 108 μm. Since the main interest was in whether raising the bar would reduce collection of the smaller particles, the smaller sizes were only used at the higher bar positions. Three bar positions were used, with the top of the bar level with the frisbee opening and then raised both 12.5 and 25 mm above this. It can be seen from the figure that with the bar present the collection efficiency still rose sharply as the particle size reduced. The increase in collection efficiency with the bar present was quite marked, for 50 μm particles it rose from about 80% with no bar to 150% at the optimum bar position. Raising the bar reduced the collection efficiency of both of the smaller particle sizes but in about the same proportion in both cases, by about 20% over the range of bar positions. The highest bar position was sufficient to block the trajectories (assuming simple trajectory paths as they

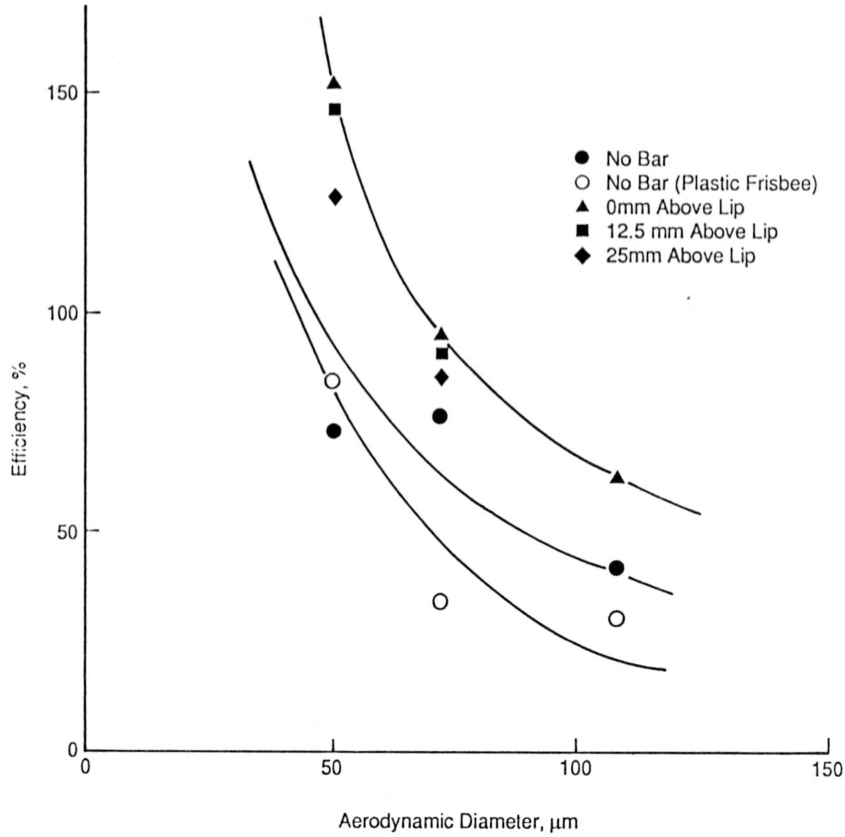

Figure 11.12 *Effect of position of eddy shedding bar on sticky metal frisbee collection efficiency: 25 mm bar, 250 mm upwind*

would occur in an undisturbed flow) of particles up to 160 μm. It seemed, therefore, that the eddy shedding bar performed far better as a collection enhancer than as a particle trap.

In view of these results the main intention of this design, which was to use the bar size selectively, partly as a collection enhancer and partly as a particle trap, did not seem practicable. With a bar arranged to improve collection efficiency by a significant amount for particle sizes around 100 μm, it was not possible to sufficiently suppress the collection efficiency of the smaller particle sizes down to levels around 100%. Investigation of this technique was not therefore continued. It may, however, be of some potential use in other gauge designs.

11.3.4 Field measurements using deposition gauges

The field measurements carried out in the Selby area by York University are reported by them in three publications (Vallack and Chadwick, 1990, 1991a, 1991b). There are as yet no formal publications of measurements by British Coal. Besides these, a number of other (unpublished) comparisons between gauge types, from a variety of sources, have been reported to the authors, mostly for short runs of data on single sites. The descriptions here will be restricted to a review of the results of the field measurements. They are, so far, restricted to deposition gauges. Also, due to the slow rate at which field data is collected (samples are normally on a monthly basis) there are only limited comparisons at the time of writing between some of the gauge modifications described here.

There are, at the time of writing, about ten sets of comparative data between the British Standard Deposit Gauge, the proposed ISO gauge and dry and sticky frisbee-type gauges available to the authors. In addition there are three sets of comparisons with frisbee-type gauges fitted with meshes and one set with a foam insert. The proposed ISO gauge is a cylindrical collector with a depth of twice its diameter and a sharp lip. Apart from the lip it is almost identical to the NILU gauge, whose lip has a moulded 45° chamfer of about 10 mm length. The results are summarized in Table 11.2, which gives the ratio of the amount collected in the different deposit gauges, factored against the British Standard. The ratios shown are from the summations of the total amount deposited into each gauge in each test series, mostly between three and ten individual measurements. At the end of the table averages of the ratios for all the test series are given.

Comparisons between the two conventional gauges, the British Standard and the proposed ISO gauges showed only small differences; on average the proposed ISO gauge collected 8% more than the British Standard. There was some variation in the comparison, the lowest ratio being 0.83 and the highest 1.5. Wind-tunnel measurements suggest that there should be little difference in the collection efficiencies of these two gauges, with the proposed ISO being a little worse. However, the proposed ISO gauge should have a much

Table 11.2 Results of deposit gauge comparisons from various sources (Figures are summed deposits for each series, ratioed against the British Standard gauge.)

			Type of Gauge⁺			
*Site**	*BS*	*ISO*	*F(D)*	*F(M)*	*F(F)*	*F(S)*
B. Coal	1	–	1.03	–	–	–
B. Coal	1	–	1.18	–	–	–
V&C 90 Barlow 87–9	1	0.99	1.1	–	–	1.15
V&C 90 Naburn 87–9	1	1.03	1.02	–	–	1.3
V&C 90 Barlow 88–9	1	0.92	1.1	–	–	1.1
V&C 90 Naburn 88–9	1	0.95	1.15	–	–	1.4
V&C 91 Barlow 89–90	1	0.95	–	0.99	–	1.16
V&C 91 Barlow 90	1	1.06	0.73	1.04	–	1.6
V&C 91 Pairs, Barlow	1	0.92	0.73	–	–	1.3
V&C 91 Pairs, Smeaton	1	–	0.99	–	–	1.6
Barlow 91	1	1.4	0.56	0.93	1.23	2.5
EOLAS	1	1.5	–	–	–	2.0
Overall means of data	1	1.08	0.96	0.99	1.23	1.47

* B. Coal = British Coal site
V&C = Vallack and Chadwick followed by year, site and dates of measurements.
Pairs = average results of gauge pairs comparisons
EOLAS = data supplied from a single intercomparison series on one site.

⁺BS = British Standard
ISO = Proposed ISO
F = Frisbee type, D = dry, M = fitted with mesh or F = fitted with foam insert, S = sticky surface coating.

higher retention than the British Standard and this is consistent with the slightly better overall performance of the proposed ISO gauge.

The dry frisbee type of gauge with no insert performed on average a few per cent worse than the British Standard. The averaged ratio was 0.96, but this difference is small by the standards of accuracy of this sort of comparison. A marked feature of all comparisons between these two types of gauge has been their high variability. All the test series have shown a high degree of scatter in the measurements and the ratio of the two gauge readings has commonly ranged by up to a factor of three either way. This does not seem to be due to any erratic qualities inherent in the gauge designs themselves, multiple measurements with either type of gauge have not shown such a degree of variation. For example, a single comparison between ten frisbee-type gauges by British Coal showed a standard deviation of only 7%. The clear implication is that both types of gauge suffer from systematic, but different, forms of sampling deficiency. The two major classes of sampling deficiency are collection and retention. From what is known of the laboratory performance of these two types of gauge it is tempting to suspect that the conventional gauge designs suffer mainly from poor collection and the shallower frisbee-type mainly from poor retention (though it must be noted that their collection characteristics, whilst very good by the standards of conventional gauges, are not perfect). It must be said against this that the frisbee-type gauges have shown better resistance to the blowout of particles in the wind-tunnel.

There is some confirmation of this hypothesis in the measurements in Table 11.2 of frisbee-type gauges modified to improve retention. There are three data sets for a frisbee-type gauge with a mesh insert, for two of which there were comparative measurements with a dry frisbee-type gauge with no insert. It is interesting that these two cases had the lowest values of collection ratio for the unmodified frisbee-type gauge and in both cases the ratio was improved markedly by the mesh insert. The mean of the ratio for the three cases was 0.99, compared with 0.96 for the unmodified gauge; however, for the two particular cases of the comparison the improvement was about 50%. The single comparison with a foam insert showed a further substantial increase in the ratio, to 1.23, which for the particular case represented a gain of about a factor of 2.5 over the unmodified gauge, but it will be some time before sufficient additional data is available to confirm the consistency or otherwise of this improvement. The final column in Table 11.2 is for the sticky frisbee-type gauge. Here all the measurements showed a significant gain over the British Standard Gauge, the ratio ranging between 1.1 and 2.5 and the average being about 1.5. This also confirms the argument that retention is a problem with the frisbee type gauge. The sticky frisbee gauge design also has the highest combination of collection and retention of those investigated.

It is clear from the comparative field data so far that the behaviour of gauges in the field is complex and not perfectly predicted by laboratory experiments alone. The latter remain vital in providing an understanding of the fundamentals of gauge performance, especially of wind effects, but must be carried out in concert with field measurements if effective new designs are to be developed.

11.4 Flux gauges

11.4.1 Summary of programme

Work on flux gauges has concentrated on development of new designs, using the wind-tunnel. The closed container type of gauge, similar to the style of the British Standard Directional Gauge, was investigated first. It proved possible to improve the performance of this type of gauge, but it also showed some fundamental limitations. Further effort was therefore devoted to through-flow designs, in which there is a flow through the gauge and it is fitted with some sort of particle trap. This latter sort of design has proved to be more promising and high collection efficiencies have been obtained from some designs. These are still under development.

Because it was desired to investigate a large number of potential gauge shapes quickly in order to identify the most promising types, it was not practicable to carry out full collection efficiency measurements over a wide range of windspeeds and particle sizes. Measurements were therefore made at four fixed conditions, two windspeeds and two particle sizes. These were windspeeds of 3 $m s^{-1}$ and 8 $m s^{-1}$, and particle sizes of 59 μm and 136 μm. These conditions can be compared in Figure 11.4 with the collection efficiency of the existing British Standard Directional Gauge. It can be seen that the lower windspeed is close to the maximum in collection efficiency that occurs at the larger particle sizes, and where there is a strong particle size dependency on collection efficiency. At the higher windspeed, the collection efficiency of this gauge is falling markedly for all particle sizes. A gauge design which can show a high and relatively uniform collection efficiency for these four conditions, should have good prospects of being effective over a wide range of windspeeds and particle sizes. The measurements for each design tried have been displayed in the form of a standard bar graph of the four measured collection efficiencies. As a convenient point of reference, the collection efficiencies of the present British Standard Directional gauge are also shown on each bar graph. These are indicated by the hatched areas of the graphs, any clear area above the hatching indicates an improved collection efficiency.

11.4.2 Investigations of closed container gauge designs

The present British Standard directional dust gauge is essentially a closed container type of flux gauge. That is, there is a closed container in which a vertical opening has been cut. With the opening facing into the wind, the flow stagnates against it, so particles are not carried into the gauge by the wind as, nominally, there is no windflow inside the gauge. Particles must enter the opening due to their inertia. As was remarked in the introduction to this paper, this form of stagnation point sampling limits the particle sizes that can be sampled, depending upon the windspeed, and the deficiency is most severe for the smallest particles and at the lower windspeeds. Also, though the flow at the entry to the gauge is nominally stagnated there is in fact a wind-driven flow inside the gauge which tends to remove particles which have entered it. Here, the deficiency is greatest at the higher windspeeds. These two causes of undersampling produce the maxima in the collection efficiency curves in Figure 11.4.

Despite these deficiencies, the closed container type of gauge has a number of practical advantages. It is simple to make and fairly straightforward to use. Once deposited into the bottom of the gauge, wind-driven removal of particles is unlikely. It is also fairly easy to recover collected particles. In view of these advantages, it was felt to be worthwhile to see to what extent this type of gauge could be further developed. The present British Standard design is clearly far from optimal in both the respects discussed above. This is considered in detail by Ralph and Hall (1989), and only briefly here. Firstly, the shape of the inlet is not ideal for stagnation sampling, as the edges of the opening are set across the flow direction. The conventional form of virtual impactor design for small particle sampling uses a sharp forward facing edge to the particle inlet for good reasons. Secondly, the shape of the container did not seem ideal for avoiding internal secondary flows which might remove particles from inside the gauge. The vertical opening is a tall and narrow slot of 340 mm height and 45 mm width, with a quite limited depth of container, 70 mm, behind it. Thus, not only is it likely that secondary flows could penetrate easily to the back of the container, but the large height of fall required of particles entering the upper part of the opening increases the likelihood of their being removed from the container by secondary flows.

The experiments on the closed container type of gauge described below were made with different shapes of container, but, except in a few stated cases, with the same shape and size of entry slot as the British Standard design, though at half scale. Only a single container was used, except in a few cases, with the opening facing forwards into the wind. The results are laid out in Figures 11.13 and 11.14. Sketches of the cross sections of the different shapes are given in Figures 11.15 and 11.16.

The first three results, (a), (b) and (c), at the top left of Figure 11.13, are of the present British Standard design. The first, (a), is for a single plain collecting container. The second, (b), for the single container lined at the back with tissue to reduce losses from particle bounce. This proved a major difficulty with this gauge in the earlier test work (Ralph and Hall, 1989), particularly as hard particles were used which bounced well. The tissue lining to the back of the gauge greatly reduced this problem, which otherwise made it difficult to assess the gauge's collection, as distinct from retention, performance. The third result, (c), was with two dummy containers arranged either side of the gauge, as with the present British Standard design. This was to check whether the aerodynamic interference of these other containers affected the particle collection. It can be seen that the tissue lining produced a small increase in collection efficiency and that the addition of the dummy containers reduced the collection again, back towards the original values. However, the differences between these three sets of measurements were relatively small, about the same as the order of accuracy of the experiment. The overall collection efficiency of the gauge was poor, around 20%, and it showed a strong dependence upon windspeed and particle size. It must also be noted that these measured collection efficiencies are distinctly lower than those in Ralph and Hall (1989) for the larger particle size. This earlier work did not cover the precise particle sizes used here, but approximately interpolated values gave collection efficiencies for the larger

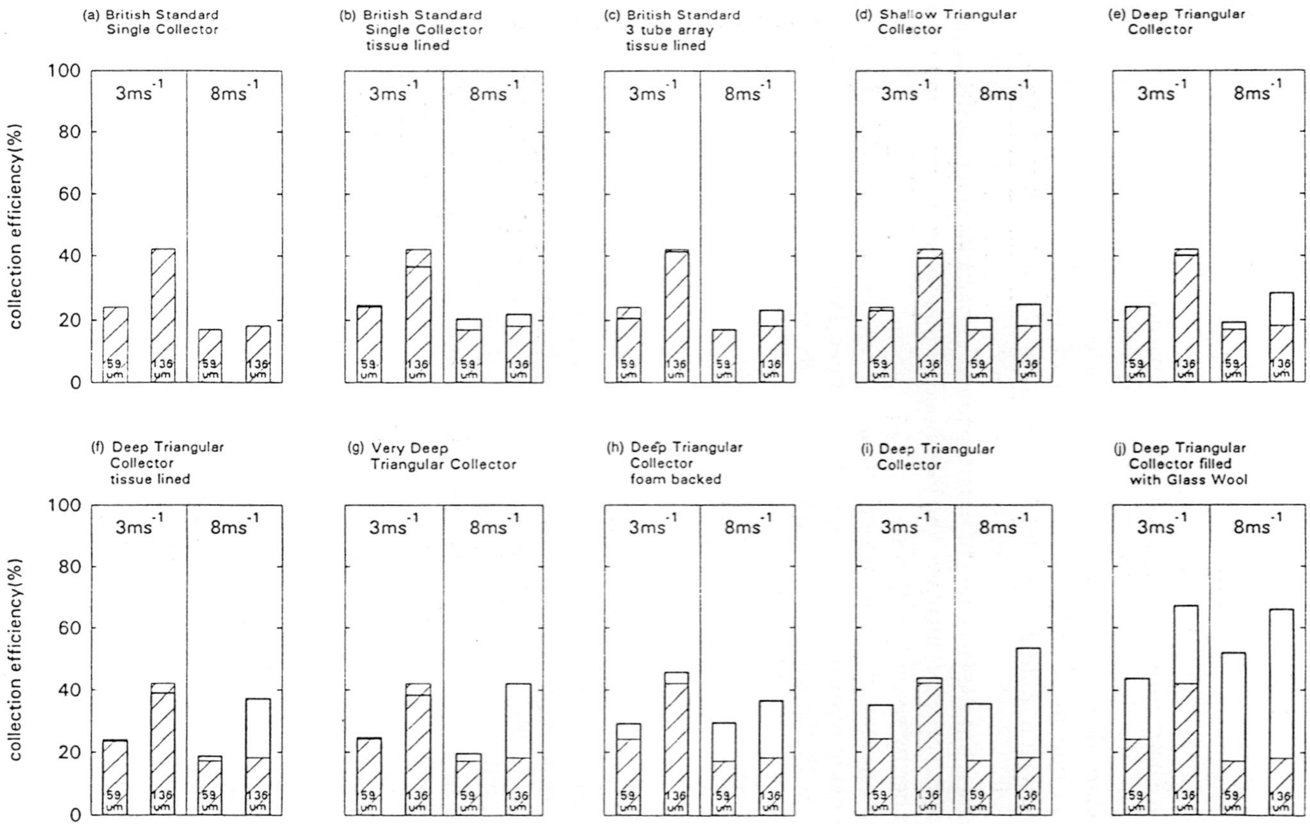

Figure 11.13 *Collection efficiencies for closed container types of flux gauge*

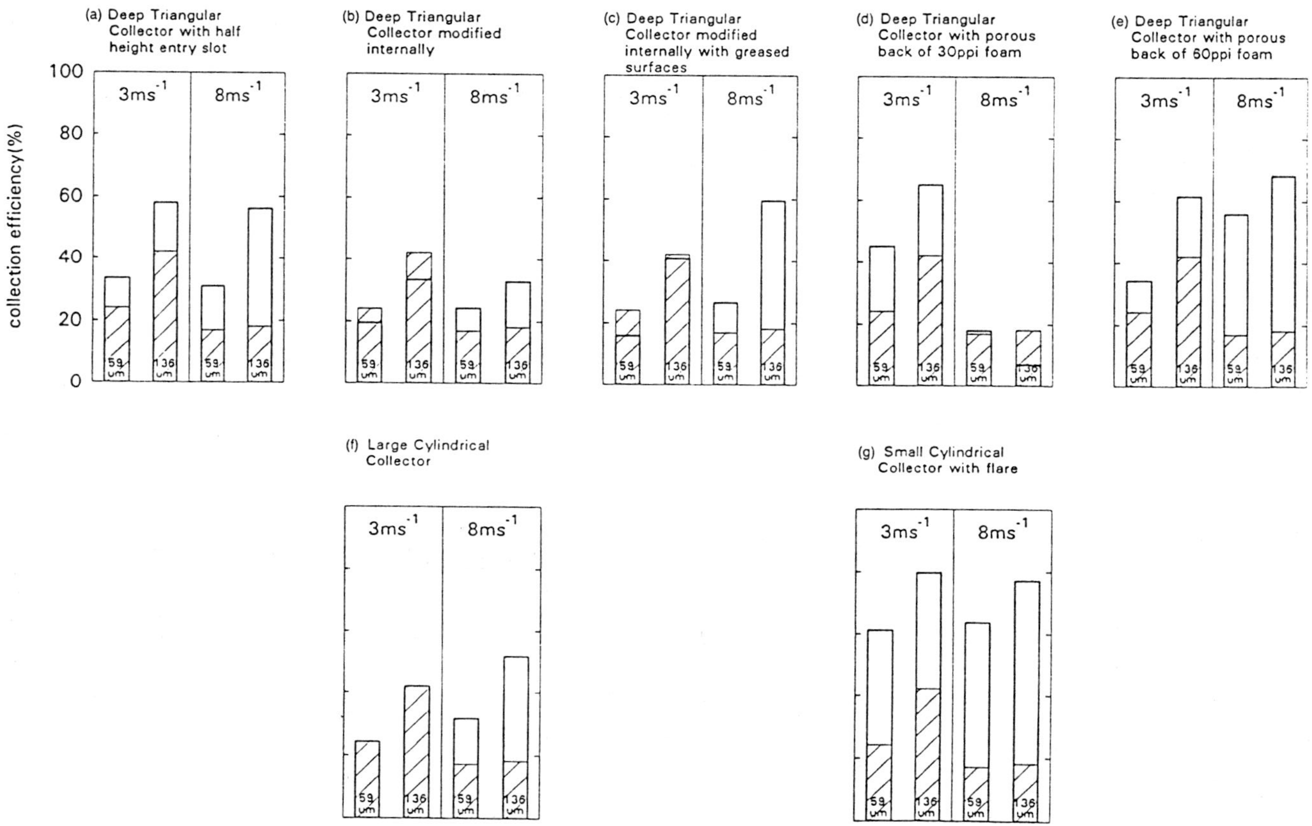

Figure 11.14 *Collection efficiencies for closed container types of flux gauge*

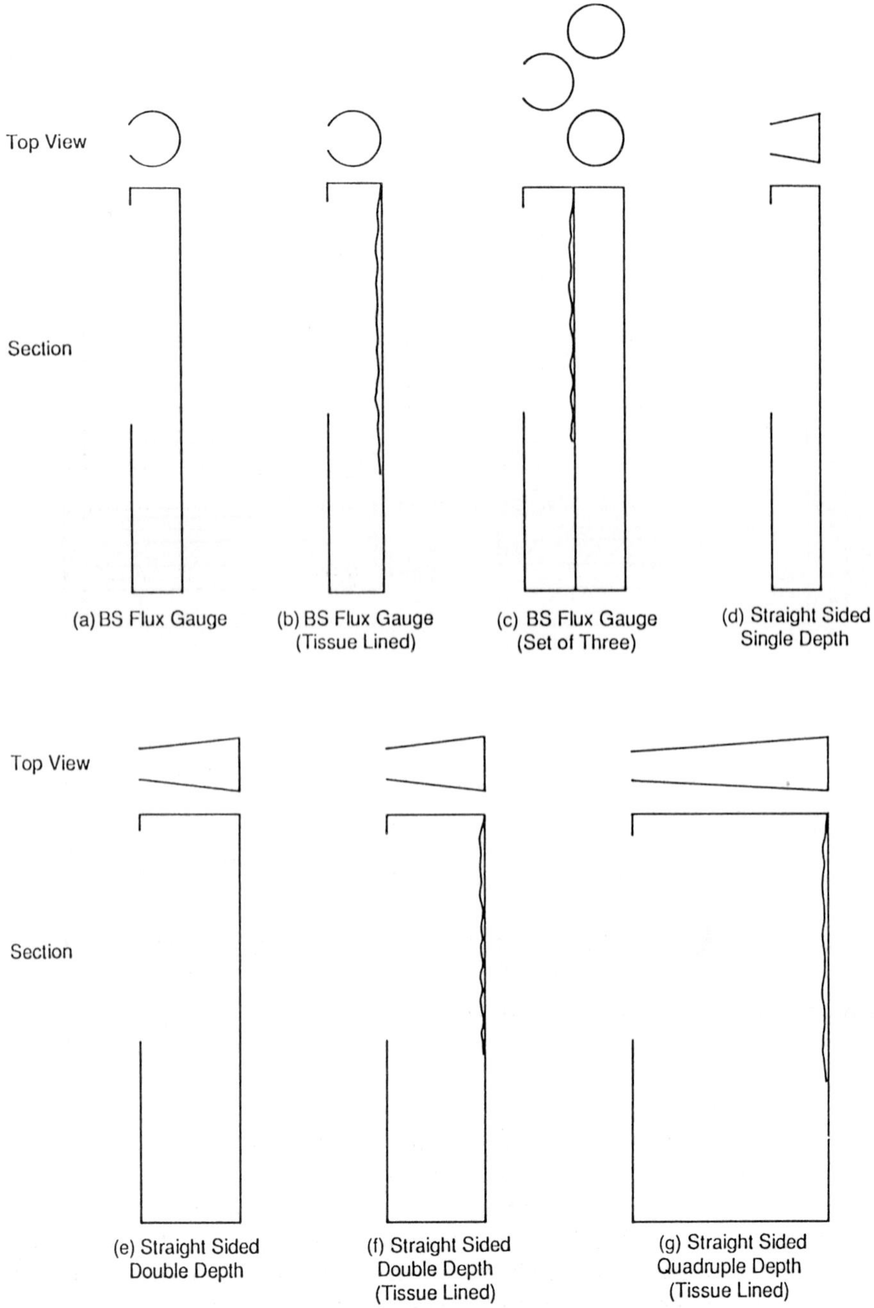

Figure 11.15 *Closed container types of flux gauge*

particle size of about 55% and 45% at 3 $m\,s^{-1}$ and 8 $m\,s^{-1}$, respectively, compared with 42% and 25% here. These latter measurements were checked and found to be quite consistent. The reasons for the differences are not clear, they may be due to scale effects (the earlier work was with one-third scale

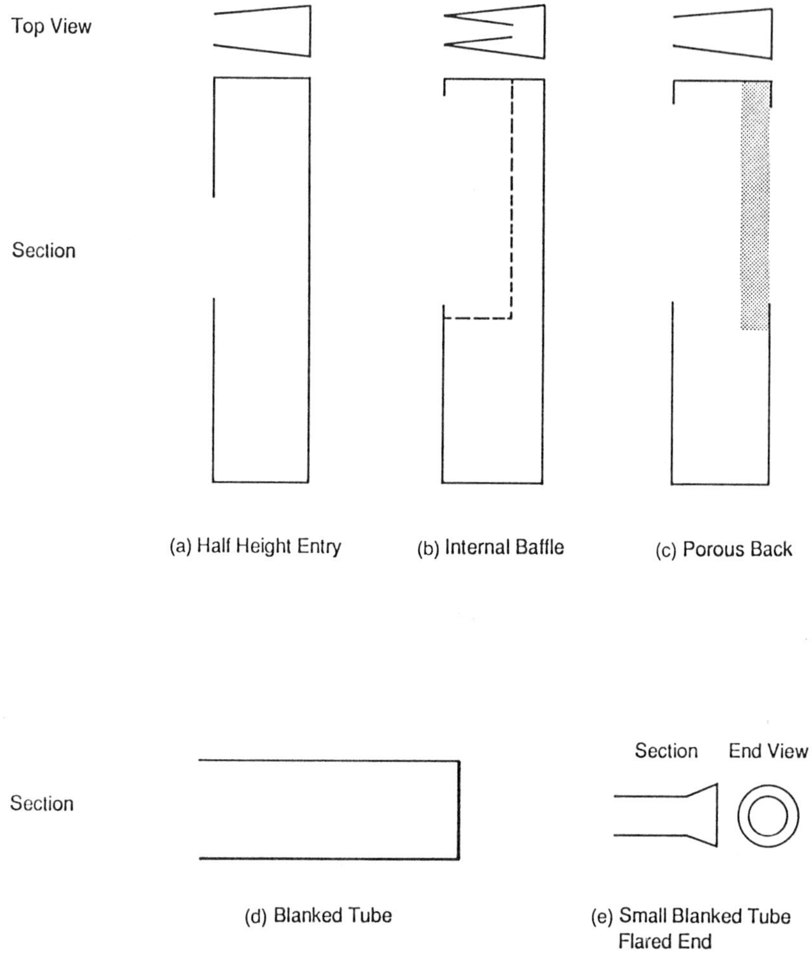

Figure 11.16 *Closed container types of flux gauge*

gauges) on either particle collection or retention.

The first modification to the gauge was to a straight-sided triangular cross-section, shown in Figure 11.15 (d), with the size of slot remaining the same. The depth and maximum width were similar to the diameter of the British Standard design. The height of the gauge was the same, with the slot in the upper half, again as with the British Standard gauge. The lower half of the modified gauge, however, retained the same triangular cross-section, there being no collecting bottle at the bottom. Thus the main change was to the edges of the slot, from being largely across the free stream flow with the original cylindrical cross section to being largely aligned with the flow. The main effect of this change, Figure 11.13 (d), was to slightly increase the collection efficiency at the higher windspeed, by about 10%, otherwise there was little difference from the original design.

The next modification was to double the depth of this triangular container, as in Figure 11.15 (e), the other dimensions, the size of slot, maximum width and depth of the container remaining the same. The intention was to try to provide a deeper region behind the inlet slot where secondary flows should be reduced in magnitude, so that particle losses from this cause would be reduced. This modification was also tried with a tissue lining at the rear of the container, intended to reduce the bouncing of collected large particles out through the opening again. The results are shown in Figure 11.13 (e) and (f). There was an improvement in the collection of large particles at the higher windspeed, especially with the tissue lining, but the overall effect on collection efficiency was relatively limited. The depth of the container was then doubled again, to four times the diameter of the British Standard gauge, see Figures 11.15 (g) and 11.13 (g). With tissue on the rear face of the container to control particle bounce, there was no improvement in collection efficiency over the double depth container. Since during the course of the experiment, particles could be seen being ejected from the gauge, the double depth container was tried with some alternative methods of particle bounce control. Figure 11.13 (h) and (i) show respectively the effects of a thin layer of foam at the rear of the container instead of tissue and of spraying the whole interior with a surface layer of silicone grease. Both modifications improved the performance of the collector by modest amounts. Finally, Figure 11.13 (j) shows the effect of loosely packing with glass wool the whole interior of the gauge. This was intended to entirely prevent particle removal from any cause, but it is not, unfortunately, a practical form of dust gauge for a number of reasons. It was used in order to find the collection efficiency of the gauge entry alone. The result was a considerable improvement in performance, the average collection efficiency of the four measurements being about 50%, double that of the British Standard. The larger particle size was collected more efficiently than the smaller, by about 30%, but there was relatively little influence of windspeed on the collection efficiency.

It appeared from these experiments that this type of inlet, though not perfect, allowed a relatively consistent collection efficiency of around 50%, provided that particles could be retained in the gauge. It also appeared that quite effective particle removal occurred even with deep gauges, within which substantial secondary flows were not expected. Further investigations were made of the flow patterns inside the gauges, using smoke flow visualization, and these revealed the reason for the poor retention of this sort of design. Inside all the gauge shapes so far discussed there was a strong circulation, driven by the external flow, over the whole depth of the opening. A sketch of this is shown in Figure 11.17. This circulation reached right to the back of the gauge, whatever its depth, and was clearly an efficient means of scouring out particles which had entered the gauge. Only particles which reached the region below the gauge opening, where there was relatively calm air, were likely to be retained. The direction of rotation of the circulation appeared to be arbitrary and it spasmodically changed its direction with a frequency of several times a minute. A curious secondary flow resulted from this main internal circulation, in the form of a pair of

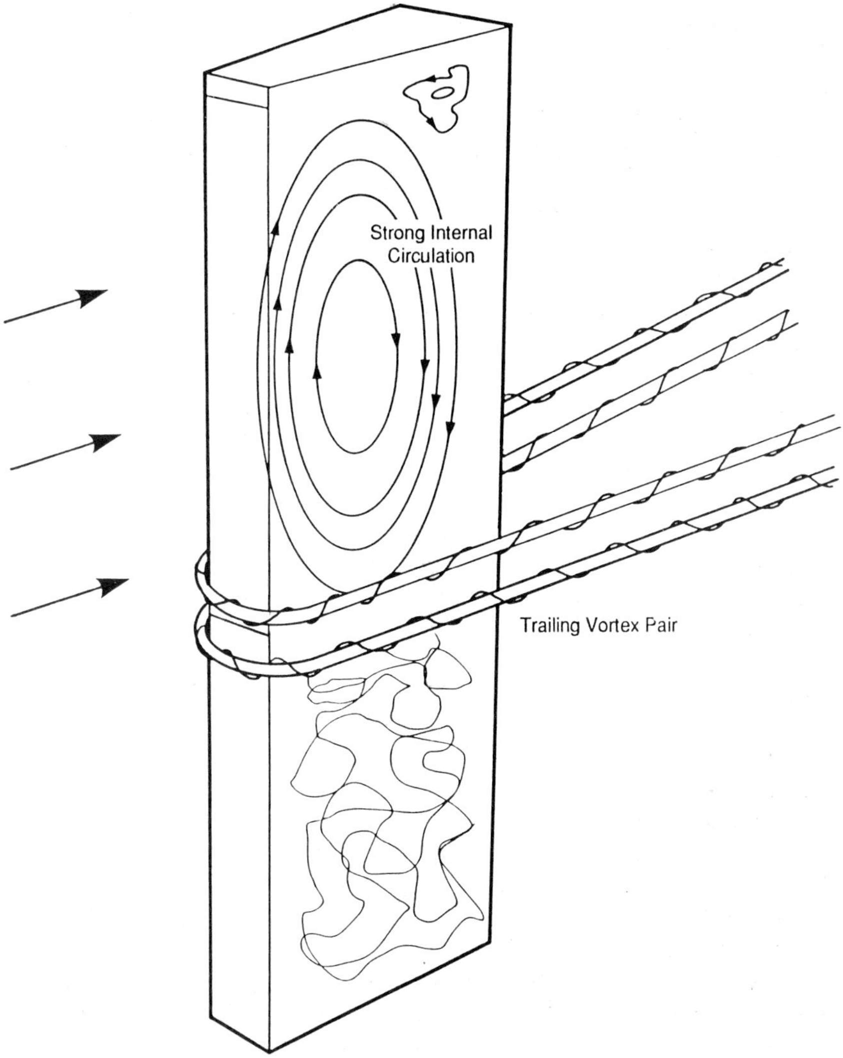

Figure 11.17 *Flow pattern inside a closed container flux gauge*

trailing vortices (rotating in opposite directions) wrapped around the front of the gauge at either the top or bottom lip of the opening, depending upon the direction of rotation of the main internal circulation. When this latter changed direction, the vortex pair would pass rapidly from one end of the collecting slot to the other.

Some further modifications of this design, intended to control the internal circulation, were investigated and some of the results are shown in Figure 11.14. None entirely eliminated the circulation. Those modifications which did were deemed unsuitable designs for a practical dust gauge. Those modifications which were both practical and seemed to reduce the internal circulation significantly were tested for collection efficiency. The result of halving the height of the opening slot, shown in Figure 11.14 (a) (the shape is shown in Figure 11.16 (a)), produced some improvement in performance, but a fairly strong dependence of collection efficiency on particle size remained. Figures 11.14 (b) and (c) show the results of an arrangement of two internal baffles (as in the diagram in Figure 11.16 (b)) intended to increase the region of calm air inside the gauge and to discourage the removal of particles by bouncing. This did improve particle retention, especially when sprayed with silicone grease, Figure 11.14 (c), but compared with Figure 11.13 (j) (for the gauge packed with glass wool, which should have had nearly complete retention) there were clearly still substantial losses from this design of gauge. Finally, Figures 11.14 (d) and (e) show the result of using a porous back to the gauge, see Figure 11.16 (c). This was intended to stabilize the flow in the gauge by allowing a small flow through it. There are a number of ways in which this might be arranged in practice without losing an excessive number of particles in the outlet flow. The method used here, fitting a porous back to the gauge, was simply the most convenient experimentally. The porous material was an open celled foam, two sizes were used, 30 pores per inch (PPI) and 60 PPI. The finer size acted as a better particle trap, but allowed a more limited flow through the gauge. The resultant windspeeds through the gauge were in the range 25-40% of the external windspeed. The coarser foam, Figure 11.14 (d), showed an improved collection at the lower windspeed but a reduced collection at the higher windspeed, probably because of a failure of the coarser foam to collect all particles impacting upon it. The finer foam, Figure 11.14 (e), showed an improved collection efficiency at all the test conditions, especially at the higher windspeed. However, flow visualization of these designs showed that despite the improved collection the internal circulation remained.

Since there seemed to be no way, within the confines required of a practicable flux gauge, of controlling the internal circulation within the closed container type of gauge with a vertical slot, it was felt that this sort of design was fundamentally unsuitable for measuring the flux of particles. It was therefore abandoned.

The only way of preventing the circulation seemed to be with a container of square or circular cross-section (looking into the wind) with a full opening at the front; that is, the container is in the form of a square or circular section tube with one end blanked off and the other facing into the wind. In principle this should not lead to a good design. An important principle of this virtual impaction type of collector is that the performance should improve as the size diminishes, because the particle inertia becomes proportionately greater. The vertical slot type of design ensures a narrow opening in one plane, which encourages collection by inertia. A square or circular opening with the same area as a slot should not perform as well. The results are less clear in practice. Figures 11.14 (f) and (g) show collection efficiency measurements of two open ended forward facing tubes. One, Figure 11.14 (f), is for a tube with the same area opening as the collecting slot of the previous work, see Figure 11.16 (d). The other, Figure 11.14 (g), having about 40% of this diameter and therefore about 14% of the area opening, see Figure 11.16 (e). Its diameter was about the same as the width of the slot in the other collectors. It can be seen that the larger tube had a similar collection performance to the British Standard design at the lower windspeed but was distinctly better at the higher windspeed. The smaller tube showed a much better performance at both windspeeds and particle sizes, the average collection efficiency being about 70%.

A practical realization of this simple forward facing tube is readily possible. Some tentative designs are shown in Figure 11.18. The depth would need to be about four or five times the diameter, perhaps with an enlarged area at the end. In order to deal with collected rainfall, the tube would need a slight downward slope (perhaps 10-15°) towards its closed end, which would need a drain into a collecting bottle. The larger tube would also probably need a mesh covering (of large open area) to its open end to prevent nesting birds, etc. The smaller tube, having a reduced collection in terms of absolute mass collected because of its smaller area opening, might need to be used in a multiple array. On an area basis, about seven of the smaller tubes would be needed, but due to its higher efficiency, three or four would collect about the same mass as the larger tube. A communal drain and collecting bottle could be used, though the drain tubes should not be too large to avoid cross flows developing between the different tubes. The tubes would need to be spaced apart by at least one tube diameter, otherwise the shape of the array would not be important. A design of this sort would be rather easier to produce than the present British Standard gauge.

11.4.3 Investigations of through-flow designs

In view of the fundamental difficulties associated with the closed container type of gauge, further efforts were devoted to the through-flow type of design. In these there is a forward facing opening with a flow through the gauge past a particle trap of some sort. In principle this sounds little different from a conventional particle sampler with a forward facing inlet. As far as the inlet is concerned, this is the case. However, it is normally expected that dust gauges should be passive devices, on the very reasonable practical grounds that there is most unlikely to be an electricity supply adequate for pumping purposes at the sampling sites which are either available or desirable. It is also desirable that there should be no moving parts to a gauge in order that it should be as reliable as possible. Thus the flow through the gauge must be driven by the wind in some way, preferably not involving any moving parts.

The main problem in designing this type of gauge is producing a wind-driven device which can generate a sufficient pressure difference through the gauge to overcome the pressure losses associated with the particle trap. It is possible to produce passive forms which can do this, but the pressure difference that can be generated is limited. In view of this, it is unfortunate that most of the effective designs of particle trap involve inertial collection

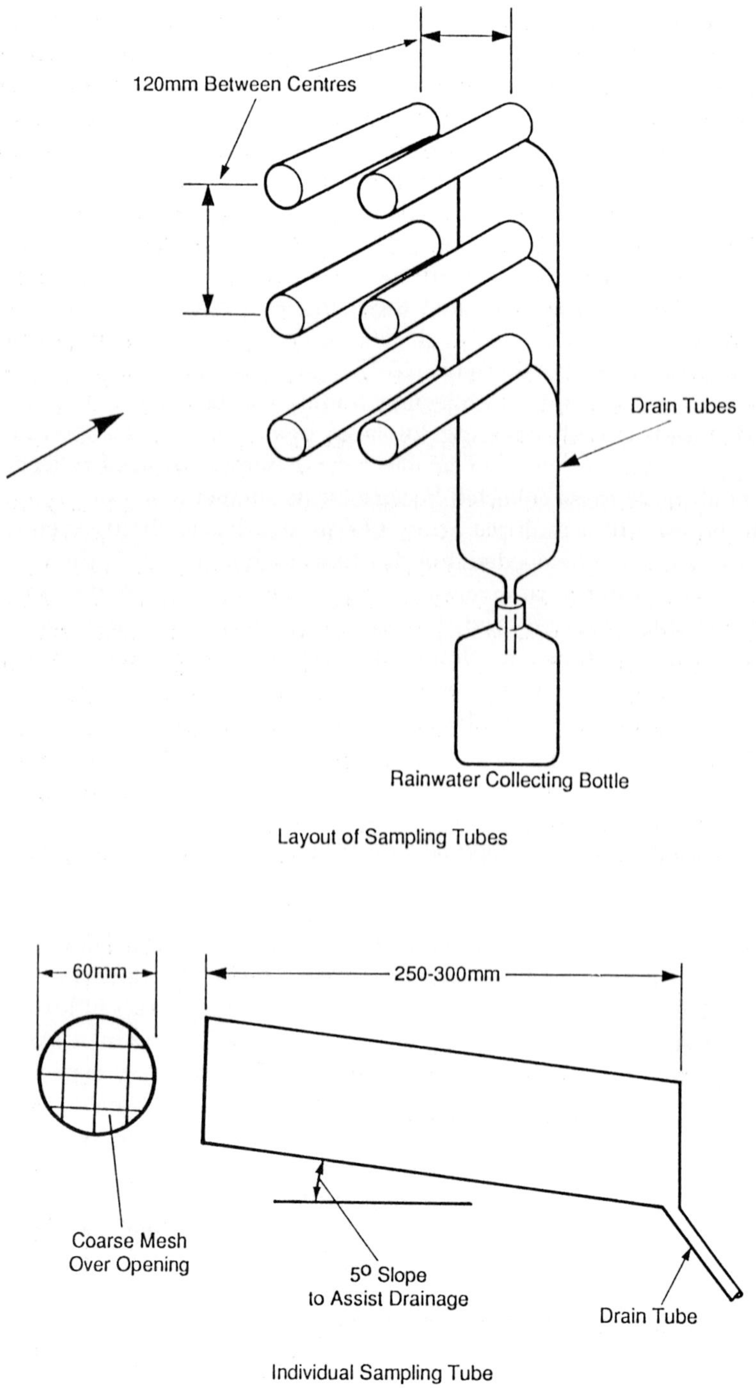

Figure 11.18 *Tentative design for a closed container flux gauge with a collection efficiency around 60–70%*

through devices which have an inherently high pressure drop. A saving feature is that isokinetic flow through the sampling inlet is not essential in order to achieve a high collection efficiency. Due to the relatively high inertia of the large particle sizes of interest, sub-isokinetic sampling can be quite effective. It will have been noted from the work described in the previous section that even a sampling inlet with a stagnated flow can have a good collection efficiency provided that the particles can be retained. Thus even a parallel tube with an efficient particle trap in its interior, which would sample sub-isokinetically due to the pressure drop in the trap, might be expected to show an improved collection efficiency over the blanked tubes described in the previous section and shown in Figures 11.14 (f) and (g).

It is possible to do better than this. A convenient way of generating a pressure drop through a device is to use a shape with a flare at the downwind end. The accelerated flow over the flare generates a reduced pressure at its base and the pressure difference so produced will assist in overcoming the pressure drop across the particle trap, thus increasing the air velocity through the gauge. Figure 11.19 shows a sketch of a tubular gauge shape of this sort, the dimensions given are of the basic shape used in most of the experiments. This technique was first applied to particle sampling by Steen (1977), who used it as an ambient aerosol sampler for fine particles; a modified form was also used as a stack sampler. The stack sampler has been used fairly widely, but the ambient sampler achieved only limited popularity. For reasons not understood by the authors this very useful method of designing through-flow samplers has largely languished. A version of it was used by Hall *et al.*, (1989) as an isokinetic sampler for measuring the concentration of large polystyrene particles (a substitute for snow) in a wind-tunnel. In this form a mesh was fitted to the end of the flare

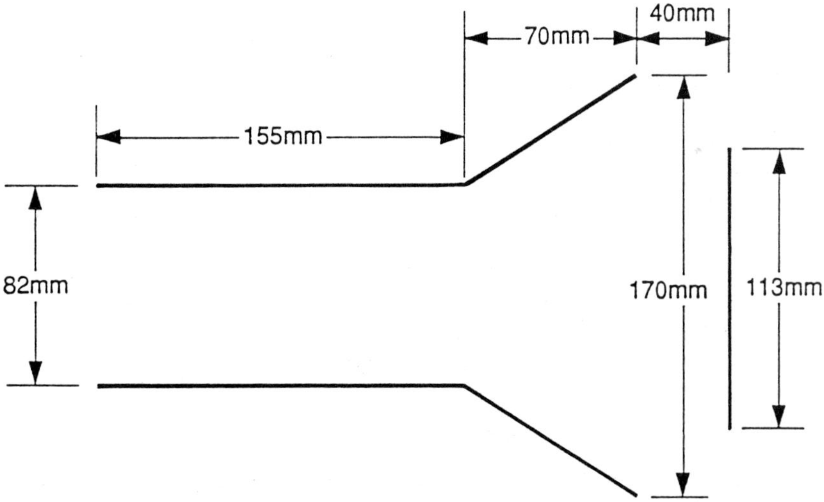

Figure 11.19 *Through-flow tubular flux gauge with flared end and fitted with baffle plate backflow preventer*

to trap the, very large, particles and it proved easily possible to obtain an isokinetic flow through the sampler. Hall and Upton (1990b) also suggested that the same design could be used for sampling much smaller particles, with a collection efficiency equal to the solid area of the mesh. This relied on the mesh collecting particles impacting on to it, and required either sticky or liquid particles, or a sticky mesh, to do this. Some later wind-tunnel tests showed that it did behave in this way and that its performance was roughly predictable. The present work has used this approach to design through-flow dust gauges, using both tubular and wedge forms. The tubular forms are similar to the sketch in Figure 11.19, the wedge forms will be discussed later.

Before considering the performance of the different designs investigated, one specific aspect of the design of through-flow flux gauges needs to be discussed. This is the prevention of backflows through the gauge. There are two modes of use of flux gauges, fixed and rotating into the wind. Fixed gauges measure the flux of particles from a given direction, while rotating gauges (usually driven by a wind vane) measure the total flux of particles from all directions. The British Standard Directional Gauge uses four fixed gauges spaced at 90° intervals, but current preferences would be for the use of both fixed and rotating gauges together. In most cases there is an interest in the flux of particles from one, or perhaps two, specific directions and this needs to be compared with the flux of particles from all other directions. This need is best served by one or two fixed gauges and one rotating gauge. With through-flow gauges of the sort sketched in Figure 11.19 there is a need for fixed gauges to sample only in forward wind directions. When the wind direction is towards the rear of the gauge there should be no wind flow through it in the reverse direction and, preferably, only minimal paths for the entry of particles into the rear of the gauge by inertia.

In principle, this could be done with a wind driven flap of some sort, but it would need to be of light construction and easily moved by light winds to be effective. This is difficult to achieve in practice where reliability in the field is needed. However, a good approximation to this requirement can be obtained simply with a baffle plate fixed slightly behind the rear of the flare. Figure 11.20 shows a sketch of this arrangement for a tubular gauge, with the flow reversed. If the size and position of the baffle plate is set so that the free shear layer from its edges just clears the edge of the flare, then there is little or no flow through the gauge in the reversed direction. There is an optimum size and position of the baffle plate. If the baffle plate was placed too close to the flare, it proved possible with some baffle plate arrangements to maintain a forward flow through the reversed gauge. As its size increases, then it must be placed closer to the flare and the size of the opening through which the normal flow through the gauge must pass is reduced. If its size is reduced, then it must be placed at greater distances from the flare and eventually the free shear layer becomes too weak to prevent a flow through the gauge. After some experiments, an optimum size and position of the baffle plate was found to be about two-thirds of the flare diameter, situated about 0.25 of the flare diameter behind its downwind edge, this arrangement

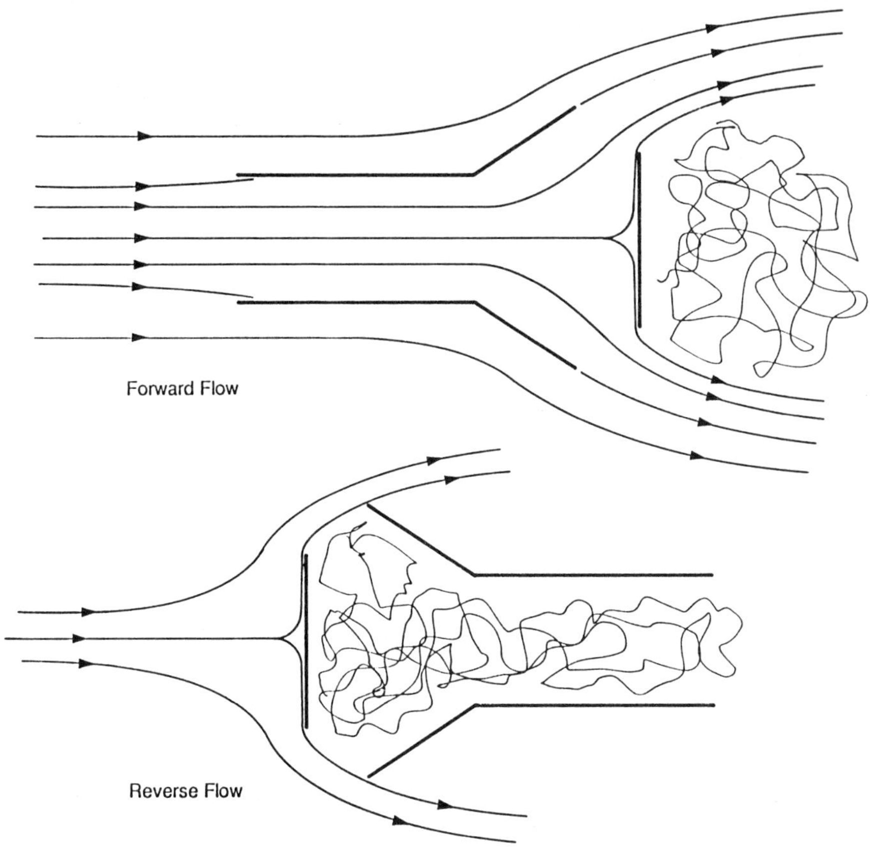

Figure 11.20 *Through-flow tubular gauge with baffle plate backflow preventer*

is shown in Figure 11.20. The baffle reduced the area of the opening around the flare by about 50%. Interestingly, this produced a slight increase in the flow through the gauge in the forward direction. The baffle plate design of Figure 11.20 was used in most of the experiments on tubular gauges, though only for forward flows. The design is provisional since experiments on it were limited to fully reversed flows, the effects of skewed wind flows, which are important, were not investigated due to time limitations.

Two sorts of particle trap were investigated, using either honeycomb or foam, usually fitted inside the parallel section of the gauge. The honeycomb had a cell size of about 8 mm and was used either in a long parallel array, as an elutriator, or sawn diagonally into slices about 30 mm deep. In the latter case, the slices of honeycomb were set normally across the flow, so that the cells were skewed to the flow direction, they then acted as small scale flow deflectors which would trap particles inertially. In principle, there ought to be a need for both types of particle trap, the elutriator working most effectively at low windspeeds and small particle sizes while the flow deflectors would be most effective with larger particles and higher windspeeds. The foam was of the open cell type discussed earlier, whose properties for particle trapping have been investigated and used in a variety of aerosol sampling devices (Brown, 1980, Gibson and Vincent, 1981). However, the present application is for significantly larger particles and higher airspeeds than in these other uses.

The designs of tubular through-flow gauge investigated are shown in Figure 11.21. All the designs were of circular cross-section except for the first two, which were of square cross-section. It was originally thought that the square cross-section might prove more convenient in use. In practice this advantage seemed small and the aerodynamic performance proved slightly inferior (the flare did not perform so well with a square as with a circular cross-section), so that the design was not persisted with. Besides the values of collection efficiency, plotted in the standard manner, the figure also shows values of the windspeed at the centre of the gauge inlet as a fraction of the undisturbed speed. Specific values are commented upon in the text, but it can be noted that in all cases sampling velocities were below isokinetic, mostly ranging between one-third and two-thirds of the normal windspeed.

Figure 11.21(a) shows the square cross-section gauge fitted with a combination of elutriator and inertial trap, a long parallel section of honeycomb followed by a short section with the cells aligned upwards at 45° to the flow. It was thought unlikely that the honeycomb trap would be very effective at retaining the hard glass spheres used in the experiments, particularly at the higher windspeed, due to particle bounce. To check this, the gauge design was first tested with the honeycomb dry, then the honeycomb was sprayed with a thin layer of silicone grease and the tests repeated. The results of the tests are shown in Figure 11.22(a) and (b). It can be seen that the particle retention was little affected at the lower windspeed, but greatly improved at the higher. With a greased trap there was a marked improvement in collection efficiency at all conditions, values around 60% being achieved. The inlet velocity to the gauge was about 45% of the windspeed. Figure 11.21(b) shows the circular cross-section gauge fitted with just an inertial trap consisting of three short sections of honeycomb with the cells aligned at 30° to the flow; the centre section was reversed so that particles had to traverse a dog-leg path to encourage trapping. The honeycomb was grease coated. This produced a much improved collection efficiency at all four test conditions, around 60–70%, being a little better at the higher windspeed. The inlet velocity to the gauge was a little above half the windspeed. No further development work was carried out on honeycomb traps at this stage, since it was desired to investigate other types of trap. However, from this limited work it was quite clear that this was a promising type of design worth further development. It appeared to be possible to obtain high collection efficiencies with only limited windspeed and particle size dependence even with significantly sub-isokinetic sampling.

The remaining results in Figure 11.22 are for measurements using foam traps of different densities. In all cases the depth of the foam was 25 mm, the thickness of the sheet in which it is normally supplied. Three different

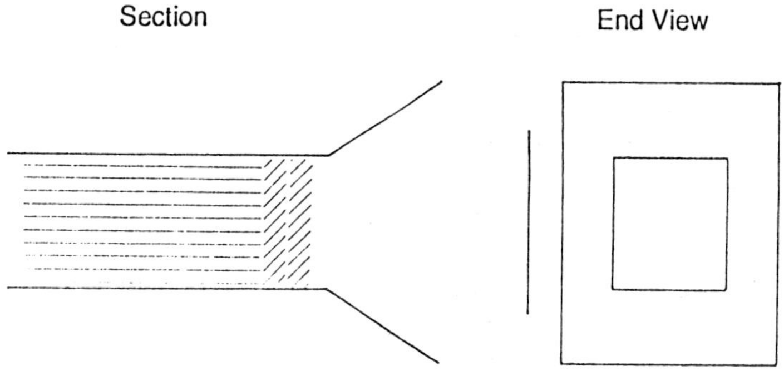

(a) Square Gauge with Honeycomb Trap

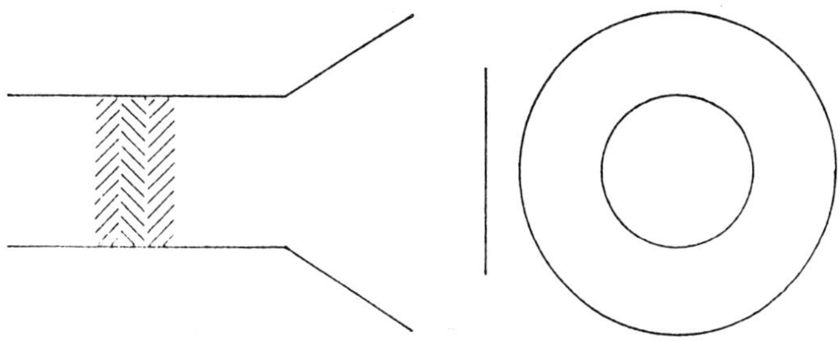

(b) Circular Gauge with Honeycomb Trap

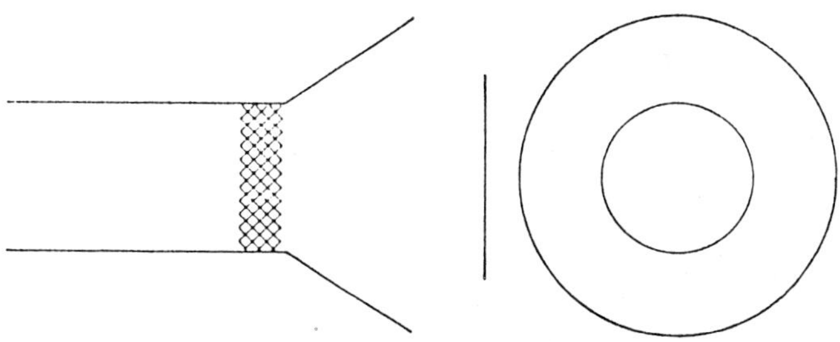

(c) Circular Gauge with Foam Trap (in Body)

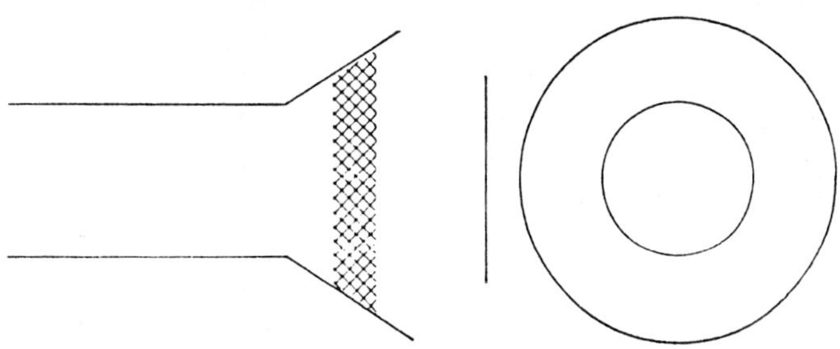

(d) Circular Gauge with foam Trap (in Flare)

Figure 11.21 *Through-flow flux gauges fitted with particle traps*

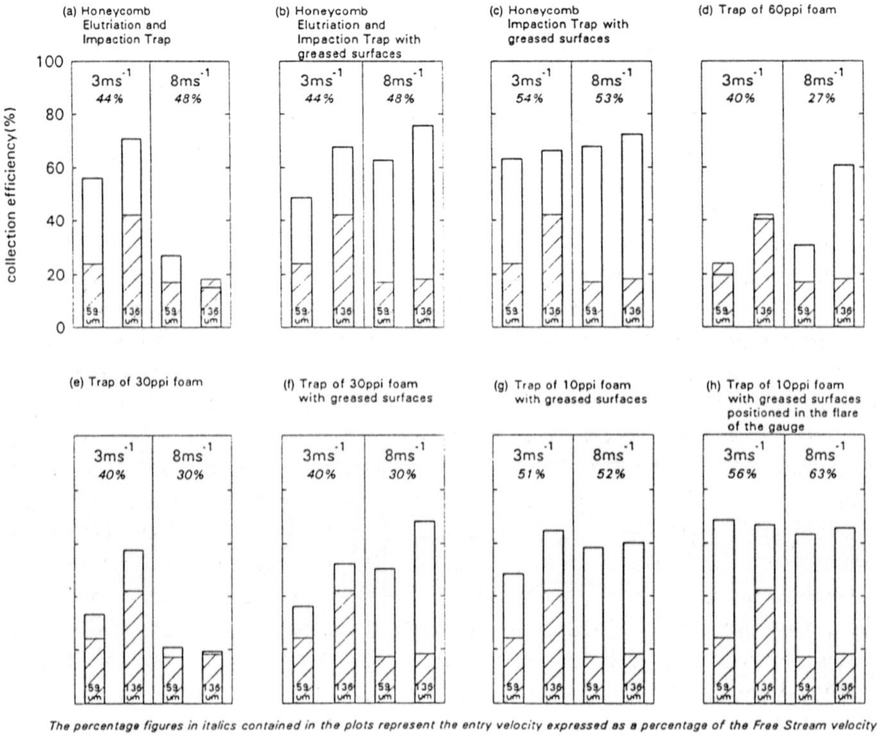

Figure 11.22 *Collection efficiencies of tubular through-flow flux gauges fitted with a variety of particle traps*

densities of foam were tried, 60, 30 and 10 PPI, normally fitted towards the end of the parallel section of the gauge, Figure 11.21 (c). The airspeed through the gauge increased as the foam became coarser, from about 30–40% of the windspeed with the finest foam, to about 60% with the coarsest. Unlike the honeycomb traps, the airspeed through the gauge also showed some windspeed dependence, being relatively greater at the lower windspeed than at the higher. For example, the inlet velocities as a proportion of the windspeed for the 30 PPI foam were 40% and 30% respectively at the lower and higher windspeeds. This behaviour is consistent with the characteristics of a number of fine filter media, which show a rapidly rising pressure loss with increasing flow rate. Figure 11.22 (d) shows the effect of a dry foam trap of 60 PPI placed at the end of the parallel portion of the gauge. This produced a small increase in collection efficiency at the lower windspeed and a substantial increase at the higher windspeed, though there was a significant particle size dependence. Spraying the foam with grease produced no significant improvement in collection efficiency. The performance was similar, though a little better, to that for the closed tube of the same diameter shown in Figure 11.14 (g), even though the airspeeds through the gauge were 30–40% of the windspeed. It appeared that this fine foam had a high particle retention, but also a high flow blockage. Figures 11.22 (e) and (f) show the effect of a 30 PPI foam, both dry and greased. The dry foam showed an improved collection efficiency over the 60 PPI foam at the lower windspeed but not at the upper, but when sprayed the collection efficiency also increased considerably at the higher windspeed. Figure 11.22 (g) shows the effect of a coarse, 10 PPI, greased foam. This produced a further increase in collection efficiency over the 30 PPI foam, mainly at the lower windspeed. There was only a limited windspeed and particle size dependence, with an overall collection efficiency around 50–60%.

Finally, Figure 11.22 (h) shows the effect of a 10 PPI foam of larger diameter placed in the flare of the gauge, Figure 11.21 (d). The intention was to place the foam in an area of larger diameter, so that the airspeed could be reduced through the foam. This resulted in a reduce pressure drop through the foam and therefore an increased airspeed through the gauge, in this case to around 60% of the windspeed. This resulted in a further gain in collection efficiency, to values around 65%, and showed only a low windspeed and particle size dependence. This is not a good position for the foam in a practical gauge (it needs to be less exposed), but was a convenient way of investigating the effect of reducing the airspeed over the foam. A practical design would need a diffuser in the body of the gauge, reducing the inlet diameter would serve the same purpose. As with the honeycomb traps, no further development work was carried out on foam traps at this stage. From the experiments carried out it appeared that foam traps had a good potential for producing effective gauges. It also appeared that the coarser foams were needed to permit an adequate flow through the gauge (perhaps with the addition of a diffuser ahead of the foam) and that they also needed a sticky coating to ensure retention of hard particles.

The wedge form of through-flow flux gauge investigated here is illustrated in Figure 11.23. It was flat sided, had a horizontal (or near-horizontal) bottom and a sloping top; the base of the wedge was finished with a vertical flat face. There was a horizontal slot entry at the front of the wedge and a similar slot exit for the airflow at the top of the blunt base. The wedge angle was 24°, the length 360 mm, the depth 200 mm and the width 200 mm (these are equivalent full-scale values). A particle trap of some sort could be fitted internally, towards the rear of the wedge. The design followed the same principle as the tubular through-flow gauges in that there was an accelerating flow up the wedge which produced a reduced pressure in the blunt base to assist in driving the flow through the gauge. The blunt base could also be arranged to provide the resistance to reversed flow required of a reversed gauge, which has been considered previously, though the shape in Figure 11.23 would not do this perfectly. Internally the wedge shape also produced a diffusing flow, so that a particle trap set towards the rear of the gauge would be in an area of reduced windspeed. The pressure drop associated with the trap would thus be reduced, increasing the airflow through the gauge.

Besides the use of a discrete particle trap, the basic form of the gauge was also designed to have inherent particle trapping qualities. By positioning the

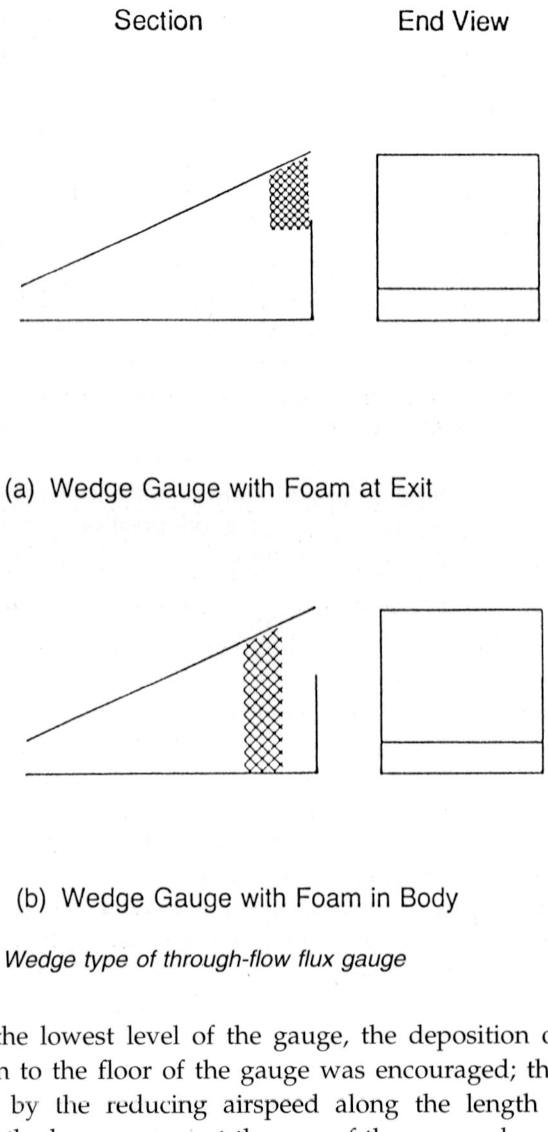

Figure 11.23 *Wedge type of through-flow flux gauge*

inlet slot at the lowest level of the gauge, the deposition of particles, by elutriation, on to the floor of the gauge was encouraged; this process was also assisted by the reducing airspeed along the length of the gauge. Furthermore, the lower corner at the rear of the gauge also acted as a trap. The airflow along the lower part of the gauge, that with the highest particle concentration due to elutriation, has to turn upwards to reach the exit slot. This, in combination with the region of separated flow in the corner, provides an inertial trap which should also have had good particle retention properties. There was also the possibility of encouraging the trapping performance of the basic gauge shape by the use of surface treatments and the fitting of internal baffles and lips.

Figure 11.24 shows three sets of measurements of collection efficiency of the wedge gauge fitted with foam traps. Figure 11.24(a) and (b) show the effect of a 30 PPI foam trap, dry and greased respectively, placed at the exit

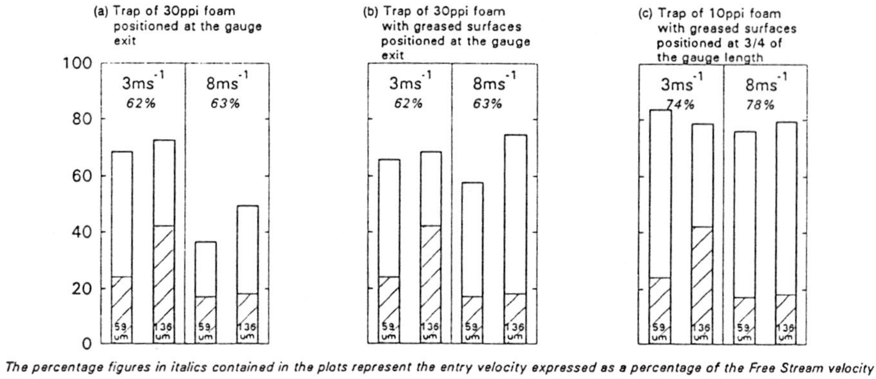

Figure 11.24 *Collection efficiencies of wedge type through-flow flux gauges fitted with a variety of particle traps*

slot of the gauge. With this configuration the airspeed at the entry slot was about 60% of the windspeed. Greasing the foam significantly improved the collection efficiency at the higher windspeed, but brought no change at the lower. With the foam greased a good collection efficiency was obtained, the average value being in excess of 60%, with only a limited variation with windspeed and particle size. Figure 11.24(c) shows a different trap arrangement, a greased foam of 10 PPI (which proved the most effective with the tubular gauges) set across the inside of the wedge about three-quarters of the way along its length. With this arrangement an airspeed in the inlet of about 75% of the free stream was achieved. The collection efficiency was high, the average being about 80%, and showed a good uniformity with windspeed and particle size.

From these limited measurements it was clear that the wedge-shaped gauge had a very good potential as a flux gauge and will be the subject of further development. This is in hand.

11.5 Conclusions

(1) Laboratory investigations of shallow, frisbee-type deposit gauges have shown that:

(a) The collection efficiency for small particles is very high, increasing with reducing particle size, and is of the same order as the enhanced deposition rates that occur at the ground due to other processes besides gravitational deposition.

(b) It is possible to enhance the collection efficiency of a deposit gauge using an eddy shedding bar around the gauge opening, but it is difficult to arrange this to vary with particle size in desired ways.

(c) Better control over splashing in rainfall can be obtained by fitting the gauge with a mesh or by covering the inner surface with a thin sheet of coarse open-celled foam. The foam is also effective in resisting the blow out of collected particles.

(2) Field comparisons between dust gauges have shown that, overall, a dry frisbee-type gauge has about the same collection efficiency as a British Standard Deposit Gauge, but that individual comparisons show a wide variation either way. If the frisbee-type gauge is made sticky and fitted with a splash preventer or a foam lining there is a considerable increase in its catch. Comparative measurements of its performance in this form are still in progress, but present evidence is that on average it collects about 50% more than the British Standard Gauge.

(3) Laboratory experiments on the British Standard Directional Gauge and on a variety of forms of this sort of closed container type of flux gauge have shown that they are fundamentally unsuited to effective dust collection. The reason is the strong internal wind driven circulation which effectively scours particles out of the gauge. There seems to be no practicable method of preventing this circulation.

(4) A very simple flux gauge made from a set of small blanked-off cylinders can achieve a collection efficiency of around 70%.

(5) Through-flow flux gauges, in which there is a flow driven (by aerodynamic means) through a passive gauge fitted with a particle trap, show much greater promise than closed container types of gauge. Initial experiments have produced a cylindrical version of a through-flow flux gauge, fitted with either honeycomb or foam traps, with collection efficiencies around 60-70%, these values being relatively insensitive to windspeed and particle size. A wedge-shaped version of a through-flow flux gauge, fitted with a greased foam trap, has shown a collection efficiency of about 80%, again with relatively little variation due to changes in particle size and windspeed. This type of gauge will be the subject of further development.

11.6 Acknowledgements

The authors wish to thank the following for their contributions to the research programme: Harry Vallack of the University of York; Mike Reynolds and Ian Addy of Selby District Council; Dave Smith and Bill Pearce of British Coal, Bretby; and Tim Hovland who assisted with the experiments at Warren Spring Laboratory.

11.7 References

Brown R C (1980). Porous foam size selectors for respirable dust samplers. *J. Aerosol Sci.*, **11**, 151-159.

Bush A W, Cross M, Gibson R D and Owst A P (1976). The collection efficiency of directional dust gauges. *Atmos. Environ.*, **10**, 997-1000.

Folland C K (1988). Numerical models of the raingauge exposure problem, field experiments and an improved collector design. *Q.J.R.Meteorol. Soc,* **114,** 1485-1516.

Gibson H and Vincent J H (1981). The penetration of dusts through porous foam filter media. *Ann. Occ. Hyg.,* **24,** 205-215.

Hall D J and Waters R A (1986). An improved, readily available dustfall gauge. *Atmos. Environ.,* **20,** No. 1, 219-222.

Hall D J, Cottrill S M, Goldsmith A L, Upton S L, Waters R A and Wright P (1988a). The Development and Field Use of a Snow Collector for Acid Precipitation Studies. Stevenage: Warren Spring Laboratory, Report No. LR585(AP).

Hall D J and Upton S L (1988b). A wind-tunnel study of the particle collection efficiency of an inverted frisbee used as a dust deposition gauge. *Atmos. Environ.,* **22,** No. 7, 1383-1394.

Hall D J, Upton S L, Campbell G W, Waters R A and Irwin J G (1989). Further Development of a Snow Collector for Use in Acid Precipitation Studies. Stevenage: Warren Spring Laboratory, Report No. LR752(AP).

Hall D J, Upton S L, Marsland G W, Douglass E J and Emmott M A (1990a). Wind-tunnel Measurements of the Particle Collection Efficiency of the Institute of Occupational Medicine Ambient Inhalable and Thoracic Particle Sampling Heads. Warren Spring Laboratory, Stevenage, Report no LR 806(PA).

Hall D J and Upton S L (1990b). A Passive Isokinetic Sampler. Paper Presented at the Aerosol Society Annual Conference, University of Surrey, Guildford, UK. 9-11 April.

Hall D J, Irwin J G, Stone B H and Upton S L (1992). Aerodynamic Considerations in Precipitation Collector Design. Paper presented at this Symposium.

Lucas D H and Moore W J (1964). The measurement in the field of pollution by dust. *Int. J. Air Wat. Poll.,* **8,** 441-453.

Ralph M O and Barrett C F (1984). A Wind-tunnel Study of the Efficiency of Three Deposit Gauges. Stevenage: Warren Spring Laboratory, Report No. LR499(AP).

Ralph M O and Hall D J (1989). Performance of the British Standard Directional Dust Gauge. Paper Presented to the Aerosol Society Annual Conference, West Bromwich.

Sehmel G (1984). Deposition and Resuspension. Atmospheric Science and Power Production (Ed. D Randerson) US Dept of Energy, DOE/TIC - 27601 (DE84005177).

Steen B (1977). A new simple isokinetic sampler for the determination of particle flux. *Atmos. Environ.,* **11,** 623-627.

Vallack H W and Chadwick M J (1990). A Two Year Survey of Airborne Dust in the Selby District, North Yorkshire. Report by the Stockholm Environment Institute at York, York University.

Vallack H W and Chadwick M J (1991a). Monitoring Airborne Dust in a High Density Coal Fired Power Station Region in North Yorkshire. Report by the Stockholm Environment Institute at York, York University.

Vallack H W and Chadwick M J (1991b). Monitoring Dust Deposition in the Vicinity of Drax Power Station, North Yorkshire, from August 1989 to December 1990. Report by the Stockholm Environment Institute at York, York University.

Vincent J H (1989). *Aerosol Sampling, Science and Practice.* Published by John Wiley and Sons Ltd, Chichester, U.K. ISBN 0 471 92175 0.

— 12 —

Dissolution procedures for the determination of lead and cadmium in airborne particulate

M G Kibblewhite, J Cox, A L Dennis and M Kendall

12.1 Abstract

Four procedures (HNO_3/H_2O_2, $HNO_3/HClO_4$, HNO_3 under pressure, and HNO_3/HCl) have been evaluated for the acid digestion of particulate collected on membrane and glass fibre filters. HNO_3 alone in a digestion bomb gave the best detection limits but this method was impracticable for the analysis of glass fibre filters. Application of the candidate procedures to filters spiked with urban particulate indicated that reflux with HNO_3/HCl or pressure digestion with HNO_3 achieves a more complete dissolution than either HNO_3/H_2O_2 or $HNO_3/HClO_4$, and this was supported by the results of analyses of metal fume reference filters.

12.2 Introduction

Lead and cadmium are recognized as hazardous to human health and in the past two decades increasingly stringent legislation has been introduced to control their release into the atmosphere. Accurate measurements of lead and cadmium in emissions are critical to the proper enforcement of controls, while reliable measurements of ambient air levels are needed to monitor the efficacy of regulations.

The great majority of instrumental techniques available for the determination of lead and cadmium require that the sample is brought into solution. This dissolution stage is at least as critical to the success of the measurement as is the effectiveness of the instrumental determination. In particular, the completeness of dissolution is critical to the accuracy of the measurement, while the variability of the procedural blank defines the method detection limit.

Many different methods can be used for the dissolution of airborne particulate collected on filters, but there is a lack of data for their comparative evaluation. The purpose of this work was to fill this gap and so provide a foundation for the definition of international standards (ISO DP9855).

The classical work of Gorsuch (1970) provides a starting point for dissolution procedure comparisons. His important observations on the risk of losses of lead and cadmium when dry ashing is used, suggests the alternative of wet digestion. Ideally the acid mixture used for dissolution should completely release all the forms of lead and cadmium present in airborne particulate – which will *inter alia* include oxides, carbonates, sulphates, silicates, sulphides and, potentially, organic compounds. Additionally, the acids employed should not form insoluble compounds with cadmium and lead and they should be volatile to assist concentration. For both these reasons sulphuric acid is not suitable. In this evaluation four acid mixtures and accompanying digestion conditions were chosen, which met the general conditions above. Variations of these have been used extensively in many laboratories.

12.2 Experimental

12.2.1 Reagents and apparatus

The acids used were analytical reagent grade HNO_3 (1.42 g ml^{-1}), $HClO_4$ (1.54 gm l^{-1}), HCl(1.18 gm6l^{-1}). H_2O_2 (100 volumes) was Aristar grade. Calibration and spiking standards were prepared from 1000 mg l^{-1} Pb and 1000 mg l^{-1} Cd solutions (both BDH Ltd). Membrane filters were Millipore Type AA (0.8 μm, 3.7 cm diameter). Glass fibre filters were Whatman QMA (3.7 cm diameter). All glassware was borosilicate. Where appropriate to the digestion procedure a thermostatically controlled hot plate or a Parr bomb (type 4744, 45 ml) was used. Determination of lead and cadmium was made using a Pye Unicam SP9 atomic absorption spectrophotometer.

12.2.2 Digestion procedures

12.2.2.1 Procedure A: dissolution with nitric acid and hydrogen peroxide

The filter is digested in a 50 ml beaker on a hot plate at 180°C with 3 ml of HNO_3 and 1 ml of H_2O_2 until almost dry. This addition of digestion reagents and their evaporation is repeated at least twice. After cooling, 1% v/v HNO_3 is used to dissolve the residue and transfer it to a volumetric flask. For membrane filters the final digestion volume is 10 ml. For glass fibre filters the digest is filtered and the filtrate, together with repeated washings, is diluted to 25 ml.

12.2.2.2 Procedure B: dissolution with nitric and perchloric acids

The filter is digested in a 50 ml beaker on a hot plate at 180°C with 10 ml of HNO_3 until almost dry. Addition and evaporation of HNO_3 is repeated until initial oxidation of the sample is complete and then the sample and beaker are allowed to cool completely. 9 ml of HNO_3 and 1 ml of $HClO_4$ are added and heating is continued on the hot plate until a white ash just appears, when 1 ml of HNO_3 is added and evaporation is continued. This final addition and evaporation of HNO_3 is continued until no white fumes of $HClO_4$ are observed. After cooling, 1% v/v HNO_3 is used to dissolve the residue and transfer it to a volumetric flask. For membrane filters the final digestion volume is 10 ml. For glass fibre filters the digest is filtered and the filtrate, together with repeated washings, is diluted to 25 ml.

12.2.2.3 Procedure C: dissolution by pressure digestion with nitric acid

The filter is placed in the PTFE inner vessel of the digestion bomb and 1 ml of HNO_3 is added. The bomb is sealed and placed in an oven at 180° for 12 hours. After complete cooling the bomb is opened and the digest transferred to a 10 ml volumetric flask with washings of 1% v/v HNO_3.

12.2.2.4 Procedure D: dissolution by reflux with nitric and hydrochloric acids

The filter is placed in a 100 ml Erlenmeyer flask and 10 ml of a solution of 1:2 HNO_3 and HCl is added. The flask is then connected to a reflux condenser and placed on a hot plate which is heated to 100°C. Reflux is continued for 2 hours. After cooling, the digest together with water washings is diluted to 50 ml in a volumetric flask.

12.3 Results and discussion

None of the digestion procedures evaluated were expected to release lead or cadmium from crystalline silicates. For this, treatment with hydrofluoric acid would be needed and it was judged that the small extra quantities of lead and cadmium released would not justify inclusion of this step.

The dissolution of particulate collected on glass fibre filters is impracticable using a digestion bomb, because the insoluble residue precludes quantitative transfer of the digest from the bomb with a suitably small volume of washings.

Detection limits for each of the digestion procedures, when combined with flame atomic absorption spectrometry (AAS), were measured by replicate analyses of unexposed filters and the results of this are presented in Table 12.1. All of the procedures provided detection limits better than 1 μg for both lead and cadmium. Factors contributing to differences in detection limits are the purity of reagents, uncorrected background interference in the flame AAS and inherent analytical precision, dependent on the number and form of analytical operations made. The use of nitric acid alone in a pressure bomb gives excellent detection limits for membrane filters because the purity of the

Table 12.1 Analysis of blank filters (Experiment 1)

	Membrane filters				*Glass fibre filters*		
Procedure	A	B	C	D	A	B	D
Results for cadmium (μg)							
Mean value	0.30	0.00	0.06	1.17	0.50	0.20	0.77
Standard deviation	0.08	0.00	0.07	0.26	0.03	0.10	0.26
Detection limit	0.24	0.00	0.21	0.78	0.09	0.30	0.78
Results for lead (μg)							
Mean value	−0.10	−0.09	−0.03	1.10	0.50	0.32	0.73
Standard deviation	0.20	0.11	0.01	0.24	0.08	0.28	0.34
Detection limit	0.60	0.33	0.03	0.72	0.24	0.84	1.02

It should be noted that the different procedures were tested separately, and that differences in the detection limits observed for these must be to some extent reflect batch to batch variations in the flame AAS precision.

digestion acid is high and opportunities for contamination are minimized. Conversely, reflux with HNO_3 and HCl gave the poorest detection limits because the HCl purity is not as good, the contact with glassware is extensive, and the final dilution volume has to be higher to accommodate sufficient washings.

It is not possible to obtain exposed air filters which all contain identical quantities of lead and cadmium. It was therefore necessary to use proxy samples and those chosen were filters spiked with aqueous standards, filters

Table 12.2 Recoveries of aqueous spikes (Experiment 2)

	Membrane filters			*Glass fibre filters*		
Procedure	A	B	D	A	B	D
Results for cadmium (μg)						
Mean value	10.1	10.0	10.3	9.73	9.89	10.1
Standard deviation	0.08	0.12	0.26	0.175	0.172	0.23
Detection limit	101	100	103	97	99	101
RSD of recovery	0.8	1.2	2.5	1.8	1.7	2.3
Results for lead (μg)						
Mean value	20.5	20.0	19.4	19.8	19.7	20.3
Standard deviation	0.16	0.16	0.67	0.32	0.27	0.56
Detection limit	102.5	100.0	97.0	99.0	97.5	101.5
RSD of recovery	0.8	0.8	3.4	1.6	1.3	2.7

spiked with an urban particulate standard reference material (SRM), and SRM reference membrane filters.

The recoveries from application of each of the digestion procedures to unexposed filters spiked with aqueous standards are presented in Table 12.2. The results are a useful indication of experimental effectiveness, but are not a real guide to dissolution capabilities because of the absence of a normal sample matrix.

Table 12.3 presents the results of application of the different digestion procedures to filters spiked with urban particulate SRM 1648. The quantities of urban particulate used were small but contained amounts of lead typically found on filters exposed in urban ambient air measurements. The recoveries observed were all in excess of 90%, but there is evidence of differences in the relative effectiveness of the procedures. Recoveries increase to 100% in the order : procedure B (HNO_3 and $HClO_4$) < procedure A (HNO_3 and H_2O_2) < procedure D (HNO_3 and HCl). Procedure C (pressure digestion with HNO_3) was only applied to membrane filters spiked with urban particulate, but for these a complete recovery of lead was obtained within the experimental error.

Table 12.3 Recoveries from urban particulate matter (Experiment 3)

	Membrane Filters				*Glass Fibre Filters*		
Procedure	A	B	C	D	A	B	D
Results for cadmium (μg)							
Mean value*	70	68	72				
Standard deviation	7.4	13.0	5.8				
Percentage recovery	93.3	90.7	96.5				
RSD of recovery	7.9	14.3	6.0				
Results for lead (% w/w of particulate matter)							
Mean value†	0.63	0.62	0.66	0.65	0.64	0.60	0.65
Standard deviation	0.013	0.021	0.018	0.015	0.027	0.014	0.020
Detection limit	96.2	94.7	100.6	99.2	97.7	91.6	101
RSD of recovery	2.1	3.4	2.7	2.3	4.2	2.3	3.0

† Note: Certified value is 75 $\pm$ 7 $\mu g g^{-1}$ of particulate
* Note: Certified value is 0.655 $\pm$ 0.008 % w/w

The concentration of cadmium in SRM 1648 is much lower than that of lead and it proved possible to only determine cadmium in the solutions prepared from the membrane filters, and then only when procedures A, B and C were used. The low precision of even these recoveries (6 to 14%) reflects the small amount of cadmium being determined. Nevertheless within the experimental errors and the confidence limits of the certified values, procedures A, B and C gave satisfactory recoveries of cadmium from the urban particulate spike.

Table 12.4 presents the results of application of digestion procedures A, B and D to sets of membrane filters containing metal fume (SRM 2676c). These indicate recoveries above 97% for both lead and cadmium, but with slightly higher results using procedure D (reflux with HNO_3 and HCl).

Table 12.4 Analysis of standard membrane filters (Experiment 4)

Procedure	*A*	*B*	*C*
Results for cadmium (μg)			
Filter lc (0.95 ± 0.01 μg)	1.0, 1.0	1.0, 0.9	0.5, 1.0
Filter llc (2.83 ± 0.04 μg)	2.8, 2.8	2.8, 2.7	3.0, 3.0
Filter lllC (10.09 ± 0.15 μg)	9.7, 9.9	9.7, 9.7	10.0, 10.5
Results for lead (μg)			
Filter lc (7.47 ± 0.11 μg)	7.0, 7.3	7.3, 7.3	7.5, 7.5
Filter llc (14.92 ± 0.22 μg)	14.6, 14.8	14.5, 14.5	15.0, 15.5
Filter lllC (29.81 ± 0.45 μg)	29.1, 29.5	29.5, 29.7	31.0, 30.5

Note: Figures in brackets are certified values.

12.4 Conclusions

Of the procedures evaluated, reflux with nitric and hydrochloric acids gave the highest recoveries of both lead and cadmium. It is therefore considered a better reference method, for samples containing greater than 1 μg of lead or cadmium, than are the methods employing nitric and perchloric acids, nitric acid and hydrogen peroxide, or nitric acid alone and under pressure.

12.5 References

ISO DP9855. Air Quality - Determination of Particulate Lead Collected on Filters from Ambient Air Atomic Absorption Spectrometry Method.
Gorsuch T (1970). *The Destruction of Organic Matter*. Pergamon Press, Oxford.

— 13 —

Directional sampling as a means of assessing the contribution of fugitive emissions to ambient pollutant levels

P Clayton and B J Davis

13.1 Abstract

Fugitive emissions from industrial processes often make significant contributions to the concentrations of pollutants in the ambient atmosphere. Standard methods of measurement which are used for air quality monitoring do not, by themselves, provide information on the source of the pollutant. Chemical fingerprinting is sometimes used to relate measured ambient concentrations to known emissions from a particular source. Meteorological data and a suitable model are necessary to interpret the data in terms of the contribution from the source in question to ambient levels at a particular monitoring site.

Two techniques have been used to quantify the contributions of fugitive emissions to ambient concentrations. The first is an empirical treatment which uses wind direction and simple ambient concentration measurements to provide a factor to relate the two and which can be used to predict concentrations based on historical wind direction data. The second, a directional sampling technique, was developed to assess directly the contribution of fugitive emissions from a works to the ground level concentrations (or deposition rates) of particular pollutants close to the works. In this technique two samples are obtained concurrently, one relating to the background levels and a second relating to periods when the wind is blowing from the works towards the sampler. Comparison of the measured levels in the two samples is an indication of the magnitude of the contribution of the works to local levels.

The current development programme, which is not described in detail in this paper, was designed to specify directional samplers for suspended and depositing particulate matter and to outline their use for the assessment of fugitive emissions.

The main factors affecting the average ground level concentration, or deposition rate at a point close to a works are the source strength and the proportion of time during which the wind blows from the direction of the works to the sampler. Thus, by correct siting of the sampling equipment and by sampling separately for periods during which the wind blows from the works and when the wind blows from other directions and knowing the lengths of those periods, it is possible to assess the contribution of the fugitive emissions of a particular pollutant.

The background to methods of assessing the contribution of fugitive emissions to ambient concentrations, the siting of sampling equipment, problems encountered in using directional samplers, repeatability of measurements and methods of interpreting the results are discussed.

13.1 Introduction

Fugitive emissions from industrial processes often make significant contributions to the concentrations of pollutants in the ambient atmosphere. For the purposes of this paper, fugitive emissions are those non-stack emissions which escape from a process without passing through an in-line emission control system. It is normally very difficult to quantify such fugitive emissions and therefore difficult to assess the effectiveness of control measures.

Standard methods of measurement which are used to monitor air quality do not on their own provide any information on the source of a pollutant although, chemical 'fingerprinting' is sometimes used to relate measured ambient concentrations to the emissions from a particular source. In the following paper two methods are described by which the contribution of fugitive emissions to ambient concentrations of a particular pollutant may be assessed. The first is an empirical method (Clayton and Ogden, 1991) which was developed by Warren Spring Laboratory (WSL) in order to predict future concentrations around a works and to attempt to estimate the improvement in fugitive emission abatement required to achieve a specific air quality standard. The second is a directional sampling technique which can be used to obtain two samples simultaneously, one relating to emissions from a works and the other relating to the background concentration of the pollutant in question. Directional sampling is applicable to both aspirated samplers for the measurement of concentrations and to passive samplers such as deposit gauges. The technique as applied to aspirated samplers is described.

13.1.1 Background

The concentration of a particular pollutant at an ambient sampling site located close to a works emitting that pollutant is dependent on the source strength and the proportion of time for which the wind blows from the source to the sampler. Thus:

$$C_t = B_0 + k_0 PE \quad (1)$$

where C_t is the measured ambient concentration of the pollutant, $\mu g\,m^{-3}$; B_0 is the concentration which would be expected if the works were not there at all, that is the absolute background concentration, $\mu g\,m^{-3}$; k_0, for a given source, is a constant which depends on such parameters as the height of emission, distance from the source, windspeed and rainfall; P is the proportion of time for which the sampling site receives pollutant from the source, i.e. the proportion of time when the wind is in the appropriate direction and E is the emission rate or source strength, $g\,s^{-1}$.

13.2 Conventional sampling

Conventional sampling techniques may be used to predict concentrations at a particular site provided that data on wind direction at that site is available over a number of years.

13.2.1 Basis of predictions

Equation (1) above shows that when $P = 0$, that is when the wind is not blowing from the works to the sampling site $k_0PE = 0$ and the site is, in effect, monitoring the background pollutant concentration B_0 including other sources of the same pollutant. It is clear therefore, that the wind direction behaviour must be a major factor in determining the ambient concentrations of a pollutant emitted from a works. The other major controlling factor is the source strength, i.e. the rate of emission of the pollutant under consideration from the works. Other factors of lesser importance include the windspeed and rainfall which could affect the source strength. Thus, provided that the source in question is the major source, that is the background contribution to the concentration of the pollutant in question is minimal, it may be assumed that:

$$C_t = k_0PE \qquad (2)$$

In fact k_0 is not a simple constant since it includes elements of windspeed, height of emission, rainfall and atmospheric stability but, for the sake of this application, it is assumed to be constant on average over the measuring period.

Rearranging equation (2) gives:

$$k_0E = C_t/P \qquad (3)$$

and, if the source strength does not change over the period under consideration,

$$C_t/P = K_1 \qquad (4)$$

where k_1 is a constant embracing the source strength and other factors.

Therefore, by calculating C_t/P for one period, the values of C_t for the following period can be predicted on the basis of historical data on the average proportion of wind from the appropriate direction for that following period which is available for most areas, that is,

$$C_p = k_1 P_a \qquad (5)$$

where C_p is the predicted concentration and P_a is the average proportion of wind from the appropriate direction obtained from historical data for the period in question. The data can also be used slightly differently since changes in k_1 indicate changes in source strength or the effectiveness of the abatement measures.

13.2.2 Prediction of concentrations

In order to utilize this method, the concentration should be measured at some point close to the works and the wind direction data for the site or somewhere close by should be available. In the following example taken from a particular investigation, the results are normalized by dividing the measured concentration by the air quality standard for the pollutant. Figure 13.1 is a scatter diagram of the normalized concentration in relation to the proportion of wind in the sector which includes the works.

In this example the measurements covered a period of over 5 years and in Figure 13.2 the 'wind factor' ($1/k_1$) is shown for each month over the period and as a running 12-month average. The value of the 'wind factor' which would be required to ensure that the annual mean would be less than the air quality standard is also shown. Because the reciprocal of k_1 is shown, all

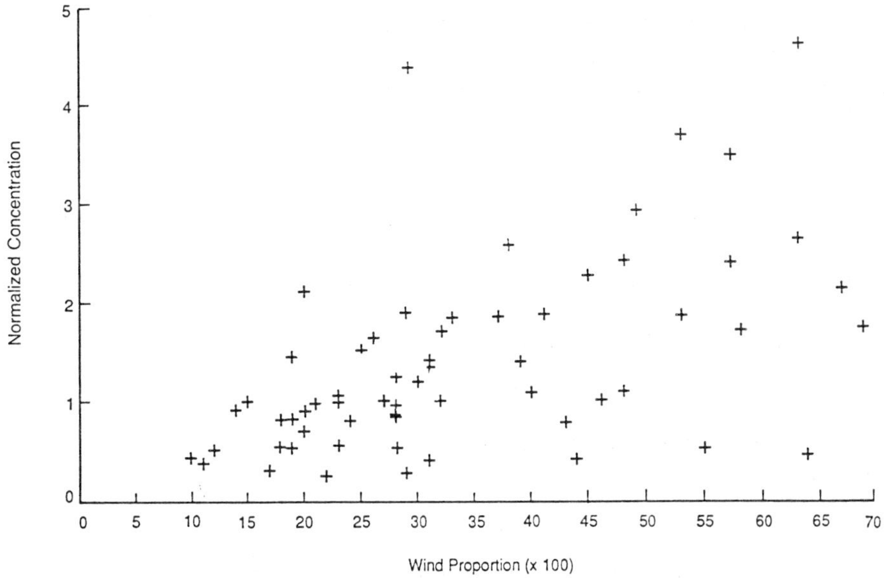

Figure 13.1 *Scatter plot of concentration and wind direction data*

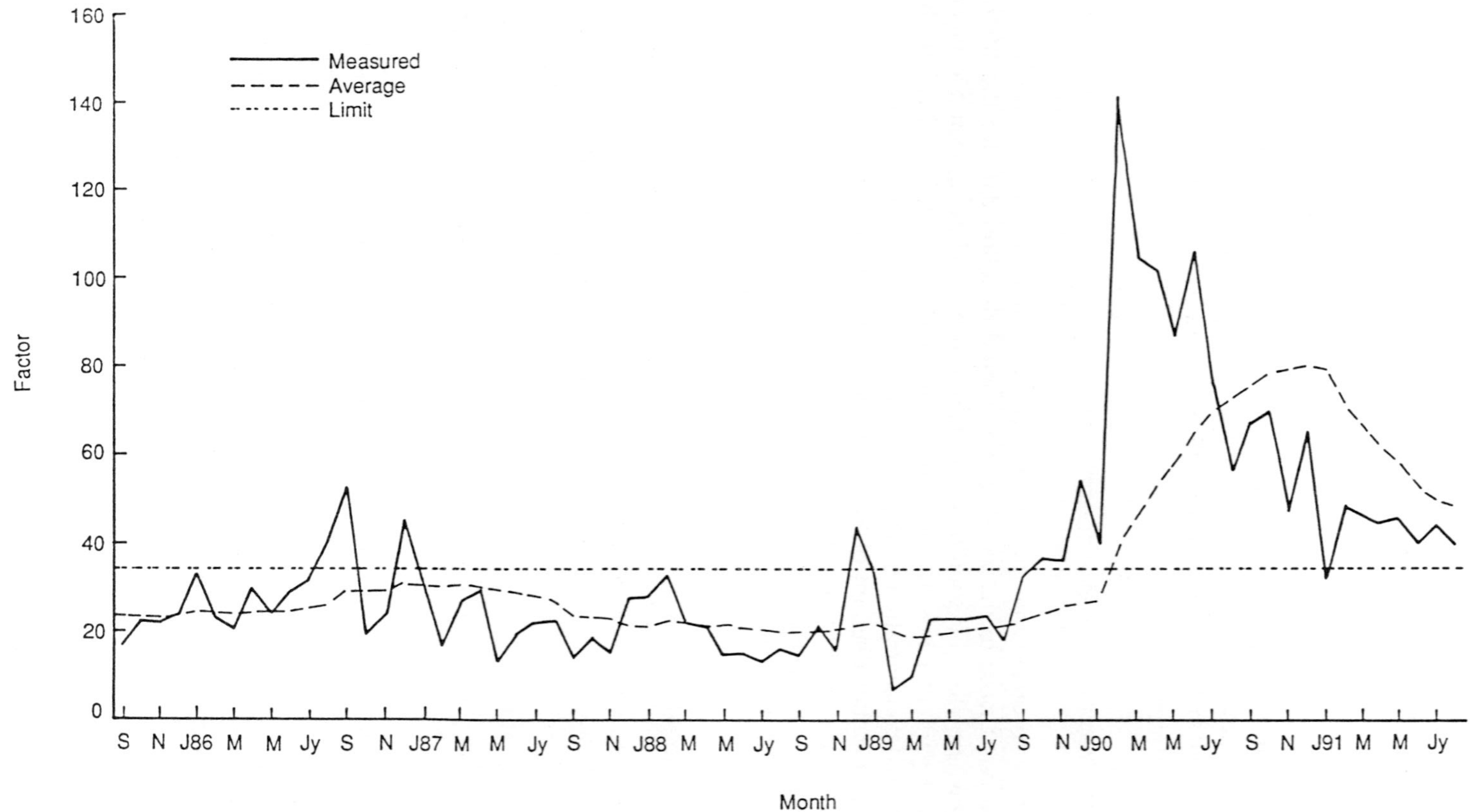

Figure 13.2 *Wind factor profile. Note: factors below the limit line will give concentrations above the air quality limit*

measurements above the datum line indicate compliance with the air quality standard. The sudden increase in the value of the factor in early 1990 coincides with the installation of new dust control measures at the works.

The running average annual mean value of the wind factor can be used to predict concentrations into the future provided that average wind direction data for the area for equivalent periods in previous years are known. In the current example, data for the 12 months up to December 1986 can be used to predict the value of the annual mean for the end of December 1987 and so on until the data for the year ending 1990 is used to predict the annual mean for December 1991, giving the results shown in Table 13.1.

Table 13.1 Measured and predicted concentrations

Year	$1/k_1$	*Normalized concentration*	
		Measured	*Predicted*
1986	23.20	1.38	1.37
1987	29.58	1.51	1.08
1988	22.06	1.79	1.45
1989	21.80	1.30	1.46
1990	26.18	1.20	1.22
1991	80.16	0.71	0.85

On this basis the concentration predicted by this empirical method is, on average, approximately 10% less than the average measured concentration. The regression line (for pre-December 1989) relating normalized concentration and wind proportion is given in Figure 13.3 and an interesting conclusion which can be drawn from this is that the air quality standard would be exceeded if the proportion of winds from the appropriate direction exceeded about 20% assuming the plant emissions remained unchanged.

13.2.3 Estimation of the reduction in emissions required to meet air quality standards

In addition to predicting the ambient concentrations at some time in the future, it is also possible by this method to determine the approximate reduction in emissions from a works which would be required to achieve a target ambient air concentration. The method is illustrated graphically in Figure 13.4 where the probability (x-axis) of achieving a particular normalized concentration (y-axis) is plotted. Provided the emission from the works remains constant, the measured concentration depends only on the proportion of time the wind blows from the direction of the source. The distribution of concentrations is therefore controlled by the distribution of the proportion of winds from the direction of the works.

Taking the results to the end of December 1989, the average concentration was 1.31 normalized units, $1/k_1$ was 26.18 and the average proportion of

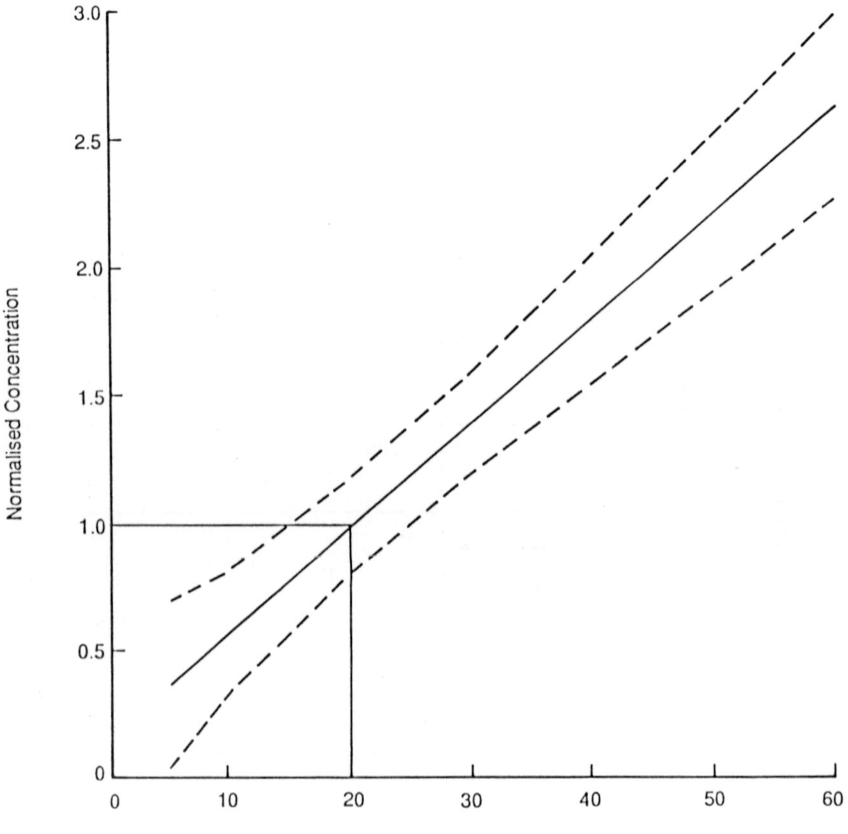

Figure 13.3 *Confidence limits for the prediction of the mean concentration*

wind from the direction of the works was 34%. The standard deviation of the proportion of wind from the direction of the works over the previous year was 16%. Thus the upper and lower 95% confidence limits for the proportion of wind from the direction of the works were 66 and 2.4% respectively. The corresponding normalized concentrations for $1/k_1 = 26.18$ are 2.5 and 0.1. Thus there is a 95% probability that the normalized concentration will be less than 2.5 and a 2.5% probability that it will be less than 0.1. This data is represented by the upper line on the graph in Figure 13.4.

The slope of the line depends on k_1; deteriorating conditions (increasing emissions) will lead to a steeper slope whilst decreasing emissions would lead to a slope which was less steep. Thus the broken line which passes through 1 at 50% probability represents the same distribution of wind data but with k_1 reduced to a level at which the air quality limit value will be met. The reduction in concentration, and therefore in emission, required to achieve the target value is thus 23%. The third, lower line, represents a value of k_1 which would give a 90% probability that the final concentration would be less than the target value by December 1990. To achieve this condition

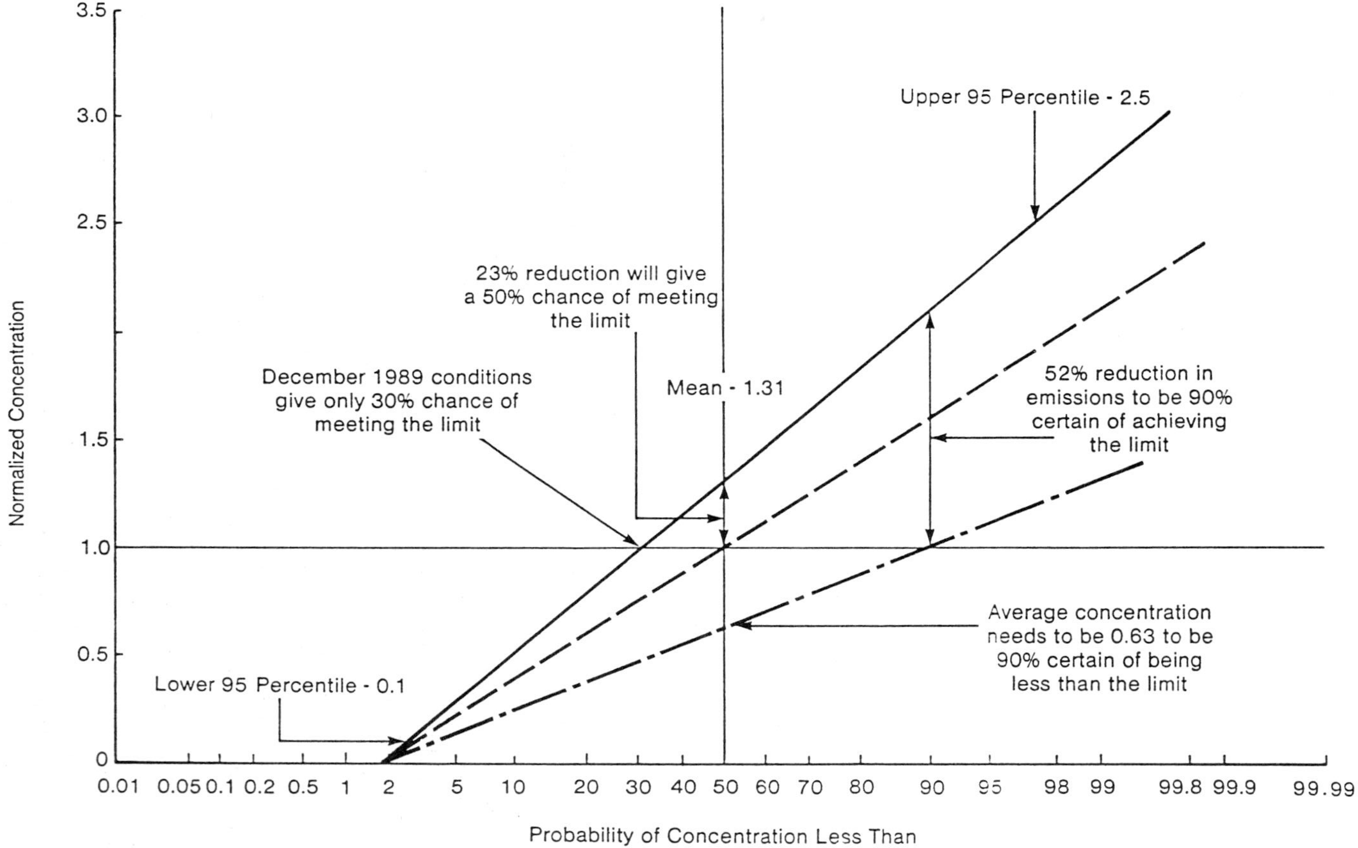

Figure 13.4 *Analysis of data up to December 1989*

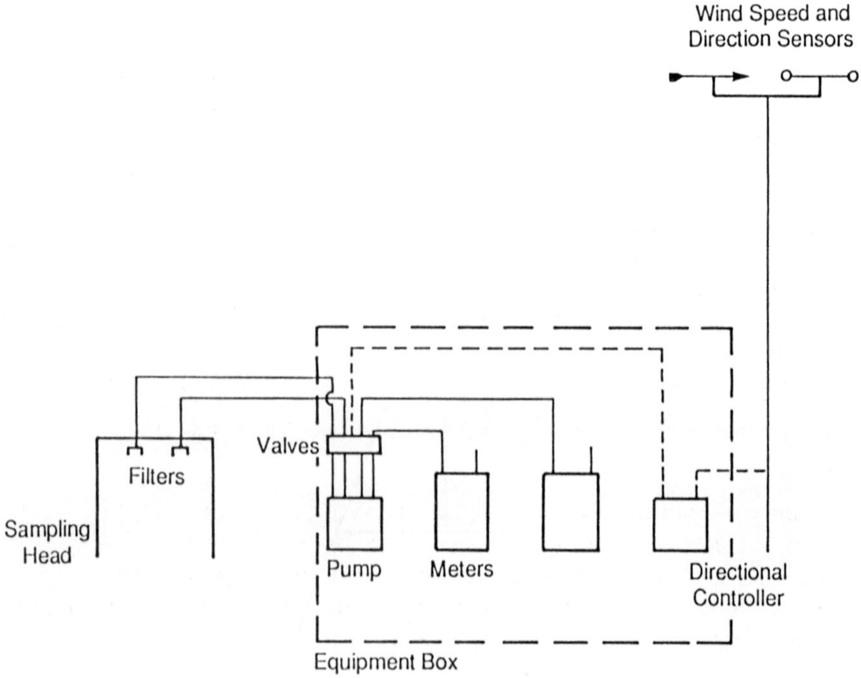

Figure 13.5 *Schematic diagram of directional M-type sampler*

emissions would need to be reduced by 52% and the target mean concentration to provide this 90% certainty is 0.63 normalized concentration units.

13.3 Directional sampling

The M-type sampler and the directional M-type sampler were developed at Warren Spring Laboratory to measure concentrations of heavy metals in the ambient atmosphere. The directional sampling technique was developed (Barnett *et al.*, 1987) to measure the contribution of a particular works to measured ambient concentrations and the equipment is shown diagrammatically in Figure 13.5.

Two samples are collected simultaneously to determine the directional concentration, C_d, which is measured when the wind blows from the direction of the works to the sampler and a background concentration, C_b, which is measured when the wind is from other directions. C_d has two components, the works contribution W_d (not to be confused with W_t, the works contribution to the overall ambient concentration) and a background concentration for the directional sampling arc B_1. The relationship between C_d, C_b and C_t is:

$$C_t = C_d P + (1 - P) C_b \tag{6}$$

where P is the proportion of the total sampling time when the wind is within the directional sampling arc.

Each filter is connected through a limiting orifice to a pump and dry gas meter. The directional sampler is controlled by an anemometer and wind vane sited close to the instrument on a mast about 5 m above the ground. The instrument is set so that the directional sampler is activated when:

1. the wind exceeds 0.5 $m\,s^{-1}$ (the operating thresholds of commercially available anemometers are variable – this is one of the areas of concern in the current programme);
2. the wind is blowing from the direction of the suspected source through an arc centred on the point of interest;
3. the wind remains within the selected arc for more than 10 seconds.

The sampler, as configured at the present time, has the following capabilities:

1. to sample when the wind is within the selected arc including the works under investigation. This may be one continuous episode or, as is most likely, several shorter episodes during the sampling period; and
2. to sample when the wind is in the major arc which does not include the point of interest.

It is not usually sufficient to use one sampler to investigate a particular works because of the possibility of other sources of the pollutants. Therefore, surveys usually include at least three sites at points around the circumference of the works as shown in Figure 13.6.

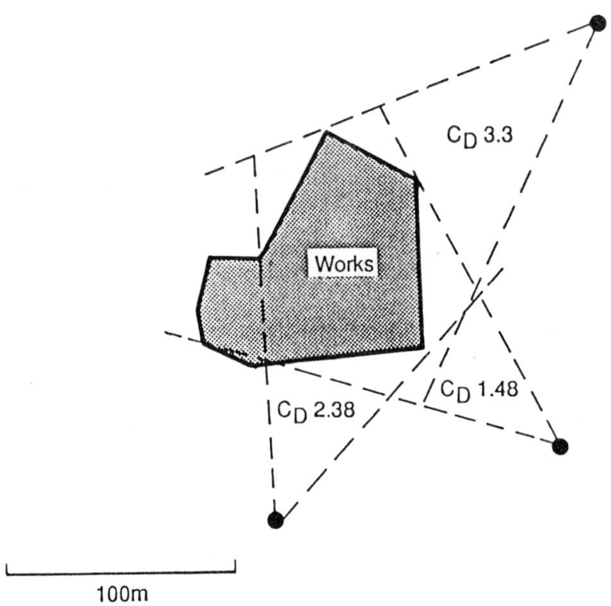

Figure 13.6 *Approximate sampling arcs and directional (C_D) lead concentrations ($\mu g\,m^{-3}$)*

Much of the current WSL programme has been concerned with the specification of the directional control system and in determining the performance characteristics of candidate anemometers and wind vanes. A second part of the current WSL programme is to provide guidelines on the siting of samplers. Perhaps the most important aspect of directional sampling is the interpretation of data; some aspects of this are presented below.

The final objective of the current WSL programme is to provide a specification for a standard method for directional sampling.

13.3.1 Determination of the repeatability of measurements

It should be appreciated that ambient field measurements cannot be expected to be absolutely reproducible if only because each instrument must be located far enough from neighbouring instruments to avoid local wind disturbance and therefore interference. Further, wind conditions (directions) close to ground level can be quite variable over short distances, especially at low windspeeds.

In order to draw up specifications for the proper functioning of an instrument suitable for making directional and background measurements of ambient air pollutants, experiments were carried out using several models of a commercially available directional M-type sampler (MD samplers). The main differences between the original WSL samplers and the commercially available samplers is in the meteorological equipment and the presentation.

13.3.1.1 Accuracy of switching between modes

Several instruments were set up at the same site (about 5 x 5 m area) in order to check the reproducibility of the electronic selection of the directional and background modes. Each instrument was set up with a 72° directional sampling angle and all were orientated towards the south west. Readings were obtained at weekly intervals of the directional and background sampling hours.

The coefficient of variation (COV = 100 x (standard deviation / mean)) for five instruments was plotted against the number of directional sampling hours during each period, see Figure 13.7. This plot shows that with very short directional sampling times (during the weekly sampling period) there is a large variation between the instruments in the recorded directional sampling hours. This suggests that for a 15% COV or less between instruments at least 6 hours of directional sampling is required and anything less should be rejected. With normal conditions it is quite usual with directional sampling for there to be less than 6 hours in the weekly directional mode and therefore this rejection is undesirable if sufficient data are to be obtained.

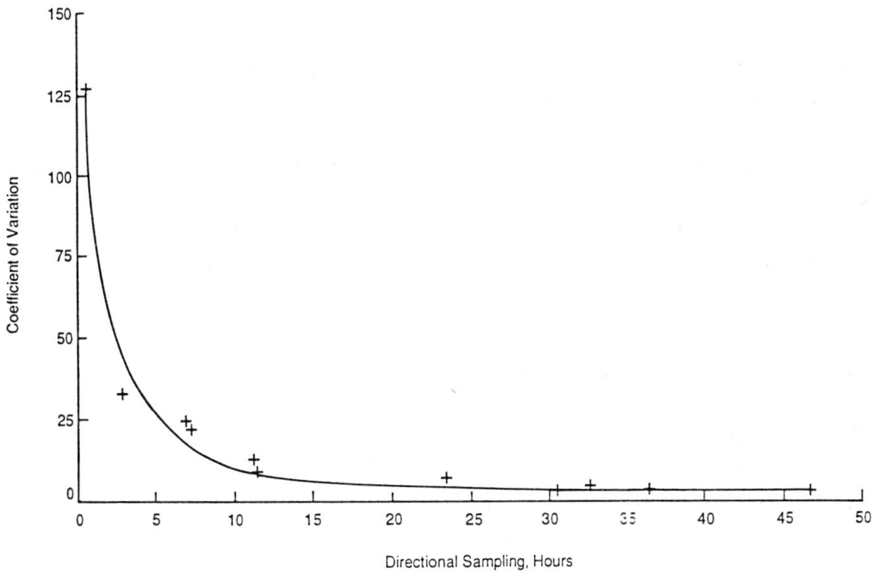

Figure 13.7 *Coefficient of variation of sampling time*

There are several possible reasons why there was this discrepancy between these instruments.

1. The wind vanes might not have been accurately aligned on the meteorological arm at the top of the masts (N on the vane furthest from the source).
2. The meteorological arm on each mast might not have been aligned correctly to the same orientation, i.e. SW or 225°.
3. The directional sampling angle might not have been set identically on each instrument.
4. There may have been a discrepancy in electronic switching in the control box.
5. The anemometer rotation threshold might not have been identical in all instruments.
6. The wind vanes might not have been indicating correctly.

Possibilities 1, 3, 4 and 6, were checked electronically and found to be satisfactory; 2, was measured at arm's length using a magnifying compass with a resolution of 2°. For 5, the anemometers were checked in the wind-tunnel at the laboratory and were found to have threshold of rotation windspeeds which varied from satisfactory to very unsatisfactory.

13.3.1.2 Trials with originally specified anemometers

Since it was known that the anemometers specified in the original WSL design were satisfactory, five control boxes were supplied to the MD instrument manufacturers for modification to accept this anemometer input. Five new anemometers of the original specification were purchased and new tests were carried out on the same site using the same test procedures as before but with the modified control boxes (Figure 13.8) and anemometers.

The results were encouraging in that a coefficient of variation (COV) of 15% or more was only exceeded for directional sampling times of 2 hours or less. Two hours in the directional mode is about 1 m^3 of sample volume (500 $l\,hr^{-1}$) and this has been used previously as the minimum quantity of aspirated air required for an acceptable sample. However there are occasions when the 15% COV is exceeded even with these modifications. It appears that this occurs when the windspeed is very low and the wind direction variable. Presently the instrument is designed to operate in the directional mode only when the windspeed exceeds 0.5 $m\,s^{-1}$. Therefore, it may be necessary to specify a somewhat higher threshold windspeed to obtain uniformity of sampling in low windspeed conditions.

13.3.1.3 Wind-tunnel measurements of anemometer thresholds

Following the results of the field trials reported in Figures 13.7 and 13.8, the manufacturers of the MD samplers supplied improved versions of their anemometers. These were tested in the small wind-tunnel at WSL and the windspeed thresholds were found to be acceptable.

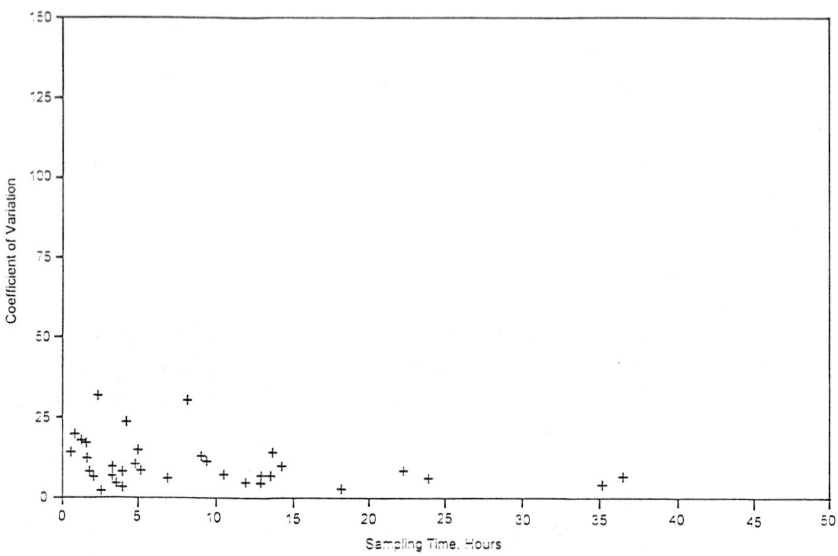

Figure 13.8 *Coefficient of variation of sampling time with improved anemometry*

13.3.2 Measurements close to an industrial source

Two samplers were set up at an industrial site. Each instrument was set to monitor directionally into a different sampling sector to assess emissions from each of two potential sources. The two sampling heads were set up at least 2 m apart in order to minimize interference.

Although the directional sectors were different, the background sectors overlapped but were not identical. However by calculating the total ambient suspended particulate pollutant concentration, C_t from the measurements made with the two instruments (C_b and C_d in equation (6)) and allowing for the fact that they were not sampling in exactly the same place the results showed in Table 13.2 are consistent. A t-test to determine if the mean of the

Table 13.2 Calculated total concentrations for two samplers at one site. Pollutant concentration $\mu g\, m^3$

Week	*Source A*	*Source B*	*Mean*
1	3.355	3.127	3.241
2	4.007	4.486	4.247
3	0.953	0.923	0.938
4	2.263	2.327	2.295
5	1.009	1.071	1.040
6	1.248	1.327	1.288
7	1.184	1.276	1.230
8	2.940	3.049	2.995
9	1.705	1.702	1.704
10	2.313	2.418	2.366
11	3.719	3.705	3.712
12	1.887	1.455	1.671
13	2.816	3.268	3.042
14	1.209	1.396	1.303
15	2.102	1.366	1.734
mean	2.181	2.193	
t-test			t = 0.158 (NS)*

* NS = not significant

differences between each pair of results differs significantly from zero shows the difference to be not significant at the 95% confidence level. These results were obtained without adjustment to the threshold windspeed and include both sampling and analytical errors.

Weeks 2, 12, 13 and 15 are not as consistent as results for other weeks; this emphasizes the need to carry out sampling over a fairly extended period in order to arrive at an overall judgement of the prevailing situation. Results in isolation or very limited surveys could give misleading results for both the nter and any regulatory authority wishing to draw conclusions exercise. These discrepancies might well have been due to low and windspeeds occurring during directional sampling periods or to the at the samplers were located about 2 m from a wall.

The overall calculated ambient concentrations for the two samplers, averaged over the 15 weekly sampling periods, were 2.18 and 2.19 $\mu g\,m^{-3}$ which agree well.

13.3.3 Optimum siting of samplers

In order to ensure that meaningful results are obtained from a survey using directional samplers, it is important that the samplers are sited to the best advantage with respect to the works under consideration. At the same time, local features which might interfere with the measurements should be avoided. In any field exercise it is difficult to satisfy these two basic requirements. This is due to the difficulty in finding an uninterrupted view of all the potential sources because of housing, roads, railways, other works or simply the general topography of the area. Therefore, it is not possible to specify arrays of sampling positions which would be applicable to all situations. Indeed every situation is different and is more or less unique. However, the following guidelines can be set and these should be followed as closely as the local topography and conditions will allow.

1. There should be a clear view of the whole potential source area.
2. The area in which the sampler is sited should be free of obstructions more than twice the height of the sampling head for 10 m from the sampler.
3. Preferably, there should be no second source of the same pollutant close by. If there is such a source, a rather more complicated array of samplers will be necessary to take the second source into account.
4. To measure fugitive emissions, as distinct from stack emissions, the sampler should be less than ten stack heights from the stack – about seven stack heights is preferable.
5. The sampler should be more than five building heights downwind of the tallest building at the source.
6. The angle subtended at the sampler by the linear aspect of the source should not be more than 70°. If, on setting up a site, it is found that this angle is too great a second sampler should be deployed in parallel so that the angle can be reduced.
7. It may be that only a small part of the works site has been identified as a source. Should this require only a small sampling angle, say 10 to 20°, it is better to move the sampler closer to the source in order to increase the sampling angle. Small sampling angles should be avoided because they lead to small directional sampling times and therefore reduce the reliability of the measurements.
8. If the site under investigation contains a number of emitters of the same

pollutant it is unlikely that setting up directional samplers to study one of those emitters would provide meaningful results because the other emitters of the same pollutant on the site would contribute the background concentration, C_b.

9. Avoid sampling at too great a distance from a small (physical size) source. The directional sampling angle will be too small and, if the angle is increased to give a greater proportion of directional sampling time, the directional component will contain a large element which should actually be included in to the background component and the results will not be representative of the real situation.

It is often very difficult to meet all these requirements because the location of the sites depends on many other factors such as site security, accessability and the provision of power supplies. It is better to select sites which are suitable from these practical aspects and then to check that the basic parameters subtended angle, distance from stack(s) and distance from buildings are within the required limits.

13.4 Interpretation of data

13.4.1 General

As stated in equation (6), the relationship between C_t, C_d and C_b can be expressed as follows:

$$C_t = PC_d + (1 - P)C_b$$

where P is the proportion of time for which there was directional sampling. The ratio C_d/C_b reflects the influence on local ambient concentrations of any source within the directional sampling sector. A value of C_d/C_b greater than unity would suggest the presence of a source within the directional arc. However, a value greater than unity could result, for example, from variations in meteorological factors such as windspeed. Thus a factor of two has been interpreted to mean that a source in the directional sector is making a significant contribution to ambient concentrations.

13.4.2 Assessing changes in works contribution using C_d and C_b

The works contribution (W_t) to the ambient concentration of a pollutant is given by:

$$W_t = P(C_d - C_b)$$

and therefore changes in the works contribution can be calculated directly for different periods. The use of this method of assessing the works contribution in the short term, for example for a single sample, is not recommended although it is the preferred method over longer periods.

The ratio C_d/C_b is also a valid estimator of the contribution of the works to measured ambient concentrations. Therefore the ratio of C_d/C_b for one period to that of another period can be taken as a measure of changes in the works contribution. The period chosen must be relatively long to take into account the normal variability of meteorological conditions. In the case of the directional sampling surveys which have been carried out in the past, a 3-month period was taken as the minimum length of time required to take into account the uncertainties associated with varying meteorological conditions.

13.5 Advantages and disadvantages of different measurement techniques

13.5.1 Measurement of ambient pollutant concentration, C_t

The measurement of C_t alone provides data for comparison with air quality standards or other relevant data. It does not, on its own, provide any information on the source of the measured pollutant or the magnitude of emissions from nearby sources.

13.5.2 Measurement of C_t *and* P

The measurement of C_t and P provides:

1. a direct measurement for comparison with air quality standards;
2. an assessment of the changes in emission from a particular source under certain circumstances;
3. a means of predicting future concentrations on the basis of wind direction statistics;
4. a means of calculating the reduction in emissions necessary to meet air quality standards.

13.5.3 Measurement of C_t, C_b *and* P

As well as having the advantages of the measurement of C_t and P the additional measurement of C_b allows the calculation of the magnitude of the contribution of a source (W_t) to the measured ambient concentration. C_d can also be calculated if required.

13.5.4 Measurement of C_d, C_b *and* P

The major advantage in measuring C_d and C_b is that the ratio C_d/C_b is a sensitive indicator of whether a works is making a significant contribution of the measured pollutant to ambient concentrations. Otherwise the advantages of measuring C_d and C_b are similar to those for measuring C_t and C_b.

13.5.5 Measurement of C_t and C_d to calculate C_b.

It might be assumed that a measurement of C_t and C_d and P would be sufficient to calculate C_b. This method has the disadvantage that C_b is usually very small compared to C_t and C_d and therefore any errors associated with the measurement of these two larger quantities will be emphasized in a subsequent calculation of C_b. In fact it is frequently possible, using this method, to calculate negative values for C_b which is obviously absurd.

13.6 Recommended use of methods

1. For assessing the significance of the works contribution to ambient concentrations – directional sampling to determine C_d/C_b.
2. For measuring the works contribution – Directional sampling to determine $(C_t - C_b)$
3. Long-term assessment of compliance with air quality standards, assessment of the works' contribution and prediction of reduction in emissions required to meet air quality standards – measurement of the ambient concentration of the pollutant and the proportion of time the wind remains in the appropriate sector.

In practice, 2 would appear to be the preferred method of measurement for most circumstances since it performs the functions required by 3 and it can be used to calculate the ratio C_d/C_b in 1.

13.7 References

Clayton P and Ogden C (1991). A Method of Predicting Trends in Ambient Pollutant Concentrations in the Vicinity of a Source. Stevenage: Warren Spring Laboratory, Report No. LR 816(PA).

Barnett J L, Clayton P and Davis B J (1987). Fugitive Emissions - A Directional Sampler for Particulates. Stevenage: Warren Spring Laboratory, 1987.

14

Stack sampling of toxic organic micropollutants

P Coleman

14.1 Abstract

The development of a specification and protocol for the determination of organic micropollutant levels in flue gas is described. Validation work was undertaken in which three sampler designs were compared at a single municipal waste incinerator. The designs tested were the EPA modified method 5, a water-cooled probe and a high volume sampler.

The measured mean concentrations varied between 5 and 25 ng TEQ m^{-3} for dioxins. The EPA modified method 5 and the water-cooled probe agreed to within 10% of the joint mean. The high volume sampler appeared to undersample.

Significant losses occurred when a flexible link was used between the probe and the filter of the EPA sampler. As a result a trial was carried out comparing the EPA modified method 5 equipment without a flexible link with a water-cooled probe with a flexible link. However losses were still large.

14.2 Introduction

This paper describes work done to establish improved procedures for extractive sampling of toxic organic micropollutants (TOMPs) from combustion and other waste gases in flues. TOMPs are compounds of very high toxicity which are present in environmental samples at low concentrations, such as the polychlorinated dibenzo-p-dioxins (PCDDs), polychlorinated dibenzofurans (PCDFs), polychlorinated biphenyls (PCBs) and polycyclic aromatic hydrocarbons (PAHs). The generic structures of PCDDs and PCDFs are shown in Figure 14.1.

There are many possible analytes of interest in these groups of compounds, for example, there are 75 PCDDs and 135 PCDFs; hence, it is not practical to routinely analyse each of these compounds. The selection of compounds of concern will depend on the aims of the sampling study but could be for legislative or toxicological reasons (WHO, 1987). For example within the PCDDs and PCDFs classifications there are seventeen 2,3,7,8-substituted compounds which are specified in regulations because they are

(a)

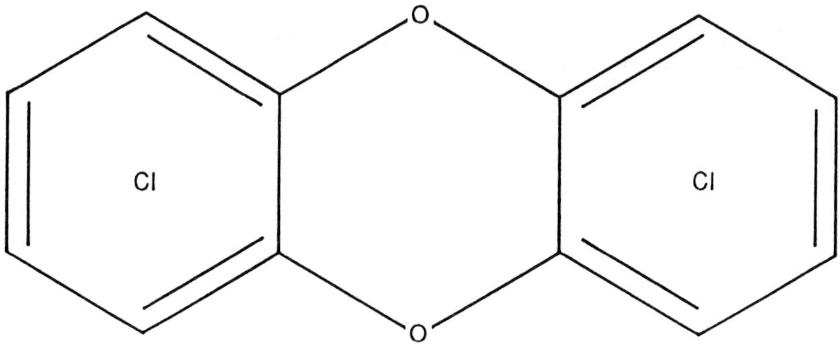

(b)

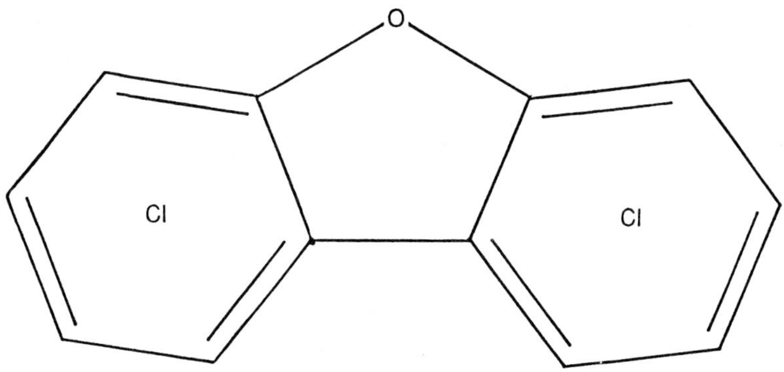

Figure 14.1 *The generic structures of (a) polychlorinated-dibenzo-p-dioxin and (b) polychlorinated dibenzofurans*

regarded as the congeners with the most significant environmental toxicity (EC, 1992, HMSO, 1992a,b,c,d).

These compounds have similar physical properties; they have limited volatility at room temperature and high volatility at typical flue gas temperatures (Eltzer and Hites, 1989; Rordorf, 1989). The dioxins, furans and PCBs are stable when adsorbed on particles at ambient temperature whereas the PAHs will degrade if exposed to light. However dioxins appear to photodegrade in the gas phase.

14.3 Sampler selection criteria and designs

14.3.1 Introduction

The purpose of the project was to develop a list of design criteria which would be sufficient to design a sampler of the required performance standard. Initially the processes involved in TOMPs sampling were investigated, then sampler designs were identified which had been used elsewhere for TOMPs sampling. Several features of sampler design and operation were identified which were thought to enable a representative sample to be obtained; these are discussed below.

14.3.2 Sampler selection criteria

14.3.2.1 Materials of construction

Surfaces which are reactive and either encourage the formation or the degradation of dioxins should not be used. Dioxin formation occurs rapidly on copper chloride and some other metal chlorides (Stieglitz and Vogg, 1988; Nestrick *et al.*, 1987); these active sites occur in most combustion fly ashes. Where steel and copper surfaces are present in samplers, these may be corroded by hydrogen chloride present in combustion flue gases to provide new active sites. In order to avoid this problem glass, quartz, PTFE or titanium should be used. It is also important that the dioxins are not irreversibly adsorbed into the sampler, hence, the use of sealing greases, rubber gaskets, etc. is discouraged.

14.3.2.2 Ability to sample isokinetically

Isokinetic sampling ensures that the sample obtained has a particle size distribution which is representative of that in the duct (Vincent, 1989). TOMPs are found in both the gas and solid phases, the proportion depending on the vapour pressure and particle concentration. The particle phase concentration of dioxins has been shown to increase with decreasing particle size in arrested electrostatic precipitator ash (Mitchell, 1989). So it is likely that the concentration of dioxins and other TOMPs on the particles in the flue will behave in a qualitatively similar way. This is perhaps caused both by the increased specific surface area of small particles and the increased specific concentration of catalytically active sites on the fly ash. Hence it is important to obtain a representative particle sample.

14.3.2.3 Temperature control

Where heating is used to prevent condensation in the sampler it should be carefully controlled. The rates of the dioxin formation reactions mentioned in Section 14.2.2.1 increase exponentially at temperatures above ambient (Hagenmaier *et al.*, 1988) up to around 350°C although below 150°C the quantities formed will generally be insignificant. However in waste incinerators fitted with advanced pollution control devices the emission concentration can be as low as 0.01 ng toxic equivalent m^{-3} (ng TEQ m^{-3}).

Hence even minute quantities of dioxin formed in the sampler becomes significant and so any heating may be significant. Throughout this paper, concentrations are expressed corrected to dry gas at 0°C, 1 atmosphere, 11% oxygen.

The activity of fly ashes varies; in some cases, at temperatures over 120°C, significant reactivity has been noticed. To avoid any sampling artefact the temperature of the sample should be maintained below 120°C, if the dew point of the gas allows it, and in any case below 130°C. This condition would mean that sampler designs in which the filter is held in the stack should not be used except at those sites where the stack temperature is always sufficiently low.

14.3.2.4 Leak-free design

It is important to avoid leaks for two reasons: firstly, the need to know accurately the volume of gas sampled from the duct; secondly, the sampler will not be operated isokinetically if the volume flow rate measured through the sampler is actually the volume flow rate through the nozzle plus an unknown leak rate. The sampler should be capable of being sealed to allow a leak rate of less than 5% of the expected flow at the maximum vacuum used.

14.3.2.5 Sufficient volume flow rate

Before a sampling campaign, the detection limit needed to be achieved in order to achieve the objectives of the study should be known. The analytical detection limit can be estimated in advance, hence, the required sample volume can then be estimated. The sampler design must allow a sufficiently high volume flow rate to achieve the desired sample volume in a reasonable time.

As an example the German emission regulations require a maximum detection limit of 5 pg m^{-3} per isomer. If the analytical detection limit is 50 pg per isomer, then 10 m^3 need to be sampled. A typical sampler operating at 1.2 $m^3 hr^{-1}$ will therefore need at least 8.5 hours of sampling time to ensure this detection limit is exceeded.

14.3.2.6 Reagent capacity

The sampler must be designed to efficiently retain the sampled TOMPs for later analysis. The filter and adsorbents need to be adequately sized for the expected sample volume and analyte concentration. However, the performance of the adsorbents commonly used are not well known and the quantities and types of adsorbents used are selected through experience.

The addition of known quantities of isotope-labelled compounds from within the group of interest, known as sampling spikes, allows losses during sampling and storage to be detected if not quantified.

14.3.3 Sampler designs

There are a wide range of samplers in use throughout the world from which 17 different designs were identified. Three of these were selected which appeared to meet the above criteria and represented different solutions to the problem.

14.3.3.1 US EPA modified method 5

The Andersen Universal sampler (Figure 14.2) is used by the majority of UK dioxin sampling teams. This sampler is made to the requirements of the US EPA Method 23 (US EPA, 1991), previously referred to as modified method 5. This is a development for dioxin sampling of method 5 (US EPA, 1985) the US particle sampling method.

The gas is sampled through a stainless steel nozzle and a glass-lined heated probe fitted with a pitot and thermocouple. The gas then passes through a glass filter holder in which a quartz fibre filter is supported on a PTFE covered gauze. The filter holder is in a heated box to prevent condensation. After leaving the heated box, the gases enter a glass condenser followed by a vessel containing XAD 2 resin. This resin is widely used as a sorbent for TOMPs however little data is available on its performance. The condenser and resin trap are both contained in a cooling water jacket. The condensate flows through the resin trap and then into a round bottom flask whilst the gas passes through a silica gel drying train to the gas control unit. This contains a pump, a gas meter and an orifice plate flow meter. The typical gas flow rate through this sampler is 1.2 $m^3 hr^{-1}$.

14.3.3.2 Water-cooled probe

A sampler with this type of probe has been developed by Stellan Marklund at the University of Umeå in Sweden (Hagenmaier and Kraft, 1987) from the impinger, polyurethane foam adsorber method of Brenner *et al.* (1984). The sampled gases pass through a glass-lined water jacketed probe (Figure 14.3) where the gas temperature falls rapidly; the resulting condensate runs out of the probe into an impinger. The particles in the sampled gas act as condensation nuclei and are collected with the condensate. The gas then flows through a second impinger containing 2-ethoxyethanol. The two impingers are held in an ice bath. Next, the gas passes through a polyurethane foam (PUF) plug to remove the gas phase TOMPs. An aerosol filter is used to trap any condensation aerosol formed in the system. Finally

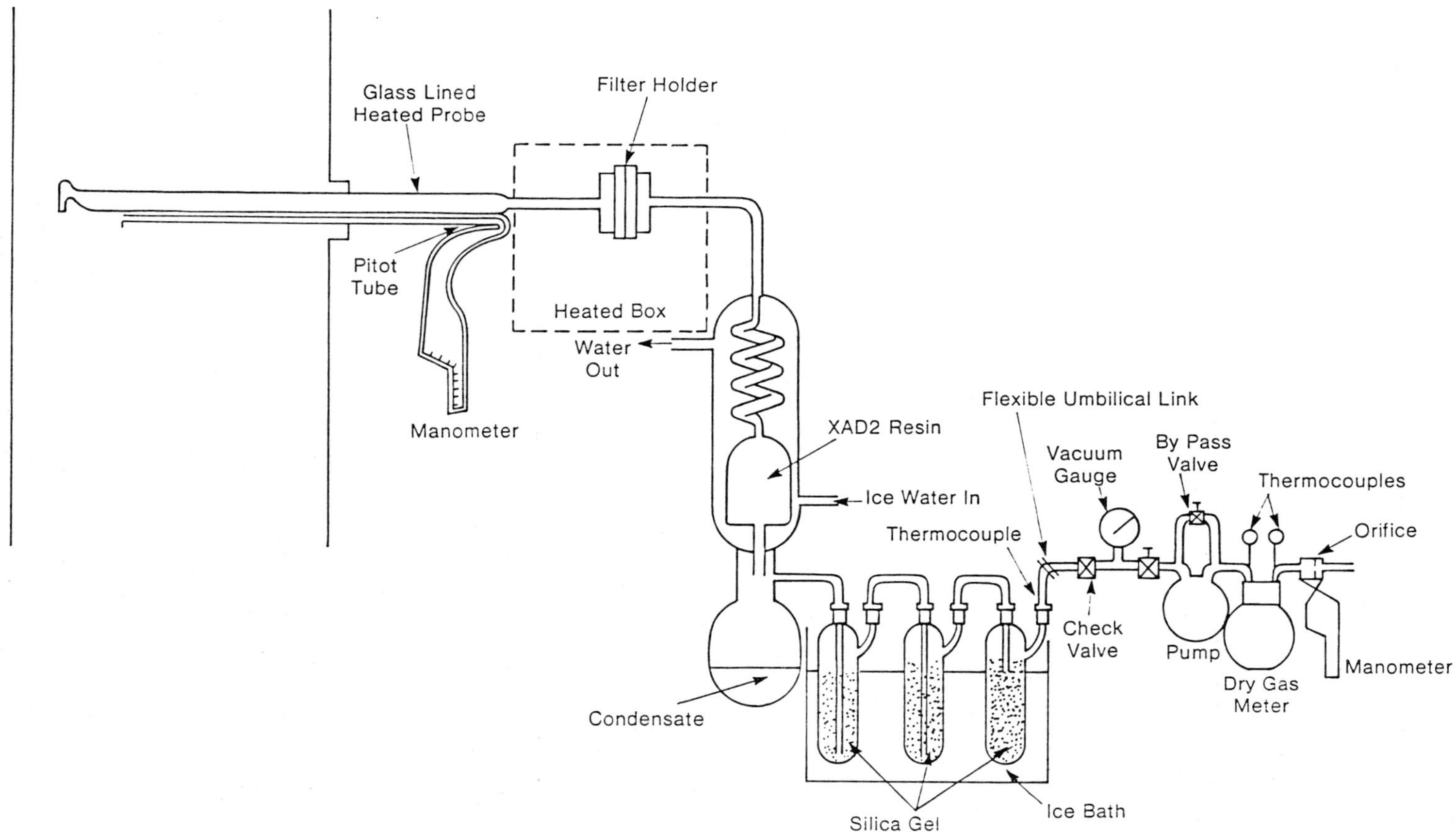

Figure 14.2 *EPA modified method V sampler for dioxin and dibenzofuran*

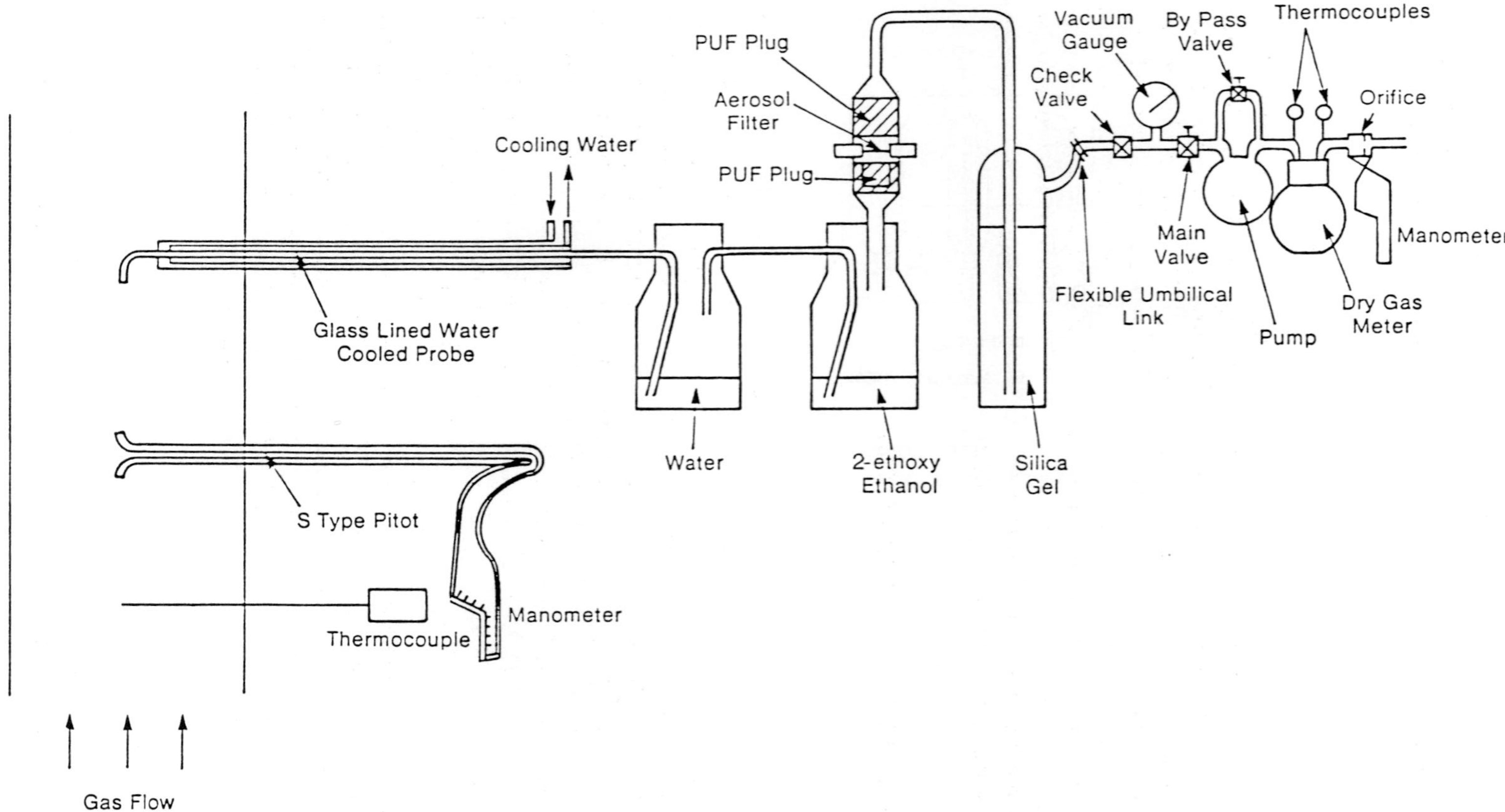

Figure 14.3 *Water-cooled probe*

the gas passes through a backup PUF plug in case there has been any volatilization from the first PUF plug or the filter. The same flow control unit is used as for the Andersen sampler hence the same volume flow rate is used.

14.3.3.3 High volume sampler

The third method used was developed at Warren Spring Laboratory (WSL) on behalf of the Department of Environment (Figure 14.4). It samples at five to ten times the flow rate of the other two designs. A glass lined heated probe is used followed by a PTFE coated cyclone and a PTFE coated filter holder containing a glass wool filter pack. The cyclone and filter are contained in a heated box. The total flow can be varied to ensure that sampling is carried out isokinetically. Following the filter a constant volume flow rate side stream is taken off to a condenser and resin trap analogous to but smaller than that described for the Andersen sampler. The gas then passes through a drying train, a pump and a gas meter.

14.4 Comparative sampling trials

14.4.1 Initial trial

In November 1990 a sampling trial was held at a UK Municipal Solid Waste Incinerator where the three samplers were compared for five runs. The samplers were operated simultaneously for 4-hour periods. Changes in gas composition across the three sampling points were taken into account by interchanging the samplers at regular intervals. Leak checking was carried out before each sampler was replaced in the duct. The sampler fractions were analysed for individual 2,3,7,8-substituted PCDDs and PCDFs, to provide information on the isomer distribution within the sampler.

In addition, the EPA modified method 5 sampler was compared with the water-cooled probe for a further two runs in which a further modified method 5 sampler was equipped with a flexible link between the probe and the filter. The normal operation of the modified method 5 sampler requires the use of a rail or support to enable the probe to be easily traversed across the duct, however, this requires a 3 m gantry. At many existing plants this is not possible because of the confined space around the duct. It is also impractical to operate the sampler vertically preventing the sampler being used for vertical sampling in a horizontal duct. The flexible link enables sampling to be carried out in these circumstances.

14.4.2 Trial with altered equipment

A further trial was then held at a municipal solid waste incinerator in which the water-cooled probe was used with a flexible link. A plain PTFE tube was

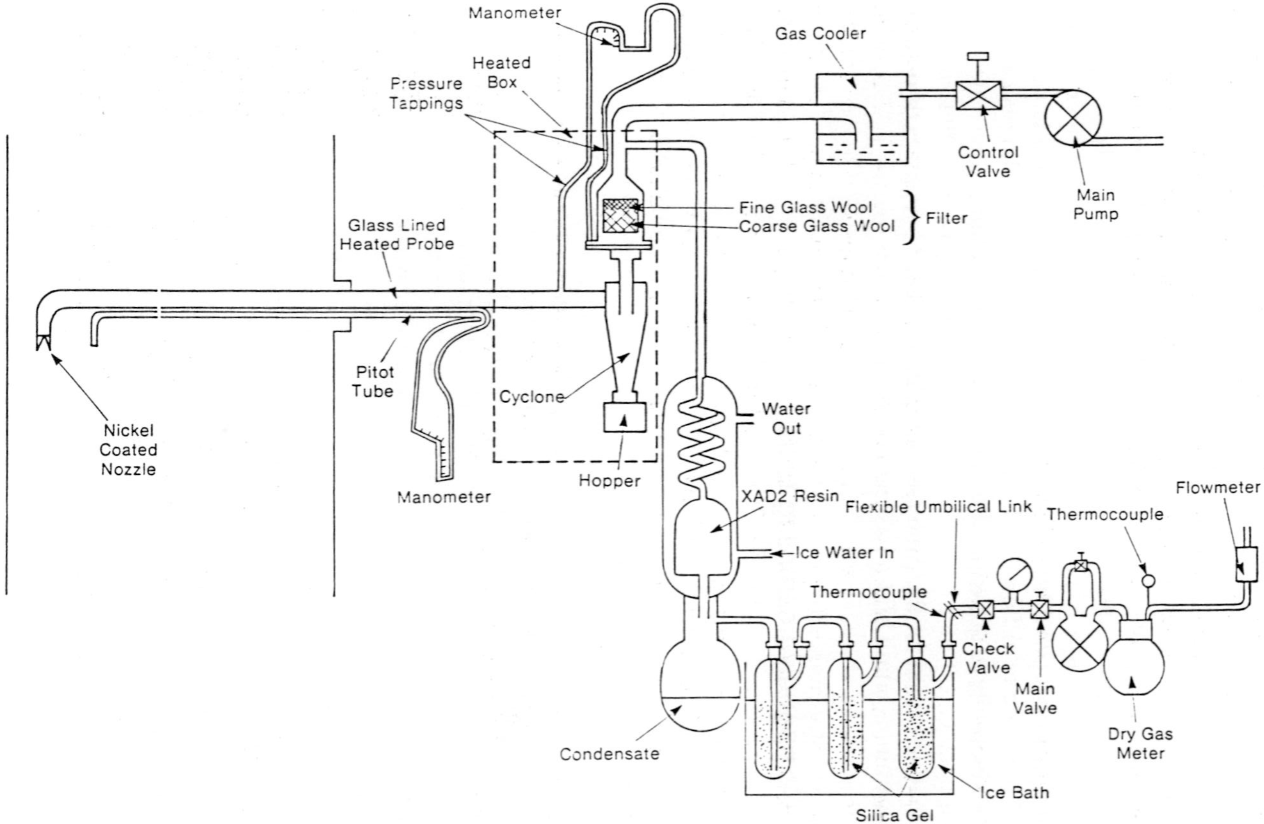

Figure 14.4 *Modified BCURA sampler for dioxin and dibenzofuran*

fitted to the end of the glass probe liner with a PTFE fitting. The end of the tube was sealed within the first impinger. By using this approach it was hoped that the sample loss problems found with the heated flexible link needed for the EPA could be avoided.

The EPA modified method 5 sampler was used as a reference method as this had compared well with the water-cooled probe in the earlier trial. Three 4-hour tests were carried out following the same procedure as had been used in the earlier trial. However the volume of acetone and toluene used to wash the water-cooled probe were doubled from 250 ml of each to 500 ml.

14.5 Results from the initial trial

14.5.1 Overall concentration

The results obtained are shown in Figure 14.5. There is good agreement between the water-cooled probe and the EPA modified method 5, with results agreeing to within 9% of their mean, although the water-cooled probe is consistently higher (Figure 14.6(a)). The high volume sampler generally reported lower concentrations than the others (Figure 14.6(b)). When the flexible link is added to the EPA modified method 5 sampler there is a significant loss of sample, see run 8 Figure 14.5. It is interesting to note that the variation in emission between different runs is greater than the variation between these two samplers even though the incinerator unit tested was in steady continuous operation.

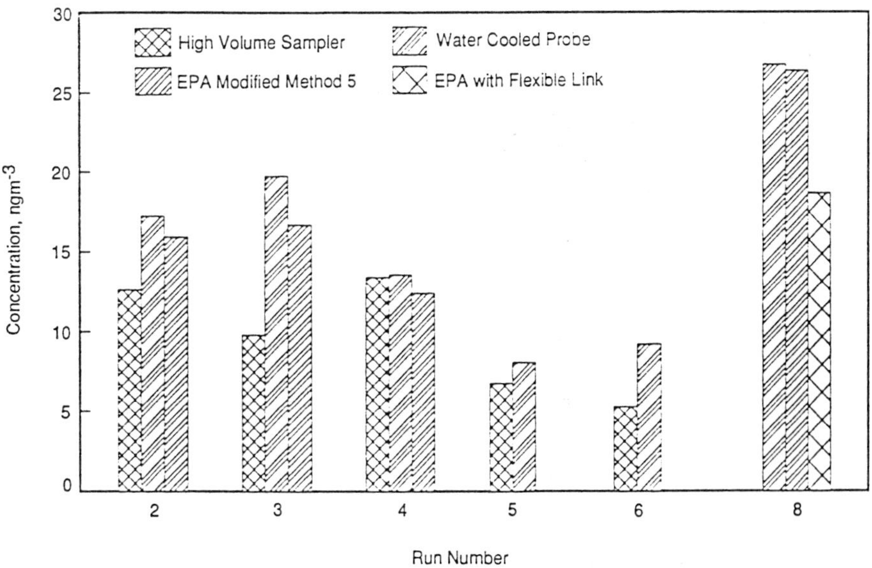

Figure 14.5 *Results from trial comparing three designs of sampler. Concentration in ng(TEQ)m^{-3} in dry air at standard temperature and pressure, containing 11% oxygen*

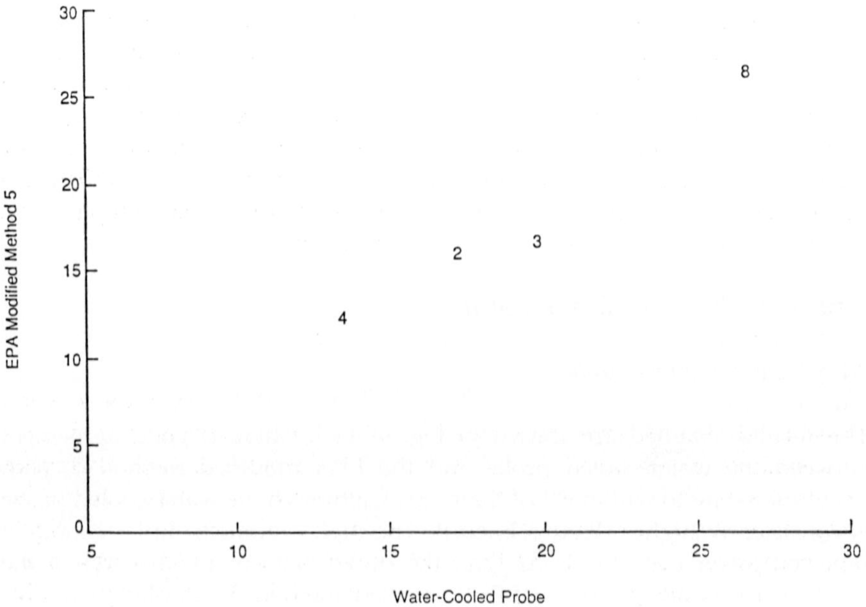

Figure 14.6a *Comparison of results from the EPA sampler with those from the water-cooled probe. Numbers are run results, axes are scaled in ng(TEQ)m^{-3}*

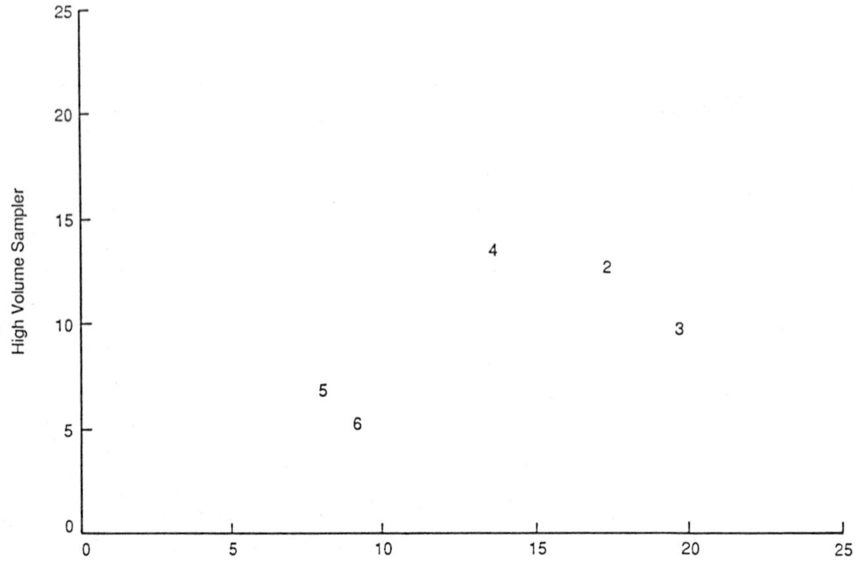

Figure 14.6b *Comparison of results from the high volume sampler with those from the water-cooled probe. Numbers are run results, axes are scaled in ng(TEQ)m^{-3}*

14.5.2 Isomer profile

As has been mentioned previously, one of the potential problems with TOMPs sampling is that reactions can take place in the sampled fly ash which may corrupt the sample. This can be checked by comparing the relative amounts of the 17 dioxin and furan isomers collected by each sampler (Figure 14.7). The results indicated good agreement except for octachloro dibenzofuran for which analytical problems occurred. There is reasonable consistency in the isomer profiles both between samplers and between runs when normalized for concentration.

14.5.3 Distribution of sample within the sampler

The individual fractions of each sampler were analysed separately where possible. The overall isomer distribution of the sample between each fraction of a sampler is discussed below. There are differences in the distribution between isomers; in general the more chlorinated the isomer the higher proportion is found near to the nozzle of the sampler. For example Figure 14.8 compares the percentage of the 17 isomers found in each compartment of the water-cooled probe. These distributions are likely to change at other incinerators with different dioxin concentrations, flue temperatures and abatement equipment.

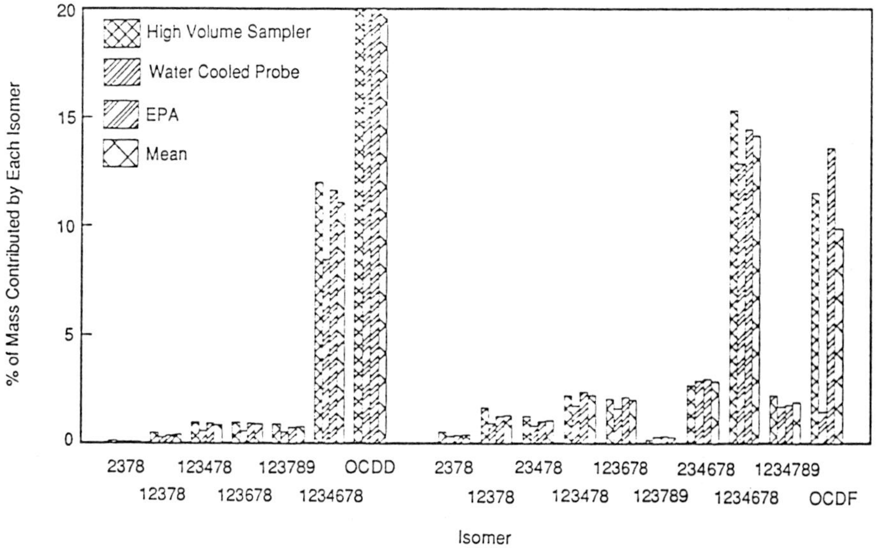

Figure 14.7 *Average isomer profile over all runs. The OCDD results were: HVS 45%, EPA 65% and WCP 45%*

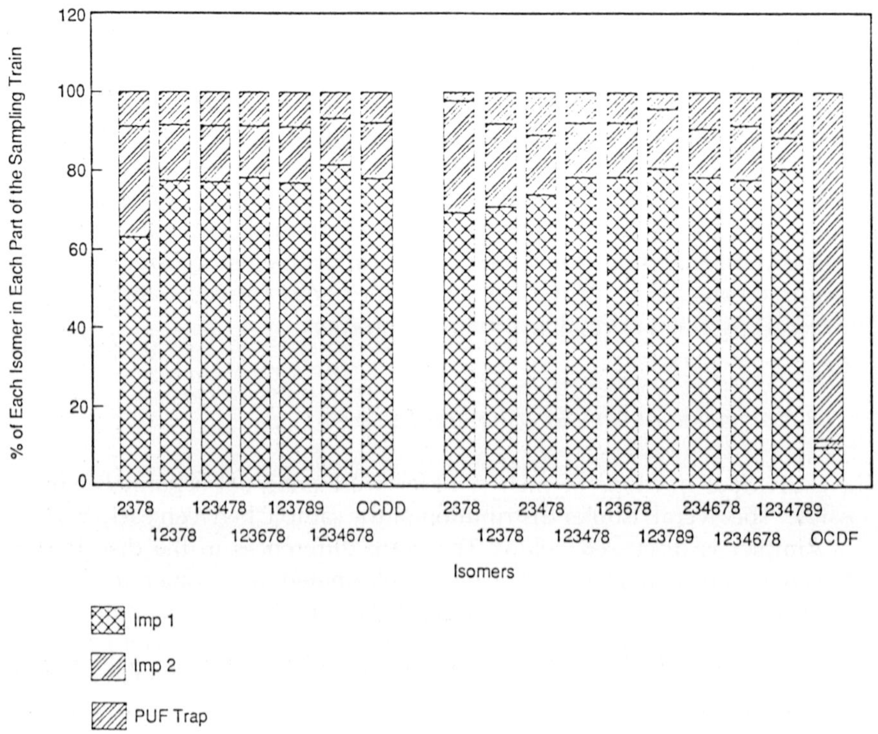

Figure 14.8 *PCDD/F distribution in each fraction for the water-cooled probe, run 2*

14.5.3.1 Water-cooled probe

The distribution of the toxic equivalent between the three sample fractions; impinger 1 containing the probe washings, impinger 2 and the PUF plug and aerosol filter is shown in Figure 14.9. The majority of the sample, 80%, is found in the first impinger, around 14% in the second impinger and the remainder in the PUF trap.

14.5.3.2 EPA modified method 5

The mean distribution of the sample through the sampler is shown in Figure 14.10. In this case the majority of the sample is caught in the resin trap. The condensate and toluene washings contain very little sample and appear in the figure as horizontal lines.

14.5.3.3 High volume sampler

The majority of the sample, expressed as concentration, was found in the sub-sample (Figure 14.11). This consists of the resin trap and condensate. The

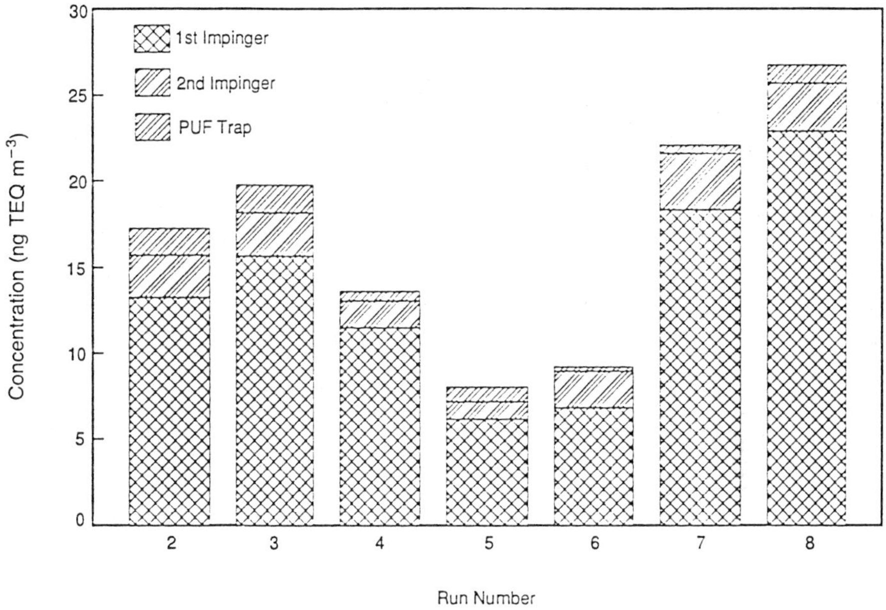

Figure 14.9 *Distribution of sample between fractions for the water-cooled probe*

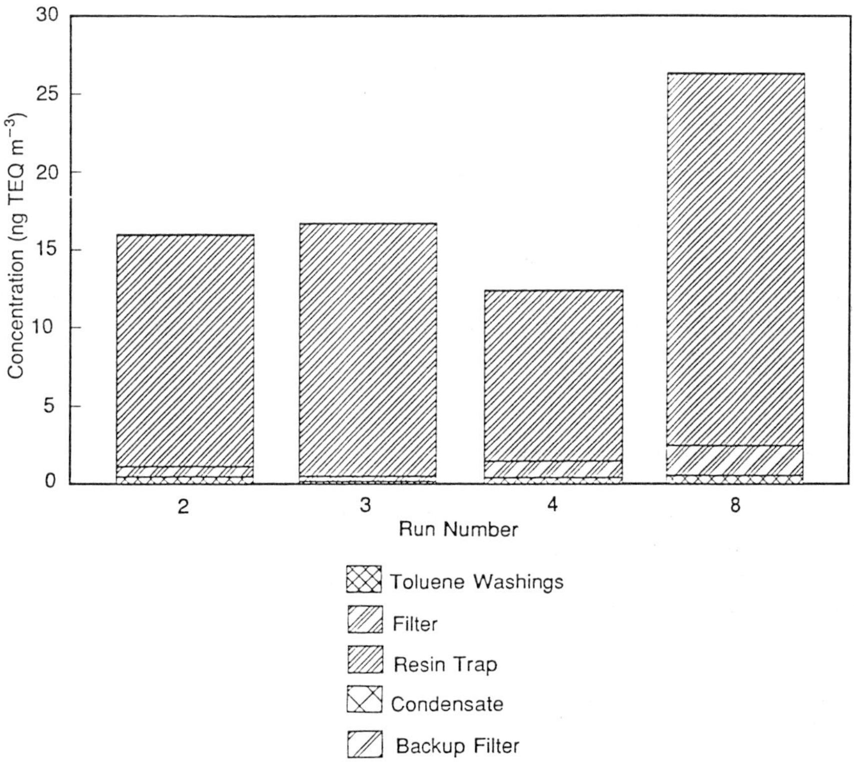

Figure 14.10 *Distribution of sample between fractions for the EPA modified method 5 sampler*

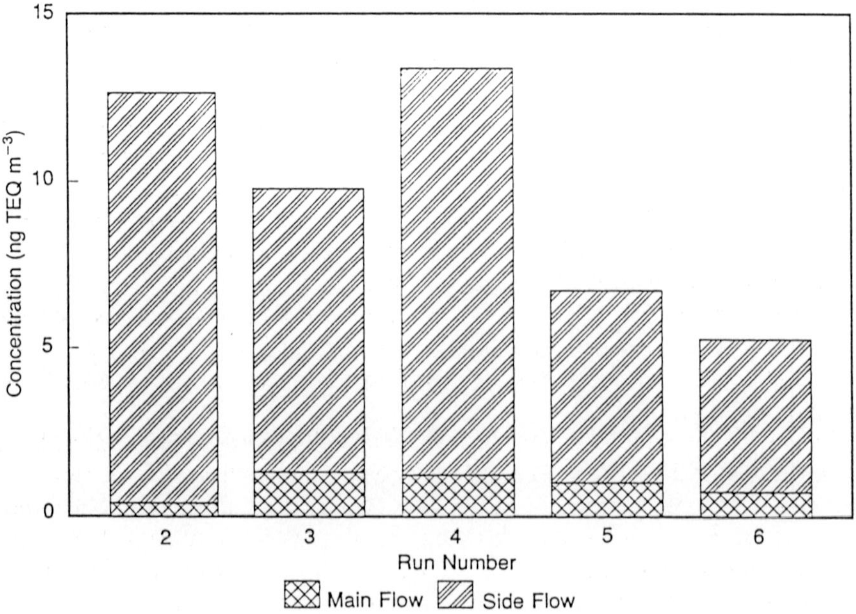

Figure 14.11 *Distribution of sample between fractions for the high volume sampler*

main stream consisting of the probe washings, hopper contents and filter, contains but a small fraction of the sample.

14.6 Results from the trial with altered equipment

This trial was based on the assumption that the EPA modified method 5 sampler would provide a sample of 100%. The results from the three tests with the water-cooled probe were –16%, +16% and –35% of the toxic equivalent collected by the EPA modified method 5. .

14.7 Discussion

The agreement between the water-cooled probe and the EPA modified method 5 was much better than suggested by the results of comparative sampling trials carried out elsewhere (Hagenmaier and Kraft, 1987). Figure 14.6(a) shows that the ratio between these two samplers does not increase consistently with concentration. The fact that the isomer profile in the samplers did not vary suggests either no dioxin formation was occurring in any of the samplers or that the formation was occurring evenly across the isomers.

If dioxin formation was occurring in the samplers, it might be expected to occur to a greater degree in the two samplers with heated filters, however, both the EPA modified method 5 and the high volume samples gave lower results. Hence significant dioxin formation did not appear to occur in this case, perhaps because the sampled particles were unreactive.

One effect of high filter temperatures may have been to volatilize the dioxins from the filter. The resin trap and the side stream in these two samplers contain the highest proportion of the sample. This was also noticeable in the behaviour of the sampling spike. The majority of the recovered ^{13}C-labelled material was found in the resin trap of the EPA modified method 5 sampler. Again in the high volume sampler the side stream contains the majority of the recovered spike when the recoveries are weighted for volume throughput. However in the water-cooled probe system the spike was mainly found in the first impinger. Hence the spike recovery in the three samplers is generally similar to the distribution of the sample.

The losses when the flexible link was used with the EPA modified method 5 sampler in the initial trial may be explained by two factors. The retention of particles in the corrugated PTFE inner surface of the flexible link and secondly it is possible that, as the flexible link is heated to avoid condensation, degradation reactions could occur on the captured fly ash.

The results of the second trial showed that despite the alternative design of flexible link losses could still occur although not consistently. This suggests that the main reason for losses of sample when flexible links are used is incomplete cleaning. Degradation reactions are unlikely to have taken place in an unheated tube.

If flexible links are to be used, and it is clear that at many existing sites there is no alternative, then cleaning procedures must be extremely thorough. Our present state of knowledge suggests that with the cleaning procedures we have investigated, significant losses can occur. However, because we have insufficient results to demonstrate that these losses are consistent, it is not possible to correct the results with a 'factor'.

We would recommend that where opportunities arise, such as in the construction of new plants, it is essential that the sampling platform is specified to enable dioxin sampling to take place without a flexible link.

14.8 Future work

A third trial took place in February 1992 at which other sampling teams were invited to compare methodology and to encourage the spread of good measurement practice in other sampling teams. This was completed successfully despite the late withdrawal of two potential participants. The results are at present being analysed.

A report is being written containing full details of the three trials carried out and the background to dioxin sampler design and operation. In addition two additional reports are being written. One will discuss the design of dioxin samplers and provide guidelines as to how to select a suitable design. The other will deal with good measurement practice.

14.9 Conclusions

1. In the range 5 to 25 ng TEQ m^{-3} the three samplers tested agree to within 40% of the mean value.
2. Two of the samplers agree to within 9% of their mean.
3. The isomer profile did not change significantly between samplers.
4. It is not necessary to specify in detail the design of a sampler for valid measurements to be made so long as the design meets certain basic criteria.
5. The use of a flexible link between the probe and the filter is to be discouraged until more information is known as to the source and consistency of losses when it is used.

14.10 References

Brenner K S, Mäder H, Steverle H, Heinrich G and Womann H (1984). Dioxin analysis in stack emission and in the wash water circuit during high temperature incineration of chlorine-containing industrial wastes. *Bull. Environ. Contam. Toxicol.*, **33**, 153-162.

EC 1992, Proposal for a Council Directive on the incineration of hazardous waste. COM (92) 9 Official Journal Reference C130 of 21 May 1992.

Eltzer B D and R A Hites (1988). Vapour pressures of chlorinated dioxins and dibenzofurans. *Environmental Science and Technology*, **22**, 1362-1364.

Hagenmaier H and Kraft M (1987). Studies towards validated sampling of PCDDs and PCDFs in stack gas. Municipal Waste Incineration Workshop, Environment Canada, Montreal, Quebec, Canada October 1-2 1987.

Hagenmaier H, Kraft M, Haag R and Brunner H (1988). Primary and secondary measures for minimising dioxin emissions from municipal waste incinerators. In: Brown A, Evemy P and Ferrero (eds) *Energy Recovery through Waste Combustion*, Elsevier Applied Science, ISBN 1-85166-285-5.

HMSO (1992a). Chief Inspector's Guidance to Inspectors, Process Guidance Note, IPR 5/1. Environmental Protection Act 1990, Waste Disposal and Recycling, Merchant and In House Chemical Waste Incineration.

HMSO (1992b). Chief Inspector's Guidance to Inspectors, Process Guidance Note, IPR 5/2. Environmental Protection Act 1990, Waste Disposal and Recycling, Clinical Waste Incineration.

HMSO (1992c). Chief Inspector's Guidance to Inspectors, Process Guidance Note, IPR 5/3. Environmental Protection Act 1990, Waste Disposal and Recycling, Municipal Waste Incineration.

HMSO (1992d). Chief Inspector's Guidance to Inspectors, Process Guidance Note, IPR 5/11. Environmental Protection Act 1990, Waste Disposal and Recycling, Sewage Sludge Incineration.

Mitchell D J (1989). An Evaluation of the Effect of Combustion Optimisation on the Size/Composition Distribution of Heavy Metals and Dioxins in Arrested Municipal Solid Waste (MSW) Incinerator Fly Ash. Stevenage: Warren Spring Laboratory, LR 815.

Nestrick T J, Lamparski L L and Crummett W B (1987). Thermolytic surface reaction of benzene and iron (III) chloride to form chlorinated dibenzo-p-dioxins and dibenzofurans. *Chemosphere,* **16**, No 4, 777-790.

Rordorf B F (1989). Prediction of vapour pressures, boiling points and enthalpies of fusion for twenty-nine halogenated dibenzo-p-dioxins and fifty-five dibenzofurans by a vapour pressure correlation method. *Chemosphere,* **18**, No 1-6, 783-788.

Stieglitz L and Vogg H (1988). Formation and decomposition of polychlorodibenzo-dioxins and furans in the fly ash of waste incineration. GIT-Supplement Umweltanalytik und Umweltschultz, 2/88, 4-11.

US EPA (1991), Method 23 - Determination of Polychlorinated Dibenzo-p-dioxins and Polychlorinated Dibenzofurans from Stationary Sources.

US EPA (1985), Method 5 Determination of Particulate Emissions from Stationary Sources. 40 Code of Federal Register, Chapter 1 Part 60, Appendix A.

Vincent J H (1989). *Aerosol Sampling, Science and Practice.* John Wiley and Sons, ISBN 0471 921720.

WHO 1987, Air Quality Guidelines for Europe: European Series No 13 Copenhagen: ISBN 92 890 1114-9.

— 15 —

Analytical methodology for the isomer specific tetra- through octa-determination of polychlorinated dibenzo-p-dioxins (PCDDs) and polychlorinated dibenzo-p-furans (PCDFs) in flue gas samples generated from coal burning

A R Fernandes, D S Wallace and B R Bushby

15.1 Abstract

A validated analytical method for the determination of all of the 2,3,7,8-chloro substituted dibenzo-p-dioxins and dibenzo-p-furans is described. The method has been successfully used in measurements involving combustion related matrices.

15.2 Introduction

The accurate determination of polychlorinated dibenzo-p-dioxins (PCDDs) and polychlorinated dibenzo-p-furans (PCDFs) at sub-nanogram levels is a demanding analytical task and depends to a great extent on the sample matrix, the levels of analyte present, and the techniques used to isolate and purify the extracted analytes.

Until recently, the requirements at Warren Spring Laboratory were generally for the analysis of arrested fly-ash samples and a good analytical method was developed towards this end (Bushby, 1989). This method allowed the measurement of the 2,3,7,8-substituted tetra-congeners as well as tetra-congener totals. The participation of the Laboratory in the Bureau of Community Reference (BCR) intercomparisons with good results was a measure of the reliability of the method used.

However, as emerging pollution abatement legislation and control technologies reduced analyte levels, and sample types changed, the method was found to be increasingly inadequate for some determinations. More recent sample types have included particulate filters and adsorbent phases for volatiles from studies of incinerator flue gases. Feedstock for the incinerators also varied from clinical waste and dried sewage sludge to various types of coal. Of these, the samples that posed the most severe problems to the determinations of PCDDs and PCDFs were those derived from the incineration of bituminous coal.

The inability of the method to clean up these types of samples is demonstrated in Figure 15.1 where (a) represents the selected ion trace for

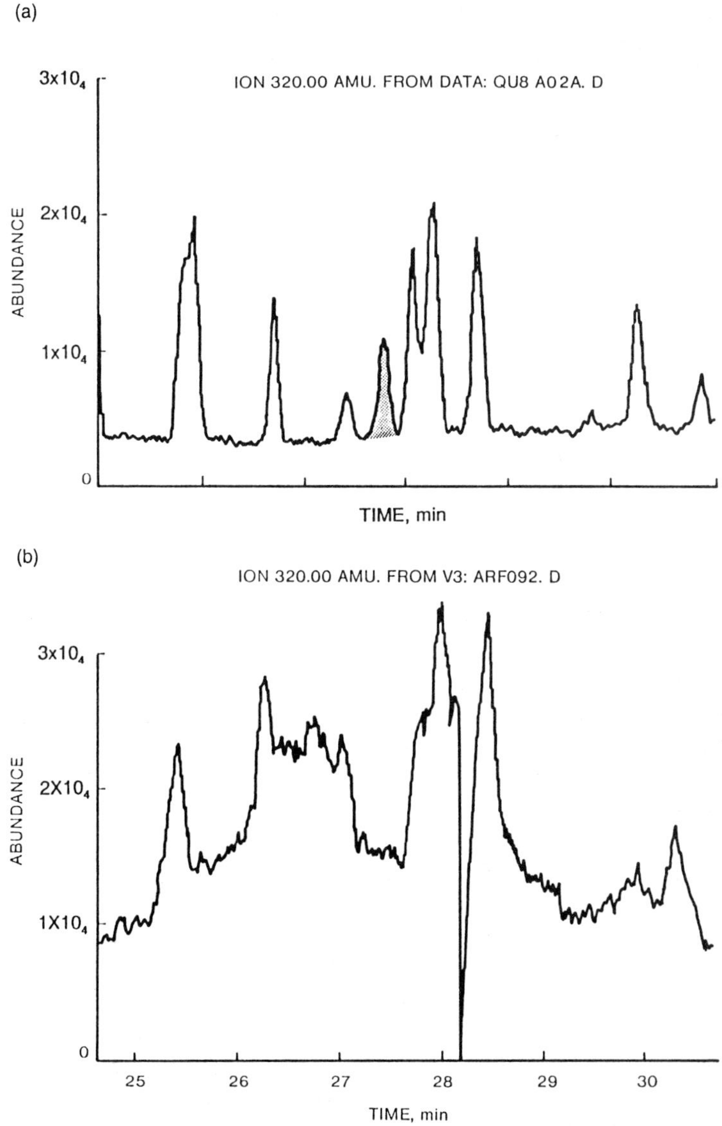

Figure 15.1 *GC/MS, selected ion response for: (a) a flyash sample; (b) flue gas from incinerated sludge*

tetra dioxins from an arrested fly-ash sample. Figure 15.1(b) shows the same trace for an incinerated sewage sludge sample. A brief study identified the areas of inadequacy of the method as:

1. the capacity of the bulk screening technique being exceeded by the greater total loading of organic matter in the more recent samples together with lower analyte concentrations;
2. the inadequate specificity of the extract purification procedure, which was failing to exclude the increased variety of organic co-extractives that survived the bulk screening.

Processed sample extracts containing co-extractives are undesirable as these not only prevent the determination of PCDDs and PCDFs, but also have a detrimental effect on the expensive gas chromatography columns used, and the sensitivity of the mass spectrometer.

Accordingly, a study was carried out to improve the method so as to provide reliable analyses of most types of combustion based samples. Simultaneously, it was also decided to extend the range of the measured analytes to include all the toxicologically significant isomers (2,3,7,8-substituted) present in the tetra- through to octa-congener groups (CDEPP, 1989). For these purposes a series of particulate filters and XAD resin traps were generated by sampling flue gases from bituminous coal incineration. This report deals with the analytical methodology developed in the course of the study. It also includes an investigation into the migration of analytes down the sampling train in the course of stack sampling (using the modified EPA 5 method).

15.3 Experimental

The reagents and apparatus used have been described in an earlier report (Bushby 1989). Additional PCDD and PCDF standards were obtained from Promochem GmbH. A list of labelled standards is given in Table 15.1.

15.3.1 Sampling

A number of particulate filters (100-125 mm) were spiked with 5 ng each of $^{13}C_{12}$-labelled 2,3,7,8 T_4CDF and submitted for stack sampling. One of these was randomly selected as the control/blank. The sampling procedure has been well documented (EPA 1988).

15.3.2 Extraction and bulk organics screening

Two exposed filters, one spiked with the sampling standard and one unspiked, were soxhlet extracted in a pre-extracted cellulose thimble for 16 hours, using 370 ml of toluene. Prior to extraction a suitable quantity (5-25 ng)

Table 15.1 Labelled internal standards used

1.	Internal sampling standard	1,2,3,4 – T_4CDD
2.	Internal quantitation / recovery standards	2,3,7,8 – T_4CDD 1,2,3,7,8 – P_5CDD 1,2,3,7,8,9 – H_6CDD 1,2,3,4,6,7,8 – H_7CDD OCDD 2,3,7,8 – T_4CDF 2,3,4,7,8 – P_5CDF
3.	Syringe standards	2,3,7,8 – T_4CDD $^{37}Cl_4$ labelled 1,2,3,7,8,9 – H_6CDF 1,2,3,4,6,7,8 – H_7CDF OCDF

All standards $^{13}C_{12}$ labelled, unless otherwise stated.

of the recovery standards was added to the filters. XAD 2 resin traps were also extracted in the same way.

Depending on the intensity of the colour of the extract, 2–5 g of H_2SO_4 modified silica (1:1) were added and the extract shaken vigorously at regular intervals over 15 minutes. The extract was filtered through a bed of base modified silica (Bushby 1989) on anhydrous sodium sulphate (Na_2SO_4) and, together with rinsings, rotary evaporated down to near dryness. Approximately 2 ml of hexane were added and the extract blown down just to the point of dryness, then immediately taken up again in 1–2 ml hexane.

15.3.3 Open column chromatography

The extract was applied together with rinsings to a multi-layered column (20 mm id x 30 cm) loaded from the bottom, as shown in Figure 15.2, with:

0.5 g silica (Biosil A)
3.0 g base modified silica (Bushby 1989)
1.0 g silica
6.0 g acid modified silica (H_2SO_4:Silica, 1:1)
2.0 g silica.

The column was eluted with 50 ml hexane directly onto a column (8 mm id x 20 cm) containing 2.5 g Florisil^R (60–100 mesh, activated at 130°C for at least 16 hours).

On completion of the hexane elution, the multi-layered column was discarded and the Florisil^R column was successively eluted with 25 ml, 12% v/v chloroform in hexane, 15 ml 50% v/v chloroform in hexane and 45 ml of dichloromethane. The final 45 ml fraction was carefully evaporated to less

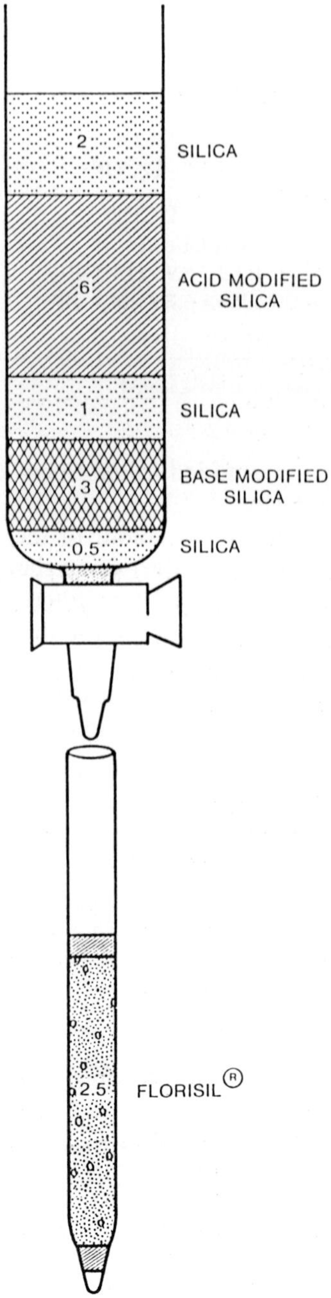

Figure 15.2 *Column for chromatographic purification*

than 1 ml and transferred to a tapered vial together with washings. 20 μl of the syringe standard in the keeper solvent (nonane or dodecane) was added to the vial and the rest of the dichloromethane blown off. This final extract was used for gas chromatography/mass spectrometry (GC/MS) analysis.

A procedural blank was included for each batch of samples analysed.

15.3.4 Gas chromatography/mass spectrometry

An HP 5890A GC/5970B MSD system driven by a 59970C chemstation was used for the GC/MS analyses. Gas chromatographic conditions were as follows:

Column conditions: 0.22 mm id x 60 m SP2331 GC column, using a head pressure of 25 psi helium.

Temperature: Splitless injection at 300°C; GC/MS interface at 255°C; the column oven programmed from 140°C for 1 min to 228°C at 17°C min^{-1} for 4 min, then 0.8°C min^{-1} to 245°C, then 1°C min^{-1} to 260°C, then 3°C min^{-1} to 275°C for 19 mins.

The column was direct-interfaced to the mass spectrometer which was held at ~250°C and 10^{-5} Torr. The mass spectrometer was operated in selected ion monitoring mode and PCDDs and PCDFs were determined in a single run. Four groups of ions were monitored, corresponding to tetra- and penta-CDDs and CDFs, hexa-CDDs and CDFs, hepta-CDDs and CDFs and octa-CDDs and CDFs, as seen in Table 15.2.

Table 15.2 GC/MS – run structure

	Group	*Ions monitored*	*Approximate retention time*
1.	T_4-CDD, CDF ^{13}C T_4-CDD, CDF P_5-CDD, CDF ^{13}C P_5-CDD, CDF $^{37}C14$-T_4CDD	320, 322, 304, 306, 332, 334, 316, 318, 356, 358, 340, 342, 368, 370, 352, 354, 328	20–38 minutes
2.	P_5-CDD, CDF ^{13}C P_5-CDD, CDF H_6-CDD, CDF ^{13}C H_6-CDD, CDF	356, 358, 340, 342, 368, 370, 352, 354, 390, 392, 374, 376, 402, 404, 386, 388	38–50 minutes
3.	H_6CDF, ^{13}C-H_6CDF, H_7-CDD, CDF ^{13}C H_7-CDD, CDF	374, 376, 386, 388, 424, 426, 408, 410, 436, 438, 420, 422	50–61 minutes
4.	OCDD, OCDF, ^{13}C OCDD	458, 460, 442, 444, 470, 472	61–71 minutes

Quantitation was based on isotope dilution principles and response factors for each of the seventeen 2,3,7,8-substituted isomers, relative to ^{13}C-labelled stable isotope standards.

15.4 Results and discussion

15.4.1 Experimental considerations

Among the variety of combustion related samples received at Warren Spring Laboratory, some of the most difficult to satisfactorily analyse for PCDDs and PCDFs have been those derived from the burning of bituminous coal.

Prior to the development of the present method, final GC/MS analyses of the processed extracts would reveal a number of co-elutants at levels which obscured the target analyte signals. As with other samples which contained a high proportion of co-extracted organic compounds the multicolumn used in the original method failed to completely remove these interfering compounds, while consequent overloading of the alumina column reduced the inherent selectivity of the purification.

In the present method the initial post-extraction bulk screening stage ensures sufficient removal of the co-extractives. The amount of acid modified silica used is regulated according to the intensity of colour of the extract and vigorous shaking over a period of 10–15 minutes ensures an adequate reaction. Further removal of co-extractives is effected by the solvent exchange from toluene to hexane, resulting in a dark insoluble residue and a clear hexane fraction containing the analytes.

The larger capacity of the modified multicolumn completes the exclusion of mainly polar and less reactive (Clement and Tosine, 1988) survivors of the screening stages and direct elution of the florisil column supplements the process, as relatively non-polar halogenated micropollutants like polychlorinated byphenyls (PCBs) are not retained by hexane elution (Creaser *et al.*, 1990). Direct elution has the added advantages of shorter analysis time and minimal analyte loss as it precludes the need for evaporation and transfer.

15.4.2 Chromatography

The mechanism of adsorption on florisil is slightly different from that on alumina. Florisil, which is composed of 84% silica, 15% magnesia and 1% sodium sulphate, is characterized by a number of adsorption sites which act as weak Lewis acids. The oxygen atoms on the dioxin molecules act as weak Lewis bases and this forms the basis for a strong interaction or adsorption between the dioxins and the florisil. The number of chloro substituents undoubtedly plays a part in the adsorptive mechanism too, which results in an elution order of tetra- to octa-CDDs.

The mechanism of dioxin retention on alumina depends to a larger extent on the structural spacing of the adsorption sites and the planarity of 2,3,7,8-

substituted dioxins which lie flat within these sites. The 2,3,7,8-T_4CDD molecule is the best example of this among the dioxin isomers. The symmetrical molecule having no ortho-substituents assumes perfect planar configuration and, on alumina it is the most strongly retained of all the dioxins.

The retention on florisil being stronger means increasing solvent strengths can be used to dislodge interfering compounds and the resulting dioxin fraction is cleaner than that obtained from alumina, as seen in Figure 15.3.

15.4.3 Gas chromatography/mass spectrometry

Gas chromatography/mass spectrometry analyses for PCDDs and PCDFs using SP2331 type columns have been documented (Buser and Rappe, 1984) and all but two of the seventeen 2,3,7,8-substituted isomers have been resolved using this method. However, the non-bonded column suffers degradation on exposure to large quantities of polar material resulting in rapid deterioration of the resolution between such isomers as 2,3,7,8-T_4CDF and 2,3,4,8-T_4CDF. This emphasizes the need for good clean-up procedures for the sample extracts.

Ions corresponding to the various PCDD and PCDF congener groups, as well as internal standards, are monitored in groups, depending on the chromatographic retention times. The elution times range from $\sim$20 min for 1,3,6,8 T_4CDF to $\sim$70 min for OCDD.

The criteria for positive identification of dioxin isomers has been established and documented (Creaser *et al.*, 1990, Rappe 1989). The important points include the simultaneous detection ($\pm$1 scan) in the correct ratio of the ion pair being monitored, at the same retention time ($\pm$2 s) as that observed for standards. Quantitation requires a minimum of $\geq$3:1 signal to noise ratio (peak to peak).

15.4.4 Quality control and assurance

The method was validated by the replicate analysis of six sub-samples of an internal reference material. A systematic error was observed for one of the samples and results for that sample were rejected. Statistical treatment of the results for the remaining five is represented in Table 15.3.

The precision of the method as determined by the relative standard deviation (RSD) suggests the range within which sample results may lie, and must be taken into consideration during the interpretation of the data. The average recovery as determined by the use of internal recovery standards was 83% with a range of 67% to 110%.

Method blanks are run with every batch of four to seven samples and involve the random use of glassware sets. The periodic determination of a standard fly-ash sample is carried out as a check on overall method performance.

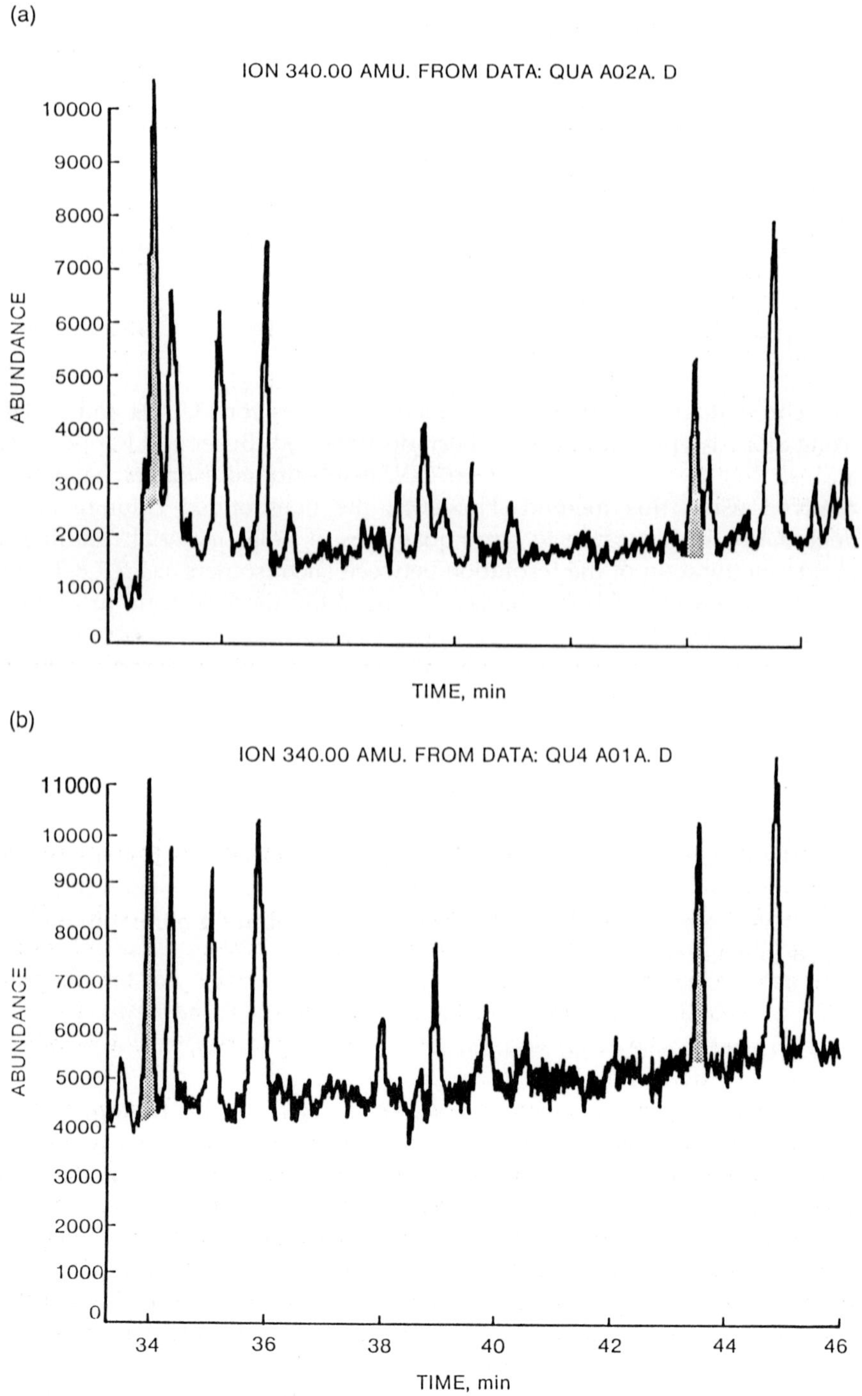

Figure 15.3 *GC/MS response demonstrating extract purification with (a) alumina, (b) Florisil*

Table 15.3 Statistical summary of precision data

Isomer	*Range* w *(ppb)*	*Mean* n	*SD* $\sigma^{(n-1)}$	*RSD* %
2378 TCDD	0.28–0.32	0.298	0.0164	5.5
2378 TCDF	0.95–1.10	1.024	0.0716	6.9
12378 PCDD	1.8–2.4	2.020	0.238	11.8
12378 PCDF	4.0–5.9	4.980	0.687	13.7
23478 PCDF	4.9–6.8	5.840	0.673	11.5
123478 H_6CDD	3.3–4.0	3.744	0.262	6.9
123678 H_6CDD	7.1–9.4	8.300	0.839	10.0
123789 H_6CDD	5.4–7.2	6.228	0.682	10.9
123478 H_6CDF	14.0–15.5	14.70	0.670	4.5
123678 H_6CDF	13.0–14.0	13.60	0.547	4.0
234678 H_6CDF	24–29	26.0	2.12	12.2
1234678 H_7CDD	72–87	77.0	6.73	8.6
1234678 H_7CDF	55–71	60.6	7.12	11.7
1234789 H_7CDF	8.5–11	9.80	0.891	9.0
OCDD	215–250	232	12.91	5.5
OCDF	79–92	86.2	5.50	6.3
		Average RSD =		8.6%

15.4.5 Dioxins from bituminous coal burning

The results of the analysis of the samples generated from the bituminous coal burn experiment are given in Table 15.4.

The table also contains the results of the stack sampling evaluation. These suggest that analyte retention on the particulate filters does not reflect the total sample loading. This is observed in the poor recovery of the sampling spike as well as the variability of the recovery figures. The results for resin trap A represent the successive use of the same trap during the course of four spiked particulate filter changes.

The migration of the sampling spike from the filters to the resin may be explained on the basis of the high temperature (~165°C) of the particulate filter housing coupled with the flow rate of the flue gases through the filter. The combination of the two parameters approximates to conditions on a gas chromatography column where the dioxin analytes move progressively.

The distribution of the levels of PCDDs and PCDFs between filters and resin reflects this migration, except for filters Y6 and Y11. It is thought that changes in furnace or sampling conditions, i.e. flow rate or temperature, may be responsible for the different levels observed.

As far as the levels of PCDDs and PCDFs in the samples are concerned the size of the data set is insufficient to yield reliable information about emissions from coal burning. However, the main observation, that PCDFs (except OCDF) are present to a greater extent than PCDDs, is substantiated

Table 15.4 Levels* of PCDD and PCDF in samples derived from bituminous coal burning

Isomer	*Filters Y4,Y9*	*Filters Y5,Y10*	*Filters Y6,Y11*	*XAD (A)*	*XAD (B)*	*Probe Wash*	*Filter House Wash*	*Det Lim*
2378 TCDD	nd	nd	nd	P	nd	nd	nd	0.05
2378 TCDF	nd	nd	nd	0.31	0.50	nd	nd	0.05
12378 PCDD	nd	nd	0.08	nd	nd	nd	nd	0.05
12378 PCDF	nd	nd	0.23	0.26	0.66	nd	nd	0.05
23478 PCDF	nd	P	0.33	0.19	0.64	nd	nd	0.05
123478 H_6CDD	nd	nd	0.08	P	P	nd	nd	0.05
123678 H_6CDD	nd	nd	0.07	P	0.27	nd	nd	0.05
123789 H_6CDD	nd	nd	nd	nd	nd	nd	nd	0.05
123478 H_6CDF	nd	P	0.36	0.19	0.57	nd	nd	0.05
123678 H_6CDF	nd	P	0.32	0.12	0.41	nd	nd	0.05
123789 H_6CDF	nd	nd	nd	nd	nd	nd	nd	0.05
234678 H_6CDF	nd	nd	0.43	0.07	0.58	nd	nd	0.05
1234678 H_7CDD	nd	nd	0.43	0.29	0.68	nd	nd	0.10
1234678 H_7CDF	nd	nd	1.02	0.31	0.97	nd	nd	0.10
1234789 H_7CDF	nd	nd	0.12	nd	nd	nd	nd	0.10
OCDD	nd	nd	nd	0.62	0.87	nd	nd	0.20
OCDF	nd	nd	nd	nd	nd	nd	nd	0.20
Sampling Spike Recovery	10%	~0%	21%	400%	NA	NA	NA	NA

nd not detected (detection limits in right hand column)
P present at, or just below detection limits
NA not applicable
* levels in ng

by reports that high PCDF to PCDD ratios have been observed for coal soot samples (Harrad *et al.*, 1991, Stieglitz and Vogg, 1987). This is probably because post-combustion conditions are more favourable to the formation of PCDFs (Stieglitz and Vogg, 1987, Harrad, 1989).

15.5 Concluding remarks

The analytical methodology developed, has been successfully used for the determination of PCDDs and PCDFs in samples which were previously found prone to interferences. The method allows for relatively large levels of co-extractives coupled with low levels of the target analytes. The method was found to be generally applicable with the successful analysis of a variety of samples ranging from arrested fly ash, volatiles/particles from flue gases, and lime used as a scavenger in an aluminium recycling process.

Better sensitivity and confirmatory analyses will be provided by the use of a high resolution mass spectrometer which will be the focus of further development of the method. This is especially relevant because stack emission levels have been reduced as a result of stricter legislation, resulting in lower analyte levels. The requirement for greater sensitivity and specificity will also hold for the future analysis of atmospheric samples.

Another area for development is the investigation of alternative GC columns such as the DB-17, which may provide better chromatographic resolution as well as greater durability.

15.6 References

Buser H and Rappe C (1984). Isomer specific separation of 2,3,7,8-substituted PCDDs by high resolution, GC-MS. *Analytical Chemistry,* **56** (3), 442.

Bushby B (1989). Analytical procedure for the determination of T_4CDD and T_4CDF in incinerator stack effluent and residual combustion products. Stevenage: Warren Spring Laboratory, Report IR 263 (CS).

CDEPP (1989). Dioxins in the environment. Report of an interdepartmental working group on PCDDs and PCDFs. Department of the Environment Central Directorate of Environmental Protection Pollution Paper No 27 HMSO.

Clement R E and Tosine H M (1988). The gas chromatography/mass spectrometry determination of chlorodibenzo-p-dioxins and dibenzofurans. *Mass Spectrometry Reviews,* **7**, 593-636.

Creaser C S, Fernandes A R, Harrad S J and Cox E A (1990). Levels and sources of PCDDs and PCDFs in urban British soils. *Chemosphere,* **21** (8), 931-939.

EPA (1988). Guidelines for stack testing of municipal waste combustion facilities. U.S. Environmental Protection Agency Report No 600/8-88-085.

Harrad S J, Fernandes A R, Creaser C S and Cox E A (1991). Domestic coal combustion as a source of PCDDs and PCDFs in the British environment. *Chemosphere,* **23** (3), 55-261.

Harrad S (1989). Levels and sources of PCDD and PCDF in UK soil. Ph.D. Thesis. University of East Anglia.

Rappe C (1989). Analytical methods and exposure assessment. *Chemosphere,* **18** (1-6), 17.

Stieglitz L and Vogg H (1987). On formation conditions of PCDD/PCDF in fly ash from municipal waste incinerators. *Chemosphere,* **16** (8/9), 1917-1922.

— 16 —

Use of principal component analysis to characterize sources for snow deposition of PCDDs and PCDFs

I Fängmark, S Marklund, M Tysklind and C Rappe

16.1 Introduction

Polychlorinated dibenzo-*p*-dioxins (PCDDs) and dibenzofurans (PCDFs) have been identified in numerous environmental samples such as ambient air, snow, soil, spruce needles and sediment samples (Rappe and Kjeller, 1987; Rappe *et al.*, 1989; Reischl *et al.*, 1989a; Eitzer and Hites, 1989; Marklund *et al.*, 1991; Reischl *et al.*, 1989b). During the last few years there has been a considerable scientific and public concern over the environmental deposition of PCDDs and PCDFs in Sweden. One reason for this is that estimates of environmental deposition from known emission sources do not agree with the actual levels found in the environment (Marklund *et al.*, 1991).

By comparing source profiles of PCDDs and PCDFs with those found in the environment it will be possible to reach a better understanding regarding the significance of different sources, and also about transport and distribution mechanisms. Today chemical data are often multivariate because it is easy to collect diverse analytical results on individual samples. The best way to extract and utilize all the available information from a large data table is by treating the data set with multivariate statistical methods such as principal component analysis (PCA) (Wold *et al.*, 1984). Snow samples have been shown to be an excellent matrix with which to estimate environmental deposition of PCDDs and PCDFs during the winter season (Marklund, 1990). In this paper we have applied PCA to snow deposition data in order to compare the congener profile of PCDDs and PCDFs with those from currently known sources.

16.2 Experimental

Snow samples were collected at three different locations in Umeå, Sweden: (1) beside two roadways in the outskirts of Umeå (Anumark and Ersmark), (2) close to a heavy traffic highway passing through the city; and (3) on the top of the water-tower in the prevailing winds from the municipal solid waste (MSW) incinerator. A background sample was collected on the ice of the Gulf of Bothnia 1000 m from the coastline 10 km SE of Umeå. The highway snow samples were collected at two locations and at a distance of 5 m and 100 m from the road. A sampling area of 0.5 x 0.5 m with a depth of 20 mm was chosen for the snow samples except for the sample collected at the water-tower. This corresponded to a two-week deposition period. At the water-tower a snow core with an area of 0.1 m^2 was collected, corresponding to a deposition period of 5 months. All samples were cleaned up and analysed on HRGC-HRMS as detailed previously (Marklund *et al.*, 1989).

16.3 Data analysis

16.3.1 Principal component analysis (PCA)

PCA is a multivariate data analysis technique which is used to investigate similarities and differences in large and complex data sets (Wold *et al.*, 1984). It is a projection method that extracts the systematic variation from a large data table and in turn describes this variation by a few descriptive dimensions. With PCA it is possible to get an overview of dominant patterns or major trends in the data.

Assume a data table with a number of samples (objects) described by n analytical data (variables). If each variable defines a coordinate axis, each object will be represented by a point in n-dimensional space. With PCA a projection is made of this swarm of points down to a space of lower dimension. The projection is made in such a way that the first principal component (PC) describes the direction through the swarm showing the largest variation in the data. The second PC is orthogonal to the first PC and shows the next largest variation, etc.

By principal component decomposition, the data will be partitioned into two parts, score values and loading values. The score values describe the position of each object when projected onto the principle components. From the score values it is possible recognize sample groupings. The loading values describe the extent to which each of the original variables contribute to the model.

Theoretically, PCA corresponds to a mathematical decomposition of the data matrix **X** into means ($\overline{X}_k$), scores (t_{ia}), loadings (ρ_{ak}) and residuals (e_{ik}). Mathematically this is expressed as:

$$x_{ik} = \overline{X}_k + \sum_{a=1}^{A} t_{ia} \cdot \rho_{ak} + e_{ik}$$

Here x_{ik} are the data elements used to describe the sample variation with respect to their content of different chlorinated dioxins and dibenzofurans. The scores (t_{ia}) specify the location of the i_{th} sample along the a_{th} principal component (PC) and the loadings (ρak) describe how much, and in what way, the k_{th} variable contributes to a certain PC. The number of significant

PCs, *A* in the equation above, is determined by a significance test called cross-validation (Wold, 1978).

All calculations were performed on an IBM PS/2 computer using the SMICA 4.3R program package (Umetri AB, S-901 24 Umeå, Sweden).

16.4 Results and discussion

16.4.1 Levels

The snow samples collected close to roadways in Umeå all showed an increased level of the tetra and penta PCDDs and PCDFs compared to those collected 100 m away from the road (Table 16.1, Figure 16.1). Earlier measurements on car emissions has shown that the exhaust from cars fuelled with leaded gasoline contains mainly the TCDF and PeCDF congeners (Marklund *et al.*, 1990). The increased level of these compounds close to the road suggests a low but measurable contribution of dibenzofurans to the environment from car emissions. The results also show that the deposition of PCDDs and PCDFs from car emissions is very local, being found only close to the roads. This agrees with results obtained by Benfenati *et al.* (1992) who found a rapid decrease of traffic associated pollutants such as Pb, PAH and PCDD, PCDF 20 m away from the road. The same tendency was also reported in a study on the accumulation of PCDDs and PCDFs in spruce needles collected at a distance of 20 m and 150 m from a highway in Germany (Reischl *et al.*, 1989b).

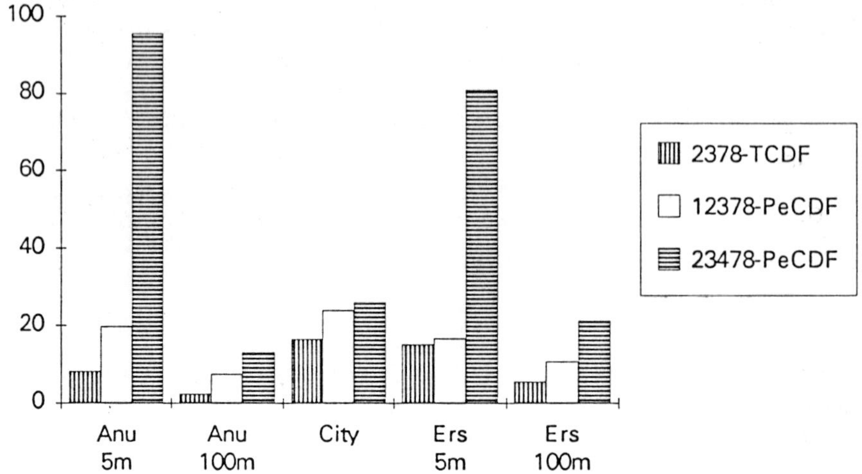

Figure 16.1 *Profile of TCDF and PeCDF in snow samples collected close to and 100 m away from roadways*

Table 16.1 Levels in snow samples (pg m^{-2}) collected 5 m and 100 m away from roadways in downtown Umeå (Ers and Anu), 5 m from a city centre roadway (city), on the ice of the Gulf of Bothnia (Bothnia) and on top of the water-tower.

Sample location distance	*Road Anu 5 m*	*Road Anu 100 m*	*Road City 5 m*	*Road Ers 5 m*	*Road Ers 100 m*	*Ice Bothnia*	*Water-tower*
2378-TCDF	8.1	2.4	16	15	5.4	6.9	91
SUM TCDF	250	99	51	98	110	170	2400
2378-TCDD	24	3.6	3.7	5.2	5.3	3.3	86
SUM TCDD	650	52	19	67	39	130	960
12378-PeCDF	20	7.4	24	17	11	40	360
23478-PeCDF	95	13	26	81	21	26	310
SUM PeCDD	920	150	250	620	170	440	8900
12378-PeCDD	9.9	<1.5	5.5	1.7	2.0	75	53
SUM PeCDD	230	46	120	260	98	300	5100
123478-HxCDF	13	8.2	20	40	6.8	30	500
123678-HxCDF	5.4	6.7	19	40	4.9	21	570
123789-HxCDF	6.1	0.9	<6	<4	1.0	5.1	150
234678-HxCDF	8.8	5.4	11	14	5.8	15	530
SUM HxCDF	120	67	170	260	60	180	12000
123478-HxCDD	380	<3	34	9.8	12	21	930
123678-HxCDD	27	<3	34	9.8	12	21	1200
123789-HxCDD	<15	<3	34	9.8	<5	18	1000
SUM HxCDD	300	370	340	98	190	310	13000
1234678-HpCDF	280	250	32	320	120	340	34000
1234789-HpCDF	<30	<15	<14	<8	<4	39	9400
SUM HpCDF	330	270	37	410	150	540	57000
1234678-HpCDD	210	190	330	200	120	140	20000
SUM HpCDD	410	340	470	370	210	260	38000
OCDF	39	47	120	<5	<6	65	13000
OCDD	600	430	1600	530	290	270	67000
TCDD-TEQ*	87	9.8	35	47	21	38	520
TCDD-TEQ* (International CCMS 1988)	126	16	42	67	24	41	1500

* TEQ = toxic equivalent.

When considering the total of PCDF and PCDD, all levels found in the snow around Umeå are low and in the same range as the background level found on the ice of the Gulf of Bothnia and in the snow samples reported earlier (Marklund *et al.*, 1989). A clear influence of the MSW incinerator can, however, be seen from the sample collected at the top of the water-tower when correcting for the sampling time, which was 5 months for the water-tower but only half a month for all other samples. The consistent levels of PCDFs in all background samples indicate that long-range transport could be a significant source for the deposition of dioxin in Sweden. Recent model calculations on the dispersion of dioxins supports the importance of long-range transport of PCDDs and PCDFs (Rappe *et al.*, 1989, van Jaarsweld and Schutter, 1992).

Known sources of PCDD and PCDF to the environment can be divided into three classes. One class is chemical manufacturing including emissions from manufacture, use, or disposal of technical products contaminated with PCDD and PCDF. Another source is combustion processes and thermal treatment. This class includes solid and chemical waste incineration, automobile emissions, steel and copper production from scrap metals, coal burning, and natural combustion. The third class includes the industrial processes using chlorine. Examples include the chlorine bleaching step in the pulp and paper industry and the production of metals, such as Mg, Al, Ni and Ti. The National Swedish Environmental Protection Agency has listed the most significant sources of PCDD and PCDF in Sweden in 1988 and made a rough estimation of the annual emissions (National Swedish Environmental Protection Board, 1988). These data are summarized in Table 16.2.

The total annual emission of PCDDs and PCDFs in Sweden in 1988 was estimated to be half a kilogram calculated by means of the toxic equivalent factors suggested by Eadon *et al.* (1986). The Swedish EPA estimates that the total national emissions will be one order of magnitude lower in 1992 due to regulations and implementation of better cleaning devices. Recent data does however indicate that small metallurgical foundries are significant contributors to the total national emissions of PCDDs and PCDFs (Lexén *et al.*, 1992).

Table 16.2 Estimated annual emissions of PCDDs and PCDFs from some of the known sources (1988).

Source	*Emissions TEQ Eadon ($g\ year^{-1}$)*
MSW and hazardous waste incinerators (flue gases)	50–100
Motor vehicles (exhaust gases)	5–15
Iron, steel and non-ferrous metal works (flue gases)	50–150
Pulp and paper mills (flue gases)	4–6
Pulp and paper mills (effluents)	15–30
Hospital furnaces (flue gases)	≅ 10

PCA was used to study the differences and similarities in isomer profiles between combustion sources (MSW-, wood- and PVC- pyrolysis, steel scrap melting and car emissions) and environmental samples (ambient air, run-off water and sludge) with the snow samples from Table 16.1. The calculations were based on all 2,3,7,8-substituted isomers and the sums of each congener group; in total 25 variables. To prevent high levels dominating the model all data were normalized against the total of all PCDDs and PCDFs in each sample. This allows for pattern recognition by isomer distribution. Since the modelling performed is scale dependent, each variable was also scaled to unit variance. Two significant principal components (PC) explain in total 67% (44% + 23%) of the variation in the data set. The resulting score-plot, PC1 versus PC2, is shown in Figure 16.2 together with variables contributing most to the directions in plot.

From the score plot, Figure 16.2, it is evident that combustion profiles and environmental samples constitute two slightly overlapping classes. The combustion related profiles are distributed over a large area depending on the type of process studied. Common for all the combustion-related samples is the tendency that the profile becomes dominated by the TCDFs to HpCDFs. Wood combustion and car emission produce a profile with a higher proportion of the lower chlorinated PCDFs (Bacher *et al.*, 1992) while laboratory experiments with PVC pyrolysis generate more of the higher chlorinated PCDFs, i.e. PeCDFs and HxCDF (Theisen *et al.*, 1989). The MSW

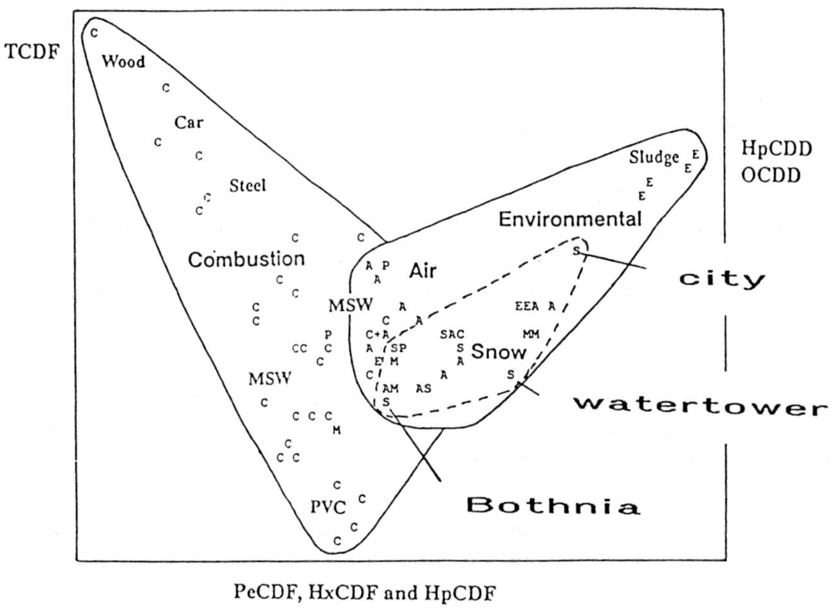

Figure 16.2 *The score plot from PCA, PC1 versus PC2, and the variables contributing to the directions in the plot: A = ambient air, C = combustion related samples, E = environmental samples, M = two overlapping objects and S = snow samples*

incineration samples are spread around the centre of the plot since those profiles contain all congener groups of PCDF and PCDD. In total there is a higher relative proportion of the PCDFs compared to the PCDDs in the MSW incinerator samples. The snow samples collected 100 m from the road as well as downtown Umeå contain a higher proportion of the PCDDs compared to the PCDFs, while the samples collected close to the road, on the ice of the Gulf of Bothnia, and on the water-tower contain about equal proportions of PCDDs and PCDFs.

The environmental samples, ambient air, run-off water, sludge, and snow are drawn towards a profile with a high relative content of HpCDD and OCDD. The same trend was also reported by Eitzer and Hites (1989) from a principal components analysis of PCDDs and PCDFs in the different compartments encountered between sources and sinks.

The snow samples (S) and the ambient air samples (A) in Figure 16.2, are located within the same area because there are only small differences in congener profiles between those two sample types. When comparing only snow samples, a minor shift towards higher chlorinated congeners of PCDD was observed for the sample collected downtown Umeå. The same applies for the sample collected on top of the water-tower, although a more incinerator-like profile would have been expected due to the fact that the water-tower is situated close to the MSW incinerator and in the main downwind direction. This suggest either a contribution from an unknown source of PCDDs or that selective deposition or transformation processes occur that have a strong influence on the profile. In order to make any statements about this further work is needed to study the full congener pattern of PCDDs and PCDFs in samples from different sources as well as in the environment. More knowledge is also needed about the vapour-particle partitioning of ambient air samples and the congener distribution on different particle sizes at different seasonal and meteorological conditions.

16.5 Conclusions

By multivariate statistical methods it is possible to make use of large data sets in order to characterize sources and sinks for environmental pollutants. A better understanding of this complex system is made since PCA takes the covariation between all the measured variables into account simultaneously instead of the traditional one variable at a time approach. PCA allows classification of samples, confirmation of source profiles, and apportionment of the ambient concentrations among the sources.

In this study we have found a discrepancy between source profiles of PCDDs and PCDFs and those found in environmental samples. This indicates either a currently unknown source of PCDDs and PCDFs or that deposition and/or transformation processes influence the observed congener profile.

16.6 References

Bacher R, Swerev M and Ballschmiter K (1992). Profile and pattern of monochloro- through octachlorodibenzodioxins and -dibenzofurans in chimney deposits from wood burning. *Environ. Sci. Technol.*, **26**, 1649-1655.

Benfenati E, Valzacchi S, Mariani G, Airoldi L and Fanelli R (1992). PCDD, PCDF, PCB, PAH, Cadmium and Lead in roadside soil: relationship between road distance and concentration. *Chemosphere*, **24**, 1077-1083.

CCMS (1988). International toxicity equivalency factor (I-TEF) method of risk assessment for complex mixtures of dioxins and related compounds. Pilot study on international information exchange on dioxins and related compounds. North Alantic Treaty Organization. Committee on the challenges of modern society (CCMS), Report number 176.

Dioxins a program for research and action (1988). National Swedish Environmental Protection Board Informs.

Eadon G, Kaminsky L, Silkworth J, Aldous K, Hilker D, O'Keefe P, Smith R, Gierthy J, Hawley J, Kim N and Decaprio A (1986). Calculation of 2,3,7,8-TCDD equivalent concentrations of complex contaminant mixtures. *Environ. Health Perspect.*, **70**, 221-227.

Eitzer B D and Hites R A (1989). Atmospheric transport and deposition of polychlorinated dibenzo-p-dioxins and dibenzofurans. *Environ. Sci Technol.*, **23**, 1396-1401.

Lexén K, de Wit C, Jansson N, Kjeller L O, Kulp S E, Ljung K, Söderström G and Rappe C (1992). Polychlorinated dibenzo-p-dioxins and dibenzofuran levels and patterns in samples from different Swedish industries analysed within the Swedish dioxin survey. Presented at 12th International Symposium on Dioxins and related Compounds, August 1992, Tampere, Finland.

Marklund S, Andersson R, Tysklind M and Rappe C (1989). Emission of PCDDs and PCDFs from a PVC-fire in Holmsund, Sweden. *Chemosphere*, **18**, 1031-1038.

Marklund S (1990). Thesis. Dioxin emission and environmental emission: A study of polychlorinated dibenzodioxins and dibenzofurans in combustion processes. University of Umeå, Umeå, Sweden.

Marklund S, Andersson R, Tysklind M, Rappe C, Egebäck K E, Björkman E and Grigoriadis V (1990). Emission of PCDDs and PCDFs in gasoline and diesel fuelled cars. *Chemosphere*, **20**, 553-561.

Marklund S, Tysklind M, Andersson R, Ljung K and Rappe C (1991). Environmental deposition of PCDDs and PCDFs as determined by the analysis of snow samples from the northern Sweden. *Chemosphere*, **23**, 1359-1364.

Rappe C and Kjeller L-O (1987). OCDDs and PCDFs in environmental samples, air particulates sediments and soil. *Chemosphere*, **18**, 1775-1780.

Rappe C, Marklund S, Kjeller L-O and Lindskog A (1989). Long-range transport of PCDDs and PCDFs on airborne particles. *Chemosphere*, **18**, 1283-1290.

Reischl A, Reissinger M, Thoma H and Hutzinger O (1989a). Accumulation of organic air constituents by plant surfaces: Part IV. *Chemosphere*, **18**, 561-568.

Reischl A, Reissinger M, Thoma H and Hutzinger O (1989b). Accumulation of organic air constituents by plant surfaces, Part IV. Plant surfaces: A sampling system for atmospheric polychlorodibenzo-p-dioxins (PCDD) and polychlorodibenzo-p-furan (PCDF). *Chemosphere*, **18**, 651-568.

Theisen J, Funke W, Balfanz E and König J (1989). Determination of PCDFs and PCDDs in fire accidents and laboratory combustion tests involving PVC-containing materials. *Chemosphere*, **19**, 423-428.

Van Jaarsweld J A and Schutter M A A (1992). Modelling the long-range transport and depositions of dioxin; first results for the North sea and surrounding countries. Presented at 12th International Symposium on Dioxins and related Compounds, August 1992, Tampere, Finland.

Wold S (1978). Cross validatory estimation of the number of components in factor and principal component analysis. *Technometrics*, **20**, 397-406.

Wold S, Albano C, Dunn W J, Edlund U, Esbensen K, Geladi P, Hellberg S, Johansson E, Lindberg W and Sjöström M (1984). Multivariate analysis in chemistry. In: B Kowalski (ed), Chemometrics, Mathematics and Statistics in Chemistry. D Riedel Publishing Company, Dordrecht, Holland, 17-95.

— 17 —

Calibration methods for sulphur dioxide and nitrogen oxides continuous emission measuring systems

D S Walker

17.1 Abstract

This paper is concerned with the main calibration methods which can be used to calibrate continuous emission measuring systems (CEMS) designed for monitoring sulphur dioxide and nitrogen oxides. In the paper the following items are briefly discussed: the meaning of calibration, the meaning of accuracy, calibration using standard reference gas mixtures, calibration by comparison with manual reference test methods employing chemical analysis, calibration by comparison with a validated CEMS, determining the precision of measurements, estimating the systematic errors of measurement. The paper is concluded by the presentation of some results obtained with a dual monitor calibration system and proposals for a standard UK calibration method.

17.2 Meaning of calibration

The calibration of a continuous measuring system for monitoring gaseous emissions is sometimes taken to mean the checking of the zero and span followed by adjustment of the response to the known value of the standard reference gas mixture. This procedure, however, is insufficient and does not comply with the British Standard definition of calibration (BS 5233). It can be argued that this type of procedure is more akin to gauging which has been defined as:

> the operation of fixing the positions of the gauge marks or scale marks of a measuring instrument (in some cases of certain principal marks only), in relation to the corresponding values of the measurand

or perhaps more appropriately to adjustment which is defined in British Standards as,

> the operation intended to bring a measuring instrument into a state of performance and freedom of bias suitable for its use. Note: Adjustment sometimes follows from a calibration.

It is, therefore, necessary to establish exactly what is meant by calibration. For the calibration of continuous emission measuring systems Warren Spring Laboratory has decided to adopt the British Standard definition.

> The set of operations which establish, under specified conditions, the relationship between values indicated by a measuring instrument or measuring system, or values represented by a material measure, and the corresponding known values of a measurand.
> Note 1. The result of a calibration permits the estimation of errors of indication of the measuring instrument, measuring systems or material measure, or the assignment of values to marks on arbitrary scales.
> Note 2. A calibration may determine other metrological properties.
> Note 3. The results of a calibration may be recorded in a document, sometimes called a calibration certificate or a calibration report.
> Note 4. The result of a calibration is sometimes expressed as a calibration factor, or as a series of calibration factors in the form of a calibration curve.
> Note 5. Calibration does not include adjustment.

It is important to note that with the BS terminology the result of a calibration permits the estimation of errors of indication of the measuring system, but it does not allow extension to include adjustment of the system to indicate the correct value. Calibration is often incorrectly understood to include adjustment.

17.3 Meaning of accuracy

Over the years there has been discussion on what the term accuracy should mean, should it include random and systematic errors and should it have only a single recognized meaning (Murphy, 1961). Only recently it was suggested that the use of 'accuracy' to refer to 'any error causing a single measurement to deviate from the true value, i.e. to encompass elements of random and systematic errors' was confusing and inappropriate, that it should be restricted to systematic errors and that it would be better if the word accuracy was never used at all (Miller and Miller, 1988). However, the consensus at present seems to suggest that the term 'accuracy' should be used as a term for overall error.

The British Standard (BS) definition defines the accuracy of a measuring instrument (BS 5233) as:

> the ability of a measuring instrument to give indications approaching the true value of a measurand.

This definition suggests that all the errors of a measuring instrument are included especially since BS definitions are also available for a variety of errors including bias error (the systematic component of the error of a measuring instrument) and repeatability error (the random component of the error of a measuring instrument) etc. The American Society for Testing and Materials (1971) has also adopted this use of the term accuracy. Therefore Warren Spring Laboratory (WSL) has decided that with respect to the calibration of CEMS the term accuracy should include and be expressed as far as is possible in terms of random and systematic errors.

Including both random and systematic errors can sometimes lead to difficulty when deciding which, of two or more systems, is the most accurate. A good example is given by Eisenhart (1968) for the precision and bias of three measurement processes in which the accuracy ranking is not immediately obvious (shown in Figure 17.1). It was concluded that process B is the most accurate and process C is the least accurate even though C is unbiased.

However, it was also pointed out that the average of a large number of measurements from process C will be more accurate than a similar average from either A or B. To summarize a procedure with a small bias and high precision can be more accurate than an unbiased procedure with low precision.

17.4 Calibration using standard reference gas mixtures

The accuracy of measurement has been defined as (BS 5233):

> the closeness of the agreement between the result of a measurement and the (conventional) true value of the measurand.

Because the 'true value' is an ideal concept and in general cannot be known exactly, the conventional true value is adopted, being defined as:

> a value of quantity which for a given purpose may be substituted for the true value.

Reference standards (BS 5233), i.e.:

> a standard generally of the highest metrological quality available at a given location; from which measurements made at that location are derived,

are often used to provide the conventional true value.

For measuring systems employing gas analysers the reference standard can take the form of a stable gas mixture of traceable origin, of known composition and known accuracy of preparation contained in a high pressure cylinder. After multi-point spanning of the measuring system is performed 'calibration' curves can be drawn. Unfortunately, when measuring air pollutant emissions, this relatively simple approach cannot, in some instances, be considered sufficient because of the unquantified presence of interfering gases in the gas sample, and in the case of extractive measuring systems the unquantified occurrence of sample losses and chemical transformations in the sampling system.

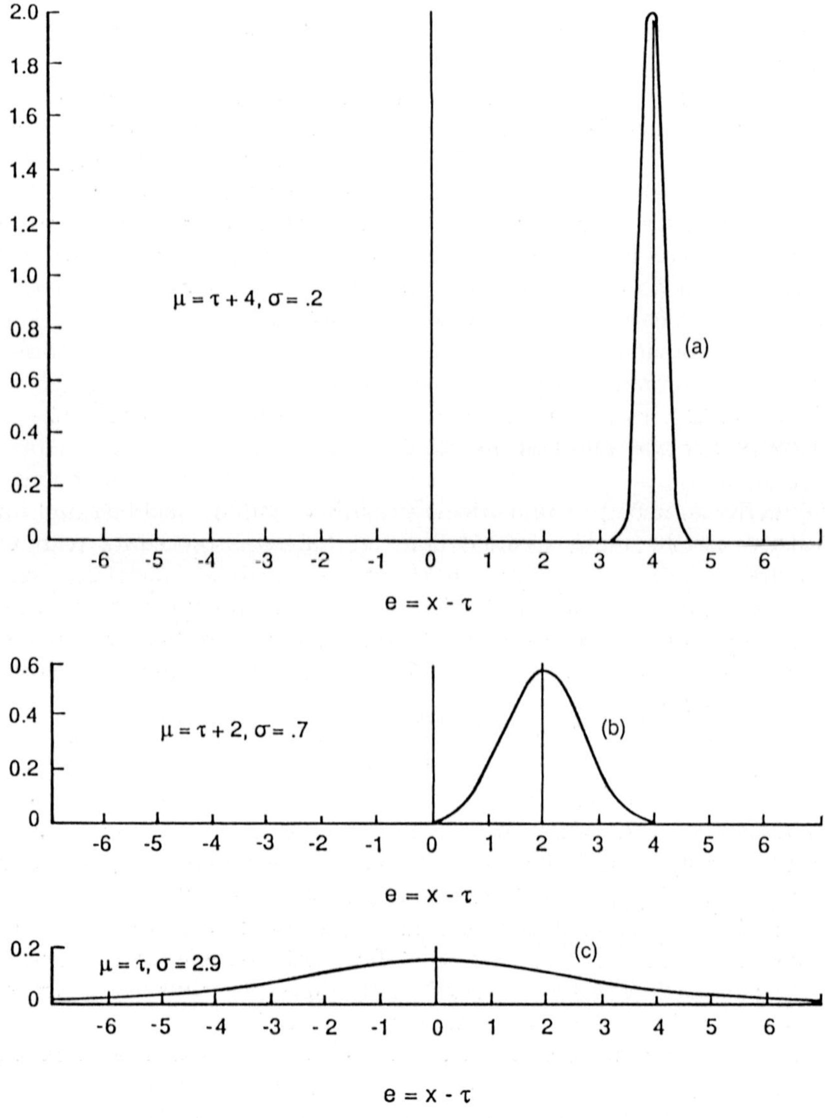

Figure 17.1 *Comparison of precision and bias of three measuring processes*

Where significant interference or losses occur the multi-point spanning procedure cannot by itself constitute a valid calibration because it does not allow an estimation of the errors of indication of the measuring system. It therefore follows that this method of calibration can only justifiably be used when the possibility of significant interference or sample losses has been excluded or if corrections are possible for the assumed systematic errors.

However, correcting for systematic errors is often not easy and is time-consuming. Furthermore even when correction factors are available and published (Tidona *et al.*, 1988), these factors are often only valid for a particular make of instrument and for measuring flue gas from a specific process type.

It is therefore not always practical or cost effective to correct for all the systematic errors. Furthermore even after an exhaustive survey of errors there is always the uncertainty that systematic errors still remain undetected. It therefore follows that another more convenient way of obtaining the conventional true value is required. For systems measuring emissions to the atmosphere this has traditionally been achieved by comparison with manual test methods which rely on chemical analysis techniques.

17.5 Calibration by comparison with manual test methods employing chemical analysis

Manual measuring methods which usually employ titrimetric chemical analysis are often used in calibration procedures in the guise of reference test methods. Titrimetric analysis has been described as a form of 'absolute analysis' (Narayanaswamy, 1989), because it is largely derived from the application of the principles of solution chemistry and requires the use of reference materials, which are usually pure chemical compounds or which can be standardized.

Reference test methods (RTM) are usually established for compliance testing and detail the sampling and analytical procedures required by the regulations. They have the advantage that they do not need calibration or adjustment with standard reference gas mixtures. Reference test methods are usually designed such that their accuracy (precision and bias) is acceptable when measuring pollutant concentrations at or near the emission release limit. However, when the concentrations measured differ from this level the accuracy can be significantly affected. Furthermore the RTM may only have been validated for some specific source categories and in some cases the precision and bias may not be known with certainty.

Reference test methods are usually wet chemical manual test methods and require a good deal of experience and skill by the on-site operator and analyst to achieve reproducible results. Suitable RTM are, however, not always available even for conventional pollutants which have been measured for many years. For example the modified Saltzman and phenoldisulphonic acid methods for measuring NO_x emissions, described in the British Standard methods are both subject to unquantified interference from sulphur dioxide (BS 1756, 1977). These methods could currently be considered to be the UK's RTM. A draft ISO standard manual test method for NO_x, currently being discussed, is also presently reputed to suffer from interference from SO_2. In addition some manual test methods for nitrogen oxides suffer from the disadvantage of being grab methods with short, about 15 s, sampling periods. This method of sampling is not the preferred choice when calibrating CEMSs, (Margeson *et al.*, 1984). A method employing integrated sampling for about 30 min is more suitable.

Manual wet chemical RTMs have also been used in the past by suppliers of standard gas mixtures (Wechter, 1975). This application could be thought ideal for RTMs because of the absence of onerous stack gas conditions and interferents. However, in an investigation concerned with the preparation of stable gas standards in cylinders the wet chemical methods were abandoned in favour of instrumental methods (Wechter, 1975). 'The wet chemical methods listed were operational early in the study but are no longer being employed. It was found that problems existed with the precision and accuracy of those methods and are probably due, at least in part, to manual techniques that were required.' Saltzman–PDS and West Gaeke methods were used in this study.

Although wet chemical RTMs provide a viable method of obtaining the conventional true value they should not, for the above-mentioned reasons, be considered as being the ideal or only acceptable method.

17.6 Calibration by comparison with a validated continuous emission measuring system

The main advantage of obtaining the conventional true value from an RTM is that the whole continuous emission measuring system is assessed via comparison with a separate manual measuring system employing: a separate probe, transport line, conditioning system and analysis using a different principle of measurement (the scientific basis of a method of measurement). This advantage of whole system assessment can be maintained if another CEMS is employed, instead of the RTM, provided a different principle of measurement is employed. It is unlikely that two independent sampling transport and conditioning systems will respond identically to sample losses or exhibit identical sample transformations, or that two different analysis systems employing different principles of measurement will respond similarly to interferents. The reference continuous emission measuring system should, however, be validated as suitable for use at each location.

Laboratory tests should be carried out to identify if any significant interferent effect, sample loss or sample transformation effect exists (that is to assess and eliminate or correct for the systematic errors).

Comparison with a validated CEMS can, therefore, provide an acceptable alternative means of calibrating an emission measuring system. Furthermore a validated CEMS also gives the advantage of real-time analysis. To summarize there are essentially three basic methods of calibrating sulphur dioxide and nitrogen oxides continuous emission measuring systems:

1. Calibration with standard reference gas mixtures.
2. Calibration by comparison with a standard reference manual wet chemical method.
3. Calibration by comparison with a validated continuous emission measuring system.

17.7 Determining the precision of the measurements

Due to the nature of gaseous emissions the measurand (a quantity subjected to measurement) is subject to ever changing variation. Hence, strictly speaking the repeatability (BS 5233) of the measurements,

> the closeness of the agreement between the results of successive measurements of the same measurand carried out subject to: the same method of measurement, the same observer, the same measuring instrument, the same location, the same conditions of use, and repetition over a short period of time,

cannot be determined. The repeatability can only be strictly determined for a standard reference gas mixture which of course lacks the characteristics of the gaseous emission.

However, if one or more additional measuring systems are used an assessment of the reproducibility (BS 5233) of the measurements,

> the closeness of the agreement between the results of measurements of the same measurand, where individual measurements are carried out changing conditions such as: method of measurement, observer, measuring instrument, location, conditions of use and time,

can be achieved provided that conditions changed are specified.

The precision, that is the repeatability and reproducibility, can both be expressed quantitatively in terms of dispersion of the results, i.e. the standard deviation can be used as an index of precision.

17.8 Estimating the systematic errors of measurement

To allow an estimation of the errors of measurement the calibration method must consider and apply to the whole of the measuring system. Calibration is sometimes mistakenly restricted to the analyser component of the system which in the case of extractive systems will ignore any bias caused by the sampling and conditioning system.

There are essentially two cases for consideration; firstly when a reference standard in the form of a stable gas mixture of traceable origin, of known composition and accuracy of preparation, contained in a high pressure gas cylinder is used, and, secondly, when comparison is made with a manual standard reference test method (SRTM) or validated CEMS.

When a reference standard is used every effort should be made to eliminate or correct for known errors. A statistical treatment of the calibration line or curve is then required to estimate the limits of error. Linear or quadratic regression analysis should be adopted which ever provides the more appropriate description of the analysis function. A review of calibration statistics was given recently (Miller, 1991).

There are several options available when comparison is made with an SRTM or CEMS. Either the previously mentioned statistical assessment, or the concept of relative error (De Wees *et al.*, 1989), or a proposed International Standards Organisation procedure based on standard deviation could be used (Proposed ISO standard method).

The concept of relative error as adopted by the USEPA is relatively simple to use. Relative error is calculated as follows (De Wees *et al.*, 1989):

$$\text{Relative error}(\%) = \frac{\underset{\text{(system being calibrated)}}{\text{measured value}} - \underset{\text{(validated CEMS)}}{\text{conventional true value}}}{\underset{\text{(validated CEMS)}}{\text{conventional true value}}} \times 100$$

A perusal of British Standard 5233 shows that this USEPA concept is fully compatible with BS definitions. Relative error is defined in BS 5233 as:

> the absolute error of measurement divided by the conventional true value of the measurand,

and the absolute error of measurement is defined as:

> the result of a measurement minus the conventional true value of the measurand.

For these reasons WSL favours the adoption in the UK of this method of estimating CEMS measurement errors.

A draft ISO document proposes the use of the formula for standard deviations to determine the working accuracy of the CEMS being calibrated against an SRTM or validated CEMS. The standard deviation of the CEMS S_A (in mg m^{-3}) is determined from $S_A = (S_D{}^2 - S_M{}^2)^{0.5}$. The standard deviation S_D is calculated from the difference in pairs of measured values, in mg m^{-3}, according to an equation for standard deviation, using the formula:

$$S_A = \left[\frac{1}{n-1}\left(\sum_{i=1}^{n} Z_i^2 - \frac{1}{n}\left(\sum_{i=1}^{n} Z_i\right)^2\right)\right]^{1/2}$$

where z_i is the difference $(x_i - y_i)$ in pairs of measured values (mg m^{-3}); x_i is the mass concentration of NO_x from the SRTM or validated CEMS; y_i is the mean mass concentration recorded by the CEMS under test and n is the number of comparative measurements. The standard deviation S_M is the standard deviation of the manual method.

This approach is claimed to give an indication of the performance characteristics of the CEMS under test and indicate if a significant systematic error is present. If the mean difference of Z lies outside its 95% confidence limits there is statistical proof that a systematic error is present. For a NO_x CEMS under test to meet the main performance specification the performance characteristic S_A should be $< \pm 5\%$ of full scale. However, it should be noted that the proposed ISO approach is still under discussion and may be amended prior to becoming an international standard.

Table 17.1 Calibration of in-stack monitor

Day	*Measuring period*	*Oxygen (%)*		*NO (mg m^{-3})*		*SO_2 (mg m^{-3})*		*Relative accuracy of monitor (%)*	
		WSL (S1)	*Op*	*WSL (S1)*	*Op*	*WSL (S1)*	*Op*	*NO*	*SO_2*
1	1420–1610	6.0	5.9	797	1141	1290	2498	43.2	93.6
2	1030–1400	5.75	5.8	832	1165	1280	2681	40.0	109.5
3	1530–1630	5.75	5.0	759	1016	1405	2019	33.9	43.7

NOTE: NO expressed as NO_2

Concentrations of NO and SO_2 expressed at 6% O_2, dry and at STP (0°C, 101.3 kPa).

Op - refers to CEMS installed at the plant and operated (including adjustment and calibration by plant personnel).

17.9 Dual monitor calibration system

When a CEMS is calibrated by comparison with a validated CEMS difficulties can arise when the two systems give very different results, yet to all intents and purposes both the systems under test and the validated CEMS have been adjusted and calibrated properly. The question concerns which system is to be believed. This is often resolved by carrying out an additional calibration via comparison with a manual SRTM. However, this takes time, costs extra and sometimes cannot be arranged at short notice.

Warren Spring Laboratory has developed the concept of comparison with one validated CEMS, to comparison with a dual calibration system comprising two separate CEMSs employing different principles of detection. Advantages include the following: the one system acts as a continuous quality control check against the other and hence 'out of control conditions' in either CEMS are immediately identified and confirmed, real-time measurement means that the calibration of the CEMS under test is achieved quickly, and systematic errors or bias are less likely to remain undetected.

Although a dual monitor calibration system (DMCS) costs extra with regard to capital equipment purchase the operating costs are only slightly higher than the costs for a calibration system comprising one CEMS.

Recently WSL carried out a calibration of a CEMS installed on a coal fired combustion plant. The results are shown in Table 17.1 and show a wide discrepancy. This calibration was carried out prior to the DMCS being fully operational therefore only three short runs are available when an additional transportable CEMS system was operated. The results comparing the two WSL measuring systems are shown in Table 17.2 and show that better agreement was obtained, especially for the NO measuring system, and provided additional evidence that the adjustment of the CEMS under test was seriously deficient.

The DMCS for nitrogen oxides and sulphur dioxide is now installed in the

Table 17.2 Relative accuracy of WSL measuring systems

Measuring period	*Oxygen (%)*		*NO (mg m^{-3})*		*SO2 (mg m^{-3})*		*Relative accuracy (%)*	
	S1	*S2*	*S1*	*S2*	*S1*	*S2*	*NO*	*SO_2*
1	5.95	5.9	812	824	1201	1030	1.5	–14.2
2	5.95	5.8	795	781	1230	1135	–1.7	–7.7
3	5.95	5.8	783	754	1258	1067	–3.7	–15.2

Note: NO expressed as NO_2.

Concentrations of NO and SO_2 expressed at 6% O_2, dry at STP (0°C, 101.3 kPa).

Table 17.3 Relative accuracy of NO dual monitor calibration system

Site	*Measuring period**	*NO (vpm) WSL (A)*	*WSL (B)*	*Relative accuracy (%) NO*
1	1	257	262	–1.9
	2	252	266	–5.3
	3	247	244	1.2
	4	237	236	0.4
	5	254	252	0.8
	6	257	240	7.1
	7	248	250	–0.8
	8	239	230	3.9
	9	252	240	5.0
Average		249	247	Range: –5.3 to 7.1
Calculated S D†		7.2	11.9	
Relative S D		2.9%	4.8%	
2	1	318	321	–0.9
	2	308	315	–2.2
	3	292	302	–3.3
	4	310	308	0.7
	5	307	312	–1.6
	6	317	338	–6.2
	7	340	367	–7.4
Average		313	323	Range: –7.4 to 0.7
Calculated S D		14.6	22.4	
Relative S D		4.7%	6.9%	
3	1	461	491	–6.1
	2	439	468	–6.2
	3	430	454	–5.3
	4	497	482	3.1
	5	483	491	–1.6
	6	414	422	–1.9
	7	483	491	–1.6
Average		458	471	Range: –6.2 to 3.1
Calculated S D		31.3	26.0	
Relative S D		6.8%	5.5%	
3	8	557	458	21.6
	9	529	435	21.6
	10	563	463	21.6
	11	507	417	21.6
	12	563	463	21.6
	13	503	518	–2.9
Average		537	459	Range: –2.9 to 21.6
Calculated S D		27.8	34.2	
Relative S D		5.2%	7.5%	
4	1	off line		
	2	509	511	–0.4
	3	505	489	3.3
	4	504	485	3.9
	5	451	441	2.3
	6	404	394	2.5
	7	524	497	5.4
	8	487	457	6.6
	9	501	468	7.1
	10	477	444	7.4
	11	503	483	4.1
	12	478	456	4.8
Average		486	466	Range: –0.4 to 7.4
Calculated S D		33.6	32.6	
Relative S D		6.9%	7.0%	

* Periods vary from about 30 mins to 2 hours in duration.
† S D = Standard deviation.

Table 17.4 Comparison of German, US and ISO calibration methods

	FRG	*USA*	*ISO*
Basic methods	1. Functional test includes checks for: gas leakage, calibration function (factor or curve) interference error, calibration drift, conveter efficiency etc. 2. Analytical test comparison of CEMS response with approved manual reference test methods	1. Relative accuracy test audit and relative accuracy audit — comparison with USEPA reference test methods (usually manual wet chemical methods) 2. Calibration gas audit and calibration drift assessment — using cylinders of standard reference gas mixtures	Three methods permitted: 1) Standard reference test gases — checks on interference, sampling losses etc. 2) Comparison with ISO approved manual method for SO_2 3) Comparison with validated instrumental measuring system
Mandated in regulations	Yes	Yes	No (may be adopted by CEN and hence included in future EC directives)
Status	Currently in use	Currently in use	Under discussion
Choice of calibration methods	No	Some choice (can now use instrumental methods for comparison)	Yes either 1, 2 or 3 (recommendations given on choice)
National or international standard method	No	No	Yes (International Standard Method)

Table 17.5 Summary of proposed UK calibration methods for NO_x and SO_2 CEMS

Item	*Proposal*	*Remarks*
Approved methods	1) Standard reference gas mixtures 2) Comparison with manual reference method 3) Comparison with validated CEMS (using different principle of detection) 4) Comparison with dual CEMS	Proposals 1, 2 and 3 are compatible with current ISO proposals Give guidance on selection (proposal 2) Item 4 is a WSL specific proposal
Precision	Adopt standard deviation as index	Compatible with BS 5233
Bias	Adopt relative error as index	Compatible with BS 5233
Statistical treatment of calibration data	Adopt linear or quadratic regression which ever best describes the analysis function	Compatible with current ISO proposal
Quality of standard reference gas mixtures including traceability	Control under the auspices of NAMAS	Similar to USEPA practice
Determination of in-service calibration curve	Include	Specify frequency
Determination of zero and span drift	Include	Specify frequency
Determination of linearity	Include for linear measuring systems	–
Determination of sampling losses	Include	For extractive measuring systems only
Determination of interference errors	Adopt VDI 3950 method	–
Determination of NO_x CEMS converter efficiency	Include and adopt ISO method if published	Only for systems employing $NO_2 \rightarrow NO$ converters
Determination of dilution ratio of CEMS employing dilution probes	Include	Only for dilution measuring systems
Determination of sample representativeness	Include	Similar to approach adopted by VDI 3950
Manual standard reference method for SO_2	Adopt BS 6069/ISO 7934	Compatible with current ISO proposals
Manual standard reference method for NO_x	Adopt USEPA method 7C if it proves suitable	Proposed ISO method has been abandoned as unsuitable
Status of calibration methods	Submit for consideration by BS	–

WSL mobile Monitoring of Emissions Laboratory which is based at East Kilbride. Since becoming operational it has been tried out at several sites. Some of the results are shown in Table 17.3 for nitric oxide. It is apparent that closer agreement of the relative accuracies is obtained at some sites than at others.

From Table 17.3 it is immediately obvious that poorer agreement of the relative accuracies was obtained at site 3 during the measuring periods 8 to 12. The consistent relative accuracy of 21.6 suggests a strong bias. It is not obvious which measuring system is at fault; for example, the precision for each measuring system is similar and did not change much from the values found for periods 1 to 7. From other measurements taken on site it is known that system A malfunctioned. When the malfunction was corrected the bias was substantially reduced. This incident demonstrates that the DMCS approach quickly identifies measuring system malfunction.

17.10 Comparison of calibration methods adopted by Germany, the US and ISO for sulphur dioxide and nitrogen oxides CEMS

A brief comparison of the methods adopted by Germany the US and ISO is given in Table 17.4. It can be seen that the draft ISO approach confers the maximum flexibility. This is possibly because the FRG and USA procedures are primarily concerned with a national regulatory function.

17.11 Proposals for UK calibration methods

A good CEMS calibration method should have the following main attributes: ensure that the measurements taken have the accuracy required to satisfy the measurement task, incur the minimum costs required to satisfy internationally recognized calibration requirements, be practicable, be as simple as possible, be flexible, and provide clear instructions and specifications. A summary of the proposals for UK calibration methods is given in Table 17.5. It can be seen from Table 17.5 that the proposals give priority to adopting international and national proposals for standards or existing standards.

17.12 References

ASTM D177-71, (1971). Use of the Terms Precision and Accuracy as Applied to Measurement of a Property of a Material. American Society for Testing and Materials, Philadelphia, PA, 1971.

British Standard BS 1756 : Part 4 : 1977. Methods for Sampling and Analysis of Flue Gases.

British Standard BS 5233 (Incorporating BS 2643). Glossary of Terms Used in Metrology.

De Wees W G, Steinsberger S C and Segall R R (1989). Recommended QA/QC for Municipal Waste Combustion Testing. For Presentation at the 82nd Annual Meeting & Exhibition, Anaheim, California, June 25-30, 1989.

Eisenhart C (1968). Realistic evaluation of the precision and accuracy of instrument calibration systems. *Journal of Research of the National Bureau of Standards - C. Engineering and Instrumentals,* **67C**, No2, April–June 1963. Reprinted with corrections, September 1968.

Margeson J H, Knoll J E, Midgett M R, Oldaker G B, Lodger K R, Grohse P M and Gutknecht W F (1984). Integrated method for determining NO_x emissions at nitric acid plants. *Anal. Chem.,* **56**, No 13, 2607-2610.

Miller J N (1991). Basic statistical methods for analytical chemistry Part 2. Calibration and regression methods: A review. *Analyst,* **116**, 3-14.

Miller J C and Miller J N (1988). Basic statistical methods for analytical chemistry Part 1. Statistics of repeated measurements. A review. *Analyst,* **113**, 1351-1356.

Murphy R B (1961) On the Meaning of Precision and Accuracy. In the ASTM Symposium on Quality of Observations. Materials Research and Standards.

Narayanaswamy R (1989). Absolute Analysis. In *Instrumentation and Analytical Science.* London: Peter Peregrinas Ltd, 1989, 259.

Proposed ISO Standard methods currently under study.

Tidona R M *et al.* (1988). Reducing interference effects in the chemiluminescent measurement of nitric oxides from combustion systems. *JAPCA,* **38**, No 6, 806-811.

Wechter S G (1975). Preparation of Stable Pollution Gas Standards using Treated Aluminium Cylinders. Paper presented at the ASTM Calibration Symposium, Boulder, Colorado, USA, August.

Questions asked during discussion session II

To D Hall

K Nicholson: Is there any concern about an upper limit to the size of particle which a gauge might collect ?

D Hall: There is no formal upper limit to particle sizes for dust collection. In practice, large particles are relatively easy to collect, so are not of such great importance in gauge design. In the field they can be important because of their disproportionate contribution to the collected mass. At any distance from a dust source, particle sizes beyond about 200 μm are unusual unless the wind-speed is high, as they deposit fairly quickly.

Questioner: Are siting factors significant as well as gauge design?

D Hall: Yes. The normal specification is for a flat area around the gauge to avoid excessive disturbances in the windflow approaching it. In practice this is very hard to find and there is a good case for trying to design gauges to tolerate disturbed flow conditions. At the moment it is difficult enough designing gauges to cope with the normal requirements.

Questioner: For organic micropollutants, small samplers can be used, or they can be scaled up if a larger sample is required. Can this be done with dust gauges?

D Hall: No. The problem is that, unlike small particle samplers, passive dust gauge performance is strongly dependent on the ratio of the gauge size to the particle stop distance. If the gauge is made larger, this ratio changes and the collection efficiency usually reduces. The best performance ought to come from the smallest gauges. However, reduced gauge size is also allied to a reduced sample size, so this advantage is not so great in practice.

D Hart: One thing missing is design criteria for designing samplers – what are we trying to collect and why?

D Hall: Deposit gauges are meant to sample particles in proportion to their deposition to the ground, while flux gauges are meant to sample the flux of particulate past the sampling station. The original requirement, which still remains, was to assess nuisance, though deposit gauge samples were often analysed for lead. However, there is a growing interest in the toxic content of deposited material as a secondary source of pollutant. It is also now appreciated that these large particles can be ingested, so they can also be a primary pollution source. Particle samplers currently in use, like the PM10 sampler and the M-type sampler for lead, collect only the small, thoracic, size fraction and do not account for these larger particle collected by dust gauges. At the moment we do not have an effective large particle sampler to match these small particle samplers.

C Schofield: Why are we measuring flux?

D Hall: It is a useful indicator of the rate of transport of dust, especially near sources where it gives some idea of the source term. Deposit gauges measure deposition locally, but give no idea of transport. Flux gauges are also good for indicating the source direction of dust, which is often of great importance.

To P Coleman

Questioner: Where did you include probe washings in the split?

P Coleman: For the Modified Method 5 sampler the probe washings were measured separately; for the water-cooled probe they were included with first impinger. For the high-volume sampler the main stream fraction consisted of the probe, filter housing and cyclone washings, along with the filter and the cyclone hopper contents.

Questioner: When the different solutions were analysed, did the isomer balance change?

P Coleman: Yes. The relative proportion of the lighter isomers increases as the sampled gas moves through the different stages of the sampler represented by the individual fractions.

To S Marklund

Questioner:	Can you work out dry deposition velocities from snow samples?
S Marklund:	We have not done this but we think it would probably be easier to do this on air samples.

To D Walker

J Scawin:	BOC prepare primary standards that are carefully analysed and compared with external SRM; agreement must be within 0.05%. This is expensive, so it would be a good idea if somewhere like NPL made primary standards and UK protocols were devised for calibration gases.
D Walker:	Yes this should be done, but would probably be expensive. However, all calibration gases used for monitoring systems should have to comply with a mandatory protocol of preparation and quality.

Index